Invitation to Nonlinear Algebra

GRADUATE STUDIES IN MATHEMATICS 211

Invitation to Nonlinear Algebra

Mateusz Michałek
Bernd Sturmfels

Providence, Rhode Island

2020 *Mathematics Subject Classification.* Primary 05E40, 13A50, 13P15, 14N07, 14Q15, 14T10, 15A69, 20G05, 52B20, 90C22.

For additional information and updates on this book, visit
www.ams.org/bookpages/gsm-211

Library of Congress Cataloging-in-Publication Data

Names: Michałek, Mateusz, 1986- author. | Sturmfels, Bernd, 1962- author.
Title: Invitation to nonlinear algebra / Mateusz Michałek, Bernd Sturmfels.
Description: Providence, Rhode Island : American Mathematical Society, 2021. | Series: Graduate studies in mathematics, 1065-7339 ; 211 | Includes bibliographical references and index.
Identifiers: LCCN 2020032321 | ISBN 9781470453671 (hardcover) | ISBN 9781470465513 (paperback) | ISBN 9781470463083 (ebook)
Subjects: LCSH: Nonlinear theories. | Mathematical analysis. | AMS: Combinatorics – Algebraic combinatorics – Combinatorial aspects of commutative algebra. | Commutative algebra – General commutative ring theory – Actions of groups on commutative rings; invariant theory. | Commutative algebra – Computational aspects and applications – Solving polynomial systems; resultants. | Algebraic geometry – Computational aspects in algebraic geometry – Higher-dimensional varieties. | Linear and multilinear algebra; matrix theory – Basic linear algebra – Multilinear algebra, tensor products. | Group theory and generalizations – Linear algebraic groups and related topics. | Convex and discrete geometry – Polytopes and polyhedra – Lattice polytopes (including relations with commutative algebra and algebraic geometry). | Operations research, mathematical programming – Mathematical programming – Semidefinite programming.
Classification: LCC QA427 .M53 2021 | DDC 512/.25–dc23
LC record available at https://lccn.loc.gov/2020032321

♾ The paper used in this book is acid-free and falls within the guidelines
established to ensure permanence and durability.
Visit the AMS home page at `https://www.ams.org/`

10 9 8 7 6 5 4 3 2 1 26 25 24 23 22 21

To Gabi and Hyungsook

Contents

Preface

This book grew out of the lecture notes for a graduate course we taught during the summer semester of 2018 at the Max-Planck Institute (MPI) for Mathematics in the Sciences in Leipzig, Germany. This was part of the general lecture series (called *Ringvorlesung* in German) offered biannually by the International Max-Planck Research School (IMPRS). The aim of our course was to introduce the theme of *Nonlinear Algebra*, which is also the name of the research group that started at MPI Leipzig in early 2017.

Linear algebra is the foundation of much of mathematics, particularly applied mathematics. Numerical linear algebra is the basis of scientific computing, and its importance for the sciences and engineering can hardly be overestimated. The ubiquity of linear algebra has overshadowed the fairly recent growth in the use of nonlinear models across the mathematical sciences. There has been a proliferation of methods based on systems of multivariate polynomial equations and inequalities. This expansion is fueled by recent theoretical advances, development of efficient software, and an increased awareness of these tools. At the heart of this growing area lies algebraic geometry, but there are links to many other branches of mathematics, such as combinatorics, algebraic topology, commutative algebra, convex and discrete geometry, tensors and multilinear algebra, number theory, representation theory, and symbolic and numerical computation. Application areas include optimization, statistics, and complexity theory, among many others.

Nonlinear algebra is not simply a rebranding of algebraic geometry. It represents a recognition that a focus on computation and applications, and the theoretical underpinnings that this requires, results in a body of inquiry that is complementary to the existing curriculum. The term nonlinear algebra is intended to capture these trends, and to be more friendly to applied scientists. A special research semester with that title, held in the fall of 2018 at the Institute for Computational and Experimental Research in Mathematics (ICERM) in Providence, Rhode Island, explored the theoretical and computational challenges that have arisen and charted a course for the future. This book supports this effort by offering students and researchers a warm welcome to the theme of nonlinear algebra.

Our presentation is structured into 13 chapters, one for each week in a semester. Many of the chapters are rather ambitious in that they promise a first introduction to an area of mathematics that would normally be covered in a full-year course. But what we offer is really just an invitation. Readers are encouraged to go further in their studies by exploring other sources. We think that students will enjoy our presentation. We hope that nonlinear algebra will encourage them to think critically and deeply, and to question the historic boundaries between "pure" and "applied" mathematics.

Mateusz Michałek and Bernd Sturmfels

Acknowledgments

The term nonlinear algebra was introduced in the setting of theoretical physics two decades ago. The second author is grateful to Shamil Shakirov for introducing him to that literature. We wish to cite the book by Dolotin and Morozov [**18**], which presents a perspective that aligns well with ours.

A two-week intense course based on our own manuscript took place at MPI Leipzig in June 2019. We are grateful to the following participants for their lectures and comments: Zachary Adams, Yulia Alexandr, Tobias Boege, Marie-Charlotte Brandenburg, Madeline Brandt, Türkü Çelik, Rodica Dinu, Eliana Duarte, Yassine El Maazouz, Yuhan Jiang, Paul Görlach, Alex Heaton, Nidhi Kaihnsa, Max Kölbl, András Lörincz, Orlando Marigliano, Milo Orlich, Yue Ren, Jose Samper, Mahsa Sayyary, Emre Sertöz, Tim Seynnaeve, Isabelle Shankar, Stefana Sorea, Martin Vodicka, and Maddie Weinstein.

We continued to give lectures on topics from this book in the winter semester of 2019/20 at MPI Leipzig and in the spring semester of 2020 at UC Berkeley. We thank our students in these courses for giving us "bug reports". Contributors include Rida Ait El Manssour, Claudia Fevola, Eugene Huang, Kyle Huang, Yelena Mandelshtam, Mathew McBride, Chiara Meroni, Max Pfeffer, Kemal Rose, Paul Vater, and Nathan Wenger.

We are grateful for valuable comments we received from the following additional readers: Sara Billey, Christian Ikenmeyer, Annachiara Korchmaros, Olga Kuznetsova, Kaie Kubjas, Michał Lasoń, Stark Ledbetter, Diane Maclagan, Rafael Mohr, Anna-Laura Sattelberger, Luca Sodomaco, and Mima Stanojkovski.

We thank Thomas Endler for help with creating the figures.

Chapter 1

Polynomial Rings

"Algebra is but written geometry", Sophie Germain

After a course in linear algebra one often encounters abstract algebra. In that course one studies algebraic structures such as fields, rings and ideals. In this first chapter we introduce basics, with a focus on polynomials and Gröbner bases. We show how to use these for computing invariants of a polynomial ideal, such as the dimension or degree. The formalism we develop now will be applied to geometric situations in later chapters.

1.1. Ideals

Our most basic algebraic structure is that of a *field*. The elements of the field serve as numbers, also called scalars. We can add, subtract, multiply and divide them. It is customary to denote fields by the letter K, for the German word *Körper*. Our favorite field is the set $K = \mathbb{Q}$ of rational numbers. Another important field is the set $K = \mathbb{R}$ of real numbers. In practice, these two fields are very different. Numbers in $\mathbb{Q}$ can be manipulated by exact *symbolic computation*, whereas numbers in $\mathbb{R}$ are approximated by floating point representations and manipulated by *numerical computation*.

Other widely used fields are the set of complex numbers $\mathbb{C}$ and the finite field $\mathbb{F}_q$ with q elements. If K is not algebraically closed then we write $\overline{K}$ for its algebraic closure. This is the smallest field in which every nonconstant polynomial with coefficients in K has a root. For instance, $\overline{\mathbb{Q}}$ and $\overline{\mathbb{F}_q}$ are the algebraic closures of the two fields above. Another important example is the field of rational functions $\mathbb{Q}(t)$. Its algebraic closure $\overline{\mathbb{Q}(t)}$ is contained in the field of *Puiseux series*, denoted by $\mathbb{C}\{\{t\}\}$, which is also algebraically closed. The elements of $\mathbb{C}\{\{t\}\}$ are formal expressions $\sum_{a=a_0}^{\infty} c_a t^{\frac{a}{m}}$, where m is a fixed positive integer, a_0 is an integer and $c_a \in \mathbb{C}$. The fields $\overline{\mathbb{Q}(t)}$ and $\mathbb{C}\{\{t\}\}$ may be unfamiliar to many of our readers. Their importance will be seen in Chapter 7, when we pass from classical algebra to tropical algebra.

In this section we study the ring of polynomials in n variables $x_1, \ldots, x_n$ with coefficients in our field K. This polynomial ring is denoted by $K[\mathbf{x}] = K[x_1, \ldots, x_n]$. If the number n is small, then we typically use letters without indices to denote the variables. For instance, we often write $K[x], K[x, y]$, or $K[x, y, z]$ for the polynomial ring when $n = 1, 2$, or 3.

Many of the constructions we present work not just for the polynomial ring $K[\mathbf{x}]$ but also for an arbitrary commutative ring R with unit 1. We allow $1 = 0$, i.e. R as a set may contain just one element 0. For the most part, the reader may assume $R = K[\mathbf{x}]$. But it would not hurt to peruse a standard textbook on *abstract algebra* and look up the axioms of a *ring* and the formal definitions of *commutative* and *unit.* Important examples of commutative rings are the integers $\mathbb{Z}$, the polynomial ring over the integers $\mathbb{Z}[\mathbf{x}]$, and the quotient of a polynomial ring by an ideal. The latter will be discussed soon.

The polynomial ring $K[\mathbf{x}]$ is an infinite-dimensional K-vector space. A distinguished basis of this vector space consists of the monomials $\mathbf{x}^{\mathbf{a}} = x_1^{a_1} x_2^{a_2} \cdots x_n^{a_n}$. There is one monomial for each nonnegative integer vector $\mathbf{a} = (a_1, a_2, \ldots, a_n) \in \mathbb{N}^n$. Every polynomial $f \in K[\mathbf{x}]$ is written uniquely as a finite K-linear combination of monomials:

$$f = \sum_{\mathbf{a}} c_{\mathbf{a}} \mathbf{x}^{\mathbf{a}}.$$

The *degree* of f is the maximum of the quantities $|\mathbf{a}| = a_1 + \cdots + a_n$ where $c_{\mathbf{a}} \neq 0$. For polynomials of degree $1, 2, 3, 4, 5$ and 6 we use the words *linear*, *quadratic*, *cubic*, *quartic*, *quintic* and *sextic*. These can be adjectives or nouns. It is also common to use the term *quadric* for a quadratic polynomial.

For example, the following is a cubic polynomial in $n = 3$ variables:

$$(1.1) \qquad f = \det \begin{pmatrix} 1 & x & y \\ x & 1 & z \\ y & z & 1 \end{pmatrix} = 2xyz - x^2 - y^2 - z^2 + 1.$$

The zero set of f is a surface in $\mathbb{R}^3$. It consists of all points at which the rank of the 3×3 matrix in (1.1) decreases. It has four singular points, namely the points $(1, 1, 1)$, $(1, -1, -1)$, $(-1, 1, -1)$, and $(-1, -1, 1)$. These points are the common zeros in $\mathbb{R}^3$ of the cubic f and its three partial derivatives

$$\frac{\partial f}{\partial x} = 2yz - 2x, \quad \frac{\partial f}{\partial y} = 2xz - 2y, \quad \frac{\partial f}{\partial z} = 2xy - 2z.$$

These are the points at which the rank of the 3×3 matrix in (1.1) equals 1. The formal definition of singular points will appear at the end of Section 2.1.

Figure 1.1. A cubic surface with four singular points.

Figure 1.1 illustrates the cubic surface $\{f = 0\}$. However, the picture is drawn in a different coordinate system. Namely, we divide each of the three coordinates by $\frac{1}{3}(1 - x - y - z)$ and clear denominators to get $g = 8x^3 + 6x^2y + 6x^2z + \cdots - 3z + 1$. Figure 1.1 shows the part of the surface $\{g = 0\}$ that lies in the box $-1.5 < x, y, z < 1.5$. This change of coordinates amounts to applying a *projective transformation* to our surface. From the vantage point of projective geometry, to be adopted in Section 2.2, it is natural to regard two varieties as being the same if they differ by a projective transformation. We therefore assert, from now on, that Figure 1.1 shows the surface $\{f = 0\}$. This will serve as a running example throughout the book.

Definition 1.1. All rings in this book are commutative and have a unit 1. An *ideal* in such a ring R is a nonempty subset I of R such that

(a) if $f \in R$ and $g \in I$, then $fg \in I$;

(b) if $f, g \in I$, then $f + g \in I$.

If $R = K[\mathbf{x}]$ then an ideal I is a nonempty subset of $K[\mathbf{x}]$ that is closed under taking linear combinations with polynomial coefficients. An alternative definition is as follows: A subset I of a ring R is an ideal if and only if there exists a ring homomorphism $\phi : R \to S$ whose kernel $\ker \phi = \phi^{-1}(0)$ is equal to I. For instance, if $R = \mathbb{Z}$ then the set I of even integers is an ideal. It is the kernel of the ring homomorphism $\mathbb{Z} \to \mathbb{Z}/2\mathbb{Z} = \{0, 1\}$ that takes an integer to either 0 or 1, depending on its parity.

Ideals in a ring play the same role as normal subgroups in a group. They are the subobjects used to define quotients. Consider the quotient of abelian groups R/I. Its elements are the congruence classes $f + I$ modulo I. The axioms (a) and (b) in Definition 1.1 ensure that the following identities hold:

$$(1.2) \quad (f+I)+(g+I) = (f+g)+I \quad \text{and} \quad (f+I)(g+I) = fg+I.$$

Proposition 1.2. *If $I \subset R$ is an ideal, then the quotient R/I is a ring.*

Given any subset $\mathcal{F}$ of a ring R, we write $\langle \mathcal{F} \rangle$ for the smallest ideal containing $\mathcal{F}$. This is the *ideal generated by* $\mathcal{F}$. If $R = K[\mathbf{x}]$ then the ideal $\langle \mathcal{F} \rangle$ is the set of all polynomial linear combinations of finite subsets of $\mathcal{F}$.

Proposition 1.3. *If I and J are ideals in a ring R, then the following subsets of R are ideals as well: the sum $I + J$, the intersection $I \cap J$, the product IJ, and the quotient $(I : J)$. The latter two subsets of R are defined as follows:*

$$IJ = \langle\, fg : f \in I, g \in J \,\rangle \quad \text{and} \quad (I : J) = \big\{ f \in R : fJ \subseteq I \big\}.$$

Proof. The product IJ is an ideal by definition. For the others one checks that conditions (a) and (b) hold. We shall carry this out for the ideal quotient $(I : J)$. To show (a), suppose that $f \in R$ and $g \in (I : J)$. We have

$$(fg)J = f(gJ) \subset fI \subset I.$$

For (b), suppose f and g are in $(I : J)$. We have

$$(f+g)J \subset fJ + gJ \subset I + I = I.$$

This implies $f + g \in (I : J)$. We have shown that $(I : J)$ is an ideal. □

The *Euclidean algorithm* works in the polynomial ring $K[x]$ in one variable x over a field K. This implies that $K[x]$ is a *principal ideal domain* (PID), i.e. every ideal I in $K[x]$ is generated by one element. That generator can be uniquely factored into irreducible polynomials.

Unique factorization of polynomials also holds when the number of variables satisfies $n \geq 2$. We say that the polynomial ring $K[\mathbf{x}]$ is a *unique factorization domain* (UFD). However, $K[\mathbf{x}]$ is not a PID when $n \geq 2$. For instance, for $n = 2$, the ideal $\langle x_1, x_2 \rangle$ is not principal. But let's first go back to the univariate case in order to illustrate the operations in Proposition 1.3.

Example 1.4 ($n = 1$)**.** Consider the following two ideals in $\mathbb{Q}[x]$:

$$I = \langle\, x^3 + 6x^2 + 12x + 8 \,\rangle \quad \text{and} \quad J = \langle\, x^2 + x - 2 \,\rangle.$$

We compute the four ideals in Proposition 1.3. For this, it helps to factor:

$$I = \langle\, (x+2)^3 \,\rangle \qquad \text{and} \qquad J = \langle\, (x-1)(x+2) \,\rangle.$$

The four new ideals are

$$\begin{aligned} I \cap J &= \langle (x-1)(x+2)^3 \rangle, & IJ &= \langle (x-1)(x+2)^4 \rangle, \\ I + J &= \langle x+2 \rangle, & I : J &= \langle (x+2)^2 \rangle. \end{aligned}$$

We see that arithmetic in $\mathbb{Q}[x]$ is just like arithmetic in the ring of integers $\mathbb{Z}$.

A nonzero element f in a ring R is called

- a *nilpotent* if $f^m = 0$ for some positive integer m;
- a *zero divisor* if there exists $0 \neq g \in R$ such that $gf = 0$.

A ring R is called an *integral domain* if it has no zero divisors and $1 \neq 0$ in R. For instance, the set $\{0\}$ is a ring but not an integral domain.

We examine these properties for the quotient ring R/I where I is an ideal in R. Properties of the ideal I correspond to properties of the ring R/I. This correspondence is summarized in the following table:

Property	**Definition**	**The quotient ring R/I**
I is *maximal*	No other proper ideal contains I	is a *field*
I is *prime*	$fg \in I \Rightarrow f \in I$ or $g \in I$	is an *integral domain*
I is *radical*	$(\exists s : f^s \in I) \Rightarrow f \in I$	has no *nilpotent* elements
I is *primary*	$fg \in I$ and $g \notin I \Rightarrow (\exists s : f^s \in I)$	*zero divisors* are nilpotent

Maximal, prime and primary ideals are proper. In other words, the ring R itself is an ideal in R, but it is neither maximal, nor prime, nor primary.

Example 1.5. The ideal $I = \langle x^2 + 10x + 34, 3y - 2x - 13 \rangle$ is maximal in the polynomial ring $\mathbb{R}[x, y]$. The field $\mathbb{R}[x, y]/I$ is isomorphic to the field of complex numbers $\mathbb{C} = \mathbb{R}[i]/\langle i^2 + 1 \rangle$. One isomorphism is obtained by sending $i = \sqrt{-1}$ to $\frac{1}{13}(x + 5y)$. The square of that expression is -1 mod I. The principal ideal $J = \langle x^2 + 10x + 34 \rangle$ is prime, but it is not maximal. The quotient $\mathbb{R}[x, y]/J$ is an integral domain. It is isomorphic to $\mathbb{C}[y]$.

Examples for the other two classes of ideals are given in the next proof.

Proposition 1.6. *We have the following implications for an ideal I in R:*

$$I \textit{ maximal} \Rightarrow I \textit{ prime} \begin{array}{l} \Rightarrow I \textit{ radical}, \\ \Rightarrow I \textit{ primary}. \end{array}$$

None of these implications is reversible. However, every ideal that is both radical and primary is prime. Every intersection of prime ideals is radical.

Proof. The first implication holds because there are no zero divisors in a field. To see that prime implies radical, we take $g = f^{s-1}$ and use induction on s. That prime implies primary is clear. To prove that every radical primary ideal is prime, assume $fg \in I$ and $f \notin I$. Then, as I is primary, we have $g^s \in I$ for some $s \in \mathbb{N}$. As I is radical, we conclude that $g \in I$.

To see that no implication is reversible, we consider the following three ideals in the polynomial ring $\mathbb{R}[x, y]$ with $n = 2$ variables:

- $I = \langle x^2 \rangle$ is primary but not radical;
- $I = \langle x(x-1) \rangle$ is radical but not primary;
- $I = \langle x \rangle$ is prime but not maximal.

The last statement holds since intersections of radical ideals are radical. □

We now revisit the surface in Figure 1.1 from the perspective of ideals.

Example 1.7 ($n = 3$). We consider the ideal generated by the partial derivatives of the cubic $f = 2xyz - x^2 - y^2 - z^2 + 1$. This is the ideal

$$I = \left\langle \frac{\partial f}{\partial x}, \frac{\partial f}{\partial y}, \frac{\partial f}{\partial z} \right\rangle = \langle yz - x,\ xz - y,\ xy - z \rangle \subset \mathbb{R}[x, y, z].$$

The cubic f is not in this ideal because every polynomial in I has zero constant term. The ideal I is radical because we can write it as the intersection of five maximal ideals. Namely, using a computer algebra system, we find

$$\begin{aligned} I = \langle x, y, z \rangle \cap \langle x-1, y-1, z-1 \rangle \cap \langle x-1, y+1, z+1 \rangle \\ \cap\ \langle x+1, y-1, z+1 \rangle \cap \langle x+1, y+1, z-1 \rangle. \end{aligned} \tag{1.3}$$

This is a primary decomposition of I, as discussed in detail in Chapter 3.

The cubic f lies in the last four maximal ideals. Their intersection equals $I + \langle f \rangle$. The zero set of the radical ideal $I + \langle f \rangle$ consists of the four singular points on the surface seen in Figure 1.1. The *Chinese Remainder Theorem* for rings implies that the quotient ring is a product of fields. Namely, we have an isomorphism $\mathbb{R}[x, y, z]/I \simeq \mathbb{R} \times \mathbb{R} \times \mathbb{R} \times \mathbb{R} \times \mathbb{R}$. It takes each polynomial modulo I to its residue classes modulo the intersectands in (1.3).

1.2. Gröbner Bases

Every ideal has many different generating sets. There is no canonical notion of a basis for an ideal. For instance, the set $\mathcal{F} = \{x^6 - 1, x^{10} - 1, x^{15} - 1\}$ minimally generates the ideal $\langle x - 1 \rangle$ in the polynomial ring $\mathbb{Q}[x]$ in one variable. Of course, the singleton $\{x - 1\}$ is a preferable generating set for that ideal. Recall that every ideal in $\mathbb{Q}[x]$ is *principal*, since we here have $n = 1$. The *Euclidean algorithm* transforms the set $\mathcal{F}$ into the set $\{x - 1\}$.

Here is a certificate for the fact that $x - 1$ is in the ideal generated by $\mathcal{F}$:

$$x^5 \cdot (x^6 - 1) \ - \ (x^5 + x) \cdot (x^{10} - 1) \ + \ 1 \cdot (x^{15} - 1) \ = \ x - 1.$$

Such identities can be found with the *extended Euclidean algorithm*. Please google this. Finding certificates for ideal membership when $n \geq 2$ is a harder problem. This topic comes up when we discuss Hilbert's Nullstellensatz in Chapter 6. In this section we introduce the basics of computing with ideals.

Gaussian elimination is familiar from linear algebra. It gives a process for manipulating ideals that are generated by linear polynomials. For example, the following two ideals are identical in the polynomial ring $\mathbb{Q}[x, y, z]$:

$$\begin{aligned} &\langle\, 2x + 3y + 5z + 7,\ 11x + 13y + 17z + 19,\ 23x + 29y + 31z + 37\,\rangle \\ &\qquad = \langle\, 7x - 16,\ 7y + 12,\ 7z + 9\,\rangle. \end{aligned}$$

Undergraduate linear algebra taught us how to transform the three generators on the left into the simpler ones on the right. This is the process of solving a system of linear equations. In our example, the three linear equations have a unique solution, namely the point $\left(\frac{16}{7}, -\frac{12}{7}, -\frac{9}{7}\right)$ in $\mathbb{R}^3$.

We next introduce Gröbner bases. The framework of Gröbner bases offers practical methods for computing with ideals in a polynomial ring $K[\mathbf{x}]$ in n variables. Here K is a field whose arithmetic we can compute. Implementations of Gröbner bases are available in many computer algebra systems. We strongly encourage readers to experiment with these tools.

Informally, we can think of computing Gröbner bases as a version of the Euclidean algorithm for polynomials in $n \geq 2$ variables, or as a version of Gaussian elimination for polynomials of degree ≥ 2. Gröbner bases for ideals are fundamental to nonlinear algebra, just like Gaussian elimination for matrices is fundamental to linear algebra. The premise of this book is that *nonlinear algebra is the next step after linear algebra.*

We identify the set $\mathbb{N}^n$ of nonnegative integer vectors with the monomial basis of the polynomial ring $K[\mathbf{x}]$. The coordinatewise partial order on $\mathbb{N}^n$ corresponds to divisibility of monomials. To be precise, we have $\mathbf{a} \leq \mathbf{b}$ in the poset $\mathbb{N}^n$ if and only if the monomial $\mathbf{x}^{\mathbf{a}}$ divides the monomial $\mathbf{x}^{\mathbf{b}}$.

Theorem 1.8 (Dickson's Lemma). *Any infinite subset of $\mathbb{N}^n$ contains a pair $\{\mathbf{a}, \mathbf{b}\}$ that satisfies $\mathbf{a} \leq \mathbf{b}$.*

Corollary 1.9. *For any nonempty set $\mathcal{M} \subset \mathbb{N}^n$, its subset of coordinatewise minimal elements is finite and nonempty.*

Proof. The fact that the subset is nonempty follows by induction on n. The subset is finite by Dickson's Lemma. □

Proof of Theorem 1.8. We proceed by induction on n. The statement is trivial for $n = 1$. Any subset of cardinality at least 2 in $\mathbb{N}$ contains a comparable pair. Suppose now that Dickson's Lemma has been proved for $n-1$, and consider an infinite subset $\mathcal{M}$ of $\mathbb{N}^n$. For each $i \in \mathbb{N}$ let $\mathcal{M}_i$ denote the set of all vectors $\mathbf{a} \in \mathbb{N}^{n-1}$ such that $(\mathbf{a}, i) \in \mathcal{M}$. If some $\mathcal{M}_i$ is infinite then we are done by the induction hypothesis. Hence we may assume that each $\mathcal{M}_i$ is a finite subset of $\mathbb{N}^{n-1}$ and that $\mathcal{M}_i \neq \emptyset$ for infinitely many i.

Consider the (possibly infinite) subset $\bigcup_{i=0}^{\infty} \mathcal{M}_i$ of $\mathbb{N}^{n-1}$. We claim that the subset of its minimal elements with respect to the coordinatewise order is finite. This is clear when $\bigcup_{i=0}^{\infty} \mathcal{M}_i$ is finite. Otherwise, by the induction hypothesis we may apply Corollary 1.9 to this subset. Hence, there is always an index j such that all minimal elements are in the finite set $\bigcup_{i=0}^{j} \mathcal{M}_i$. Pick any $\mathbf{b} \in \mathcal{M}_k$ for $k > j$. Since the element $\mathbf{b}$ must be greater than or equal to some minimal element, there exists an index i with $i \leq j < k$ and an element $\mathbf{a} \in \mathcal{M}_i$ with $\mathbf{a} \leq \mathbf{b}$. We have $(\mathbf{a}, i) \leq (\mathbf{b}, k)$ in $\mathcal{M}$. □

Definition 1.10. Consider a total ordering $\prec$ of the set $\mathbb{N}^n$. We write $\mathbf{a} \preceq \mathbf{b}$ if $\mathbf{a} \prec \mathbf{b}$ or $\mathbf{a} = \mathbf{b}$. The ordering $\prec$ is a *monomial order* if for all $\mathbf{a}, \mathbf{b}, \mathbf{c} \in \mathbb{N}^n$,

- $(0, 0, \ldots, 0) \preceq \mathbf{a}$;
- $\mathbf{a} \preceq \mathbf{b}$ implies $\mathbf{a} + \mathbf{c} \preceq \mathbf{b} + \mathbf{c}$.

This gives a total order on monomials in $K[\mathbf{x}]$. Three standard examples are

- *the lexicographic order*: $\mathbf{a} \prec_{\text{lex}} \mathbf{b}$ if the leftmost nonzero entry of $\mathbf{b} - \mathbf{a}$ is positive. This ordering is important for elimination.
- *the degree lexicographic order*: $\mathbf{a} \prec_{\text{deglex}} \mathbf{b}$ if either $|\mathbf{a}| < |\mathbf{b}|$, or $|\mathbf{a}| = |\mathbf{b}|$ and the leftmost nonzero entry of $\mathbf{b} - \mathbf{a}$ is positive.
- *the degree reverse lexicographic order*: $\mathbf{a} \prec_{\text{revlex}} \mathbf{b}$ if either $|\mathbf{a}| < |\mathbf{b}|$, or $|\mathbf{a}| = |\mathbf{b}|$ and the rightmost nonzero entry of $\mathbf{b} - \mathbf{a}$ is negative.

All three orders satisfy $x_1 \succ x_2 \succ \cdots \succ x_n$, but they differ on monomials of higher degree. We recommend that the reader list the 10 quadratic monomials for $n = 4$ in each of the three orderings above.

Throughout this book we specify a monomial order by giving the name of the order and how the variables are sorted. For instance, we might say: "let $\prec$ denote the degree lexicographic order on $K[x, y, z]$ given by $y \prec z \prec x$". Other choices of monomial orders are obtained by assigning positive weights to the variables; see [**10**, Exercise 11 in §2.4]. We also note that any monomial order is a refinement of the coordinatewise partial order on $\mathbb{N}^n$:

$$\text{if } \mathbf{x}^{\mathbf{a}} \text{ divides } \mathbf{x}^{\mathbf{b}}, \quad \text{then } \mathbf{a} \preceq \mathbf{b}.$$

Remark 1.11. Fix a monomial order $\prec$ and let $\mathcal{M}$ be any nonempty subset of $\mathbb{N}^n$. Then $\mathcal{M}$ has a unique minimal element with respect to $\prec$. To show this, we apply Dickson's Lemma as in Corollary 1.9. Our set $\mathcal{M}$ has a finite, nonempty subset of minimal elements in the componentwise order on $\mathbb{N}^n$. This finite subset is linearly ordered by $\prec$. We select its minimal element.

We now fix a monomial order $\prec$. Given any nonzero polynomial $f \in K[\mathbf{x}]$, its *initial monomial* $\text{in}_\prec(f)$ is the $\prec$-largest monomial $\mathbf{x}^{\mathbf{a}}$ among those that appear in f with nonzero coefficient. To illustrate this for the orders above, let $n = 3$ with variable order $x \succ y \succ z$: Fix the polynomial $f = x^2 + xz^2 + y^3$. Then $\text{in}_{\prec_{\text{lex}}}(f) = x^2$, $\text{in}_{\prec_{\text{deglex}}}(f) = xz^2$ and $\text{in}_{\prec_{\text{revlex}}}(f) = y^3$.

For any ideal $I \subset K[\mathbf{x}]$, we define the *initial ideal* of I with respect to a given monomial order $\prec$ as follows:

$$\mathrm{in}_\prec(I) \;=\; \langle\, \mathrm{in}_\prec(f) \,:\, f \in I\backslash\{0\}\,\rangle.$$

This is a *monomial ideal*, i.e. it is generated by a set of monomials. A priori, this generating set is infinite. However, it turns out that we can always choose a finite subset that suffices to generate this monomial ideal.

Proposition 1.12. *Fix a monomial order $\prec$. Every ideal I in the polynomial ring $K[\mathbf{x}]$ has a finite subset $\mathcal{G}$ such that*

$$\mathrm{in}_\prec(I) \;=\; \langle\, \mathrm{in}_\prec(f) \,:\, f \in \mathcal{G}\,\rangle.$$

Such a finite subset $\mathcal{G}$ of I is called a Gröbner basis *for I with respect to $\prec$.*

Proof. Suppose no such finite set $\mathcal{G}$ exists. Then we can create a list of infinitely many polynomials $f_1, f_2, f_3, \ldots$ in I such that none of the initial monomials $\mathrm{in}_\prec(f_i)$ divides any other initial monomial $\mathrm{in}_\prec(f_j)$. This would be a contradiction to Dickson's Lemma (Theorem 1.8). □

We next show that every Gröbner basis actually generates its ideal.

Theorem 1.13. *If $\mathcal{G}$ is a Gröbner basis for an ideal I in $K[\mathbf{x}]$, then $I = \langle \mathcal{G} \rangle$.*

Proof. Suppose that $\mathcal{G}$ does not generate I. Among all polynomials f in the set $I\backslash\langle\mathcal{G}\rangle$, there exists an f whose initial monomial $\mathbf{x}^{\mathbf{b}} = \mathrm{in}_\prec(f)$ is minimal with respect to $\prec$. This follows from Remark 1.11. Since $\mathbf{x}^{\mathbf{b}} \in \mathrm{in}_\prec(I)$, there exists $g \in \mathcal{G}$ whose initial monomial divides $\mathbf{x}^{\mathbf{b}}$, say $\mathbf{x}^{\mathbf{b}} = \mathbf{x}^{\mathbf{c}} \cdot \mathrm{in}_\prec(g)$. Now, $f - \mathbf{x}^{\mathbf{c}} g$ is a polynomial with strictly smaller initial monomial. It lies in I but does not lie in the ideal $\langle\mathcal{G}\rangle$. This contradicts our choice of f. □

Corollary 1.14 (Hilbert's Basis Theorem). *Every ideal I in the polynomial ring $K[\mathbf{x}]$ is finitely generated.*

Proof. Fix any monomial order $\prec$. By Proposition 1.12, the ideal I has a finite Gröbner basis $\mathcal{G}$. By Theorem 1.13, the finite set $\mathcal{G}$ generates I. □

Gröbner bases are not unique. If $\mathcal{G}$ is a Gröbner basis of an ideal I with respect to a monomial order $\prec$, then so is every other finite subset of I that contains $\mathcal{G}$. In that sense, Gröbner bases differ from the bases we know from linear algebra. The issue of minimality and uniqueness is addressed next.

Definition 1.15. Fix I and $\prec$. A Gröbner basis $\mathcal{G}$ is *reduced* if the following two conditions hold:

(a) The leading coefficient of each polynomial $g \in \mathcal{G}$ is 1.

(b) For distinct $g, h \in \mathcal{G}$, no monomial in g is a multiple of $\mathrm{in}_\prec(h)$.

In what follows we fix an ideal $I \subset K[\mathbf{x}]$ and a monomial order $\prec$.

Theorem 1.16. *The ideal I has a unique reduced Gröbner basis for $\prec$.*

Proof idea. We refer to [**10**, §2.7, Theorem 5]. The idea is as follows. We start with any Gröbner basis $\mathcal{G}$ and turn it into a reduced Gröbner basis by applying the following steps. First we divide each $g \in \mathcal{G}$ by its leading coefficient to make it monic, so that (a) holds. We then remove from $\mathcal{G}$ all elements g whose initial monomial is not a minimal generator of $\text{in}_\prec(I)$. For any pair of polynomials with the same initial monomial, we delete one of them. Next we apply the division algorithm [**10**, §2.3] to any nonleading monomial until there are no more nonleading monomials divisible by any leading monomial. The resulting set is the reduced Gröbner basis. □

Let $\mathcal{S}_\prec(I)$ be the set of all monomials $\mathbf{x}^\mathbf{b}$ that are not in the initial ideal $\text{in}_\prec(I)$. We call these $\mathbf{x}^\mathbf{b}$ the *standard monomials* of I with respect to $\prec$.

Theorem 1.17. *The set $\mathcal{S}_\prec(I)$ of standard monomials is a basis for the K-vector space $K[\mathbf{x}]/I$.*

Proof. The image of $\mathcal{S}_\prec(I)$ in $K[\mathbf{x}]/I$ is linearly independent because every nonzero polynomial f whose image is zero in $K[\mathbf{x}]/I$ lies in the ideal I. Any such f has at least one monomial, namely $\text{in}_\prec(f)$, that is not in $\mathcal{S}_\prec(I)$.

We next prove that $\mathcal{S}_\prec(I)$ spans $K[\mathbf{x}]/I$. Suppose not. Then there exists a monomial $\mathbf{x}^\mathbf{c}$ which is not in the K-span of $\mathcal{S}_\prec(I)$ modulo I. We may assume that $\mathbf{x}^\mathbf{c}$ is minimal with respect to the monomial order $\prec$. Since $\mathbf{x}^\mathbf{c}$ is not in $\mathcal{S}_\prec(I)$, it lies in the initial ideal $\text{in}_\prec(I)$. Hence there exists $h \in I$ with $\text{in}_\prec(h) = \mathbf{x}^\mathbf{c}$. Each monomial in h other than $\mathbf{x}^\mathbf{c}$ is smaller with respect to $\prec$, so it lies in the K-span of $\mathcal{S}_\prec(I)$ modulo I. Hence $\mathbf{x}^\mathbf{c}$ lies in the K-span of $\mathcal{S}_\prec(I)$ modulo I. This is a contradiction. □

The most well-known tool for computing Gröbner bases is *Buchberger's algorithm* [**10**, §2.7]. Variants of this algorithm are implemented in all major computer algebra systems. The algorithm takes as its input a monomial order $\prec$ and a finite set $\mathcal{F}$ of polynomials in $K[\mathbf{x}]$. The output is the unique reduced Gröbner basis $\mathcal{G}$ for the ideal $I = \langle \mathcal{F} \rangle$ with respect to $\prec$. Experimenting with such an implementation is strongly recommended.

In what follows we present some examples of input-output pairs $(\mathcal{F}, \mathcal{G})$ for $n = 3$. Here we take the lexicographic monomial order with $x \succ y \succ z$.

Example 1.18. A computer algebra system, such as `Maple`, `Mathematica`, `Magma`, `Macaulay2`, or `Singular`, transforms the input $\mathcal{F} \subset \mathbb{Q}[x, y, z]$ into its reduced Gröbner basis $\mathcal{G}$. The initial monomials are always underlined:

- For $n = 1$, computing the reduced Gröbner basis means computing the greatest common divisor of the input polynomials: $\mathcal{F} = \{x^3-6x^2-5x-14, 3x^3+8x^2+11x+10, 4x^4+4x^3+7x^2-x-2\}$, $\mathcal{G} = \{\underline{x^2} + x + 2\}$.
- For linear polynomials, running Buchberger's algorithm amounts to Gaussian elimination. For $\mathcal{F} = \{2x + 3y + 5z + 7,\ 11x + 13y + 17z+19,\ 23x+29y+31z+37\}$, the reduced Gröbner basis is found to be $\mathcal{G} = \{\underline{x} - \frac{16}{7},\ \underline{y} + \frac{12}{7},\ \underline{z} + \frac{9}{7}\}$.
- The input $\mathcal{F} = \{xy - z, xz - y, yz - x\}$ yields the output $\mathcal{G} = \{\underline{x} - yz, \underline{y^2} - z^2, \underline{yz^2} - y, \underline{z^3} - z\}$. There are precisely five standard monomials: $\mathcal{S}_\prec(I) = \{1, y, z, yz, z^2\}$. The number five also occurred in Example 1.7, where we saw that $\mathcal{F}$ has five zeros in $\mathbb{C}^3$.
- Let the input be the curve in the (y, z)-plane parametrized by the two cubics $(x^3 - 4x,\ x^3 + x - 1)$ in one variable x. We write this as $\mathcal{F} = \{y - x^3 + 4x, z - x^3 - x + 1\}$. The Gröbner basis has the implicit equation of this curve as its second element: $\mathcal{G} = \{\,\underline{x} + \frac{1}{5}y+\frac{1}{5}z-\frac{1}{5},\ \underline{y^3}-3y^2z-3y^2+3yz^2+6yz+28y-z^3-3z^2+97z+99\,\}$.
- Let z be the sum of $x = \sqrt[3]{7}$ and $y = \sqrt[4]{5}$. We encode this in the set $\mathcal{F} = \{x^3 - 7, y^4 - 5, z - x - y\}$. The real number $z = \sqrt[3]{7} + \sqrt[4]{5}$ is algebraic of degree 12 over $\mathbb{Q}$. Its minimal polynomial is the first element in the Gröbner basis $\mathcal{G} = \{\underline{z^{12}} - 28z^9 - 15z^8 + 294z^6 - 1680z^5 + 75z^4 - 1372z^3 - 7350z^2 - 2100z + 2276, \ldots\}$.
- The elementary symmetric polynomials $\mathcal{F} = \{x + y + z, xy + xz + yz, xyz\}$ have the reduced Gröbner basis $\mathcal{G} = \{\,\underline{x} + y + z,\ \underline{y^2} + yz + z^2,\ \underline{z^3}\,\}$. There are six standard monomials. The quotient $\mathbb{Q}[x, y, z]/I$ is the regular representation of the symmetric group S_3.

For each of the six ideals above, what is the reduced Gröbner basis for the degree lexicographic order? What are the possible initial monomial ideals?

Many such examples boil down to the fact that lexicographic Gröbner bases are useful for eliminating variables. We shall see this in Theorem 4.5.

In general, the choice of monomial order can make a huge difference in the complexity of the reduced Gröbner basis, even for two input polynomials.

Example 1.19 (Intersecting two quartic surfaces in projective 3-space $\mathbb{P}^3$). A random homogeneous polynomial of degree 4 in $n = 4$ variables has 35 monomials. Consider the ideal I generated by two such random polynomials. If $\prec$ is the degree reverse lexicographic order, then the reduced Gröbner basis

$\mathcal{G}$ consists of 5 elements of degree up to 7. If $\prec$ is the lexicographic order, then $\mathcal{G}$ consists of 150 elements of degree up to 73.

Naturally, one uses a computer to find the 150 polynomials above. Many computer algebra systems offer an implementation of Buchberger's algorithm for Gröbner bases. We reiterate that readers are strongly encouraged to experiment with a computer algebra system while studying this book.

For an introduction to Buchberger's algorithm and many further details regarding Gröbner bases, we refer to the textbooks by Cox-Little-O'Shea [**10**], Greuel-Pfister [**23**] and Kreuzer-Robbiano [**30**]. In later chapters we shall freely use concepts from this area, such as S-polynomials and Buchberger's criterion. After all, our book is nothing but an "invitation".

1.3. Dimension and Degree

The two most important invariants of an ideal I in a polynomial ring $K[\mathbf{x}]$ are its dimension and its degree. We shall define these invariants, starting with the case of monomial ideals. In this section we focus on combinatorial aspects. The geometric interpretation will be presented in Chapter 2.

Definition 1.20 (Hilbert function). Let $I \subset K[\mathbf{x}]$ be a monomial ideal. The *Hilbert function* h_I takes nonnegative integers to nonnegative integers. The value $h_I(q)$ is the number of monomials of degree q *not* belonging to I.

A convenient way to represent a function $\mathbb{N} \to \mathbb{N}$ is by its generating function. This is a formal power series with nonnegative integer coefficients. The generating function for the Hilbert function is called the *Hilbert series*.

Definition 1.21 (Hilbert series). Let $I \subset K[\mathbf{x}]$ be a monomial ideal. We fix a formal variable z. The Hilbert series of I is the generating function

$$\mathrm{HS}_I(z) \quad = \quad \sum_{q=0}^{\infty} h_I(q) z^q.$$

We begin with the zero ideal $I = \{0\}$. We count all monomials in $K[\mathbf{x}]$.

Example 1.22. The Hilbert series of the zero ideal is the rational function

$$\mathrm{HS}_{\{0\}}(z) \quad = \quad \frac{1}{(1-z)^n} \quad = \quad \sum_{q=0}^{\infty} \binom{n+q-1}{n-1} z^q.$$

The number of monomials of degree q in n variables is $h_I(q) = \binom{n+q-1}{n-1}$. Note that the Hilbert function $h_{\{0\}}(q)$ is a polynomial of degree $n-1$ in q.

We next consider the case of a principal ideal.

Example 1.23. Let $I = \langle x_1^{a_1} \cdots x_n^{a_n} \rangle$, where $\sum_{i=1}^n a_i = e$. We must count monomials of degree q that are not divisible by the generator of I. To do this, we count all monomials and then subtract those that are in I. This yields

$$\mathrm{HS}_I(z) \quad = \quad \frac{1-z^e}{(1-z)^n} \quad = \quad \sum_{q=0}^{\infty} \left[\binom{n+q-1}{n-1} - \binom{n+q-e-1}{n-1} \right] z^q.$$

The second binomial coefficient is zero when $q < e$. For all $q \geq e$, the Hilbert function $h_I(q) = \binom{n+q-1}{n-1} - \binom{n+q-e-1}{n-1}$ is a polynomial in q of degree $n-2$. The highest-order term of this polynomial is found to be $\frac{e}{(n-2)!} q^{n-2}$.

Our third example concerns ideals generated by two monomials:

Example 1.24. Fix an ideal $I = \langle m_1, m_2 \rangle$ in $K[\mathbf{x}]$, where m_i is a monomial of degree e_i for $i = 1, 2$. We count the monomials in I of degree q by

(1) computing the number of monomials divisible by m_1,

(2) adding the number of monomials divisible by m_2, and

(3) subtracting the number of monomials divisible by both m_1 and m_2.

Step (3) concerns monomials that are divisible by the least common multiple $m_{12} = \mathrm{lcm}(m_1, m_2)$. Let e_{12} denote the degree of m_{12}. The Hilbert series is

$$\mathrm{HS}_I(z) \quad = \quad \frac{1 - z^{e_1} - z^{e_2} + z^{e_{12}}}{(1-z)^n}.$$

Therefore, the Hilbert function is an alternating sum of binomial coefficients:

$$h_I(q) \;=\; \binom{n+q-1}{n-1} - \binom{n+q-e_1-1}{n-1} - \binom{n+q-e_2-1}{n-1} + \binom{n+q-e_{12}-1}{n-1}.$$

This expression agrees with a polynomial in q, provided $q \geq e_{12}$.

Theorem 1.25. *The Hilbert series of a monomial ideal $I \subset K[\mathbf{x}]$ is*

$$\mathrm{HS}_I(z) \quad = \quad \frac{\kappa_I(z)}{(1-z)^n}, \tag{1.4}$$

where $\kappa_I(z)$ is a polynomial with integer coefficients and $\kappa_I(0) = 1$. There exists a polynomial HP *in one unknown q of degree $\leq n-1$, known as the* Hilbert polynomial *of the ideal I, such that $\mathrm{HP}(q) = h_I(q)$ for all values of the integer q that are sufficiently large.*

Proof. We prove this result by counting monomials. This is done using the inclusion-exclusion principle, as hinted at in the three examples above. Let $m_1, m_2, \ldots, m_r$ be the monomials that minimally generate I. For any subset τ of the index set $\{1, 2, \ldots, r\}$, we write m_τ for the least common multiple of the set $\{m_i : i \in \tau\}$, and we set $e_\tau = \mathrm{degree}(m_\tau)$. This includes the empty set $\tau = \emptyset$, for which $m_\emptyset = 1$ and $e_\emptyset = 0$. The desired numerator polynomial (1.4) can be written as alternating sums of 2^r powers of z:

$$\kappa_I(z) \quad = \quad \sum_{\tau \subseteq \{1,2,\ldots,r\}} (-1)^{|\tau|} \cdot z^{e_\tau}.$$

The cases $r = 0, 1, 2$ were seen above. The general case is inclusion-exclusion. Note that $\kappa_I \in \mathbb{Z}[z]$ with $\kappa_I(0) = 1$. By regrouping the terms of (1.4),

$$(1.5) \qquad h_I(q) \quad = \quad \sum_{\tau \subseteq \{1,2,\ldots,r\}} (-1)^{|\tau|} \binom{n+q-e_\tau-1}{n-1}.$$

This is a polynomial for $q \gg 0$. More precisely, the Hilbert function $h_I(q)$ equals the Hilbert polynomial $\mathrm{HP}_I(q)$ for all q that exceed $e_{\{1,2,\ldots,r\}}$. This bound is the degree of the least common multiple of all generators of I. □

Remark 1.26. The inclusion-exclusion principle carried out in the proof of Theorem 1.25 is a powerful idea, but it also hints at possible simplifications. We wrote the numerator polynomial $\kappa_I(z)$ and the Hilbert polynomial $\mathrm{HP}_I(q)$ as alternating sums of 2^r terms. However, in most applications r is much larger than n, and the vast majority of terms will cancel each other. Doing the correct bookkeeping leads us to the topic of *minimal free resolutions* of monomial ideals. This is a main theme in a subject area known as *combinatorial commutative algebra*. Yes, please google this.

Example 1.27. Let $n = 2$ and consider the monomial ideal

$$I \;=\; \langle x \rangle \cap \langle y \rangle \cap \langle x, y \rangle^{r+1} \;=\; \langle x^r y, x^{r-1}y^2, x^{r-2}y^3, \ldots, x^2y^{r-1}, xy^r \rangle.$$

Our formula for κ_I involves 2^r terms. After cancellations, only $2r$ remain:

$$\kappa_I(z) \;=\; 1 - rz^{r+1} + (r-1)z^{r+2}.$$

The Hilbert polynomial is the constant $\mathrm{HP}(q) \equiv 2$. This is also the value of the Hilbert function $h_I(q)$ for $q > r$. Note that $h_I(q) = q+1$ for $q \le r$.

Definition 1.28 (Dimension and degree)**.** Let I be a monomial ideal and write

$$\mathrm{HP}_I(q) \;=\; \frac{g}{(d-1)!} q^{d-1} + \text{lower-order terms in } q.$$

If the Hilbert polynomial HP is nonzero, the *dimension* of I is d and the *degree* of I is g. Here g is a positive integer. If $\mathrm{HP}_I(q) \equiv 0$ then we say that I is 0-dimensional. In this case, $K[\mathbf{x}]/I$ is a finite-dimensional K-vector space. We define the degree of I to be the dimension of that vector space.

Remark 1.29. The fact that g is a positive integer is a result in combinatorics. The proof is omitted here, but we revisit this theme in Chapter 13. From the inclusion-exclusion formulas above, one can show that the numerator of the Hilbert series factors as $\kappa_I(z) = \lambda_I(z) \cdot (1-z)^{n-d}$, where $\lambda_I(z)$ is also a polynomial with integer coefficients. The degree of I equals $g = \lambda_I(1)$.

Remark 1.30. It may seem artificial to define dimension and degree by distinguishing between the cases where the Hilbert polynomial is zero or not. However, if the zero polynomial has degree -1, then the two definitions are compatible. Furthermore, given any ideal I in $K[\mathbf{x}]$, the degree of $\tilde{I}$ in $K[\mathbf{x}, z]$, with one extra variable z but the same generators as I, remains the same. Hence, we could equivalently define the degree of I by adding z and extracting the leading coefficient of $\mathrm{HP}_{\tilde{I}}(q)$. This operation increases the dimension by 1.

Example 1.31. Let I be a principal ideal as in Example 1.23, generated by a monomial of degree $e > 0$. Then the dimension of I is $n-1$ and the degree of I is e. This follows from the formula we gave for the Hilbert function, which reveals that the Hilbert polynomial satisfies $\mathrm{HP}(I) = \frac{e}{(n-2)!}q^{n-2} + O(q^{n-3})$.

Example 1.32. Let $n = 2m$ be even and consider the monomial ideal

$$I \;=\; \langle\, x_1x_2,\, x_3x_4,\, x_5x_6,\, \ldots\,,\, x_{2m-3}x_{2m-2},\, x_{2m-1}x_{2m}\,\rangle.$$

The dimension of I equals m and the degree of I equals 2^m. It is instructive to work out the Hilbert series and the Hilbert polynomial of I for $m = 3, 4$.

We now consider an arbitrary ideal I in $K[\mathbf{x}]$. We no longer assume that I is generated by monomials. Let $\prec$ be any *degree-compatible* monomial order. This means that $|\mathbf{a}| < |\mathbf{b}|$ implies $\mathbf{a} \prec \mathbf{b}$ for all $\mathbf{a}$ and $\mathbf{b}$.

Lemma 1.33. *The number of standard monomials of I of degree q is independent of the choice of monomial order $\prec$, provided $\prec$ is degree-compatible.*

Proof. Let $K[\mathbf{x}]_{\leq q}$ denote the vector space of polynomials of degree $\leq q$. We write $I_{\leq q} := I \cap K[\mathbf{x}]_{\leq q}$ for the subspace of polynomials that lie in the ideal I. Also, consider the set of standard monomials of degree at most q:

$$\mathcal{S}_\prec(I)_{\leq q} \;=\; \mathcal{S}_\prec(I) \cap K[\mathbf{x}]_{\leq q}.$$

We claim that $\mathcal{S}_\prec(I)_{\leq q}$ is a K-vector space basis for the quotient space $K[\mathbf{x}]_{\leq q}/I_{\leq q}$. This set is linearly independent since no K-linear combination of $\mathcal{S}_\prec(I)$ lies in I. But, given that $\prec$ is degree-compatible, it also spans. This is because taking the normal form of a polynomial modulo the Gröbner basis can never increase the total degree. $\square$

Definition 1.34. Let I be an arbitrary ideal in a polynomial ring $K[\mathbf{x}]$. The function from $\mathbb{N}$ to $\mathbb{N}$ that associates to q the dimension of the quotient space $\dim K[\mathbf{x}]_{\leq q}/I_{\leq q}$ is known as the *affine Hilbert function.* We also define the *Hilbert function* h_I of I to be the Hilbert function of the initial ideal $\mathrm{in}_\prec(I)$, where $\prec$ is any degree-compatible term order. For all $q \in \mathbb{N}$ we have

$$\begin{aligned} h_I(q) \;=\; h_{\mathrm{in}_\prec(I)}(q) \;&=\; |\mathcal{S}_\prec(I)_{\leq q}| \;-\; |\mathcal{S}_\prec(I)_{\leq q-1}| \\ &=\; \dim(K[\mathbf{x}]_{\leq q}/I_{\leq q}) - \dim(K[\mathbf{x}]_{\leq q-1}/I_{\leq q-1}). \end{aligned}$$

This is the number of standard monomials whose degree equals q. This number is independent of $\prec$, thanks to Lemma 1.33. Thus the affine Hilbert function $q \mapsto \sum_{j=0}^{q} h_I(j)$ is determined by the Hilbert function and vice versa.

We also define the *Hilbert series* and the *Hilbert polynomial* to be the series and polynomial of any degree-compatible initial monomial ideal. Namely, we set

$$(1.6) \qquad \mathrm{HS}_I(z) := \mathrm{HS}_{\mathrm{in}_\prec(I)}(z) \quad \text{and} \quad \mathrm{HP}_I(q) := \mathrm{HP}_{\mathrm{in}_\prec(I)}(q).$$

We similarly define the *affine Hilbert series* and the *affine Hilbert polynomial* of I. These can also be defined by the formulas in (1.6), assuming we take $\mathrm{in}_\prec(I)$ in a polynomial ring $K[\mathbf{x}, y]$ that has one more dummy variable y.

We define the *dimension* and *degree* of I as the dimension and degree of $\mathrm{in}_\prec(I)$. These concepts are now well-defined, thanks to Lemma 1.33.

Example 1.35. Let I be a principal ideal generated by a polynomial f of degree e in $n \geq 1$ variables. The Hilbert series of I is $\mathrm{HS}_I(q) = \frac{1-q^e}{(1-q)^n}$. The affine Hilbert series equals $\frac{1-q^e}{(1-q)^{n+1}}$. The dimension of I is $n-1$. The degree of I is e. This follows from Example 1.31 because the singleton $\{f\}$ is a Gröbner basis and its initial monomial $\mathrm{in}_\prec(f)$ has degree e in any degree-compatible monomial order $\prec$.

Remark 1.36. The prefix "affine" for the Hilbert function is important in order to distinguish affine varieties from projective varieties. We will discuss these geometric concepts in Chapter 2. Later on in the book, and elsewhere in algebraic geometry, it will usually be clear from the context whether the affine version or the projective version is meant. However, in this chapter we want to be precise and make that distinction.

What we have accomplished in this section is to give a purely combinatorial definition of the dimension and degree of an ideal I. In Chapter 2 we shall see that this notion of dimension agrees with the intuitive one for the associated algebraic variety $\mathcal{V}(I)$. Namely, a variety has dimension 0 if and only if it consists of finitely many points. The number of these points is counted by the degree of the corresponding radical ideal. Likewise, the ideal of a curve has dimension 1, the ideal of a surface has dimension 2, etc. The degree is a measure of how curvy these shapes are. One can show that a prime ideal has degree 1 if and only if it is generated by linear polynomials.

Example 1.37. Fix the polynomial ring $K[x, y, z]$ and let $f = xyz - x^2 - y^2 - z^2 + 1$ as in (1.1). The ideal $\langle f \rangle$ has dimension 2 and degree 3. Let I be the ideal generated by its partial derivatives, as in Example 1.7. Then I has dimension 0 and degree 5. The ideal $I + \langle f \rangle$, whose zeros are the four singular points of the surface in Figure 1.1, has dimension 0 and degree 4.

Exercises

(1) Prove that an ideal in a polynomial ring $K[\mathbf{x}]$ is principal if and only if its reduced Gröbner basis is a singleton.

(2) Draw the plane curve $\{f = 0\}$ that is defined by polynomial $f = 5x^3 - 25x^2y + 25y^3 + 15xy - 50y^2 - 5x + 25y - 1$. What do you observe?

(3) For $n = 2$, define a monomial order $\prec$ such that $(2,3) \prec (4,2) \prec (1,4)$.

(4) Let $n = 2$ and fix the monomial ideals $I = \langle x, y^2 \rangle$ and $J = \langle x^2, y \rangle$. Compute the ideals $I + J$, $I \cap J$, IJ and $I^3J^4 = IIIJJJJ$. How many minimal generators does the ideal $I^{123}J^{234}$ have?

(5) The *radical* $\mathrm{rad}(I)$ of an ideal I in a ring R is the smallest radical ideal containing I. Prove that the radical of a primary ideal is prime.

(6) For ideals in a polynomial ring $K[\mathbf{x}]$, prove that
- the radical of a principal ideal is principal;
- the radical of a monomial ideal is a monomial ideal.

(7) Show that the following inclusions always hold and are strict in general:

$$\mathrm{rad}(I)\,\mathrm{rad}(J) \subseteq \mathrm{rad}(IJ) \quad \text{and} \quad \mathrm{in}_\prec(\mathrm{rad}(I)) \subseteq \mathrm{rad}(\mathrm{in}_\prec(I)).$$

(8) Using Gröbner bases, find the minimal polynomials of $\sqrt[5]{6} + \sqrt[7]{8}$ and $\sqrt[5]{6} - \sqrt[7]{8}$. This is analogous to the fifth item in Example 1.18.

(9) Find the implicit equation of the curve $\{(x^5 - 6, x^7 - 8) \in \mathbb{R}^2 : x \in \mathbb{R}\}$.

(10) Study the ideal $I = \langle x^3 - yz, y^3 - xz, z^3 - xy \rangle$. Is it radical? If not, find $\mathrm{rad}(I)$. Regarding I as a system of three equations, what are its solutions in $\mathbb{R}^3$?

(11) For the ideals I and $\mathrm{rad}(I)$ in the previous exercise, determine the Hilbert function, Hilbert series, Hilbert polynomial, dimension, and degree. Find these same objects and quantities preceded by the adjective "affine" where possible.

(12) Find an ideal in $\mathbb{Q}[x, y]$ whose reduced Gröbner basis (in lexicographic order) has cardinality 5 and there are precisely 19 standard monomials.

(13) Let I be the ideal generated by the n elementary symmetric polynomials in $x_1, \ldots, x_n$. Pick a monomial order and find the initial ideal $\mathrm{in}_\prec(I)$.

(14) Let X be a 2×2 matrix whose entries are variables. Let I_s be the ideal generated by the entries of the matrix power X^s for $s = 2, 3, 4, \ldots$. Investigate these ideals. What are the dimension and the degree of I_s?

(15) A symmetric 3×3 matrix has seven principal minors: three of size 1×1, three of size 2×2, and one of size 3×3. Does there exist an algebraic relation between these minors? Hint: Use the lexicographic Gröbner basis.

(16) Prove that if $\text{in}_\prec(I)$ is radical then I is radical. Does the converse hold?

(17) Determine all straight lines that lie on the cubic surface in Figure 1.1.

(18) Identify maximal, prime, radical and primary ideals in the ring $R = \mathbb{Z}$.

(19) Let I be the ideal generated by all 2×2 minors of a $2 \times n$ matrix filled with $2n$ variables. What are the degree and dimension of I for $n = 2, 3, 4$?

(20) Find a prime ideal I of degree 3 and dimension 1 in n variables for $n = 2$ and $n = 3$. In the latter case, we require further that $h_I(1) = 3$.

(21) Compute the dimension and degree of the ideal generated by two random homogeneous polynomials of degree 4 in $n = 4$ variables, as in Example 1.19. Next drop the hypothesis "homogeneous" and redo.

Chapter 2

Varieties

"*Geometry is but drawn algebra*", Sophie Germain

A *variety* is the set of solutions to a system of polynomial equations in several unknowns. Varieties are the main objects of study in algebraic geometry. They are the geometric counterparts of ideals in a polynomial ring. The latter live on the algebraic side. We distinguish between *affine varieties* and *projective varieties.* The former arise from arbitrary polynomials, while the latter are the zero sets of systems of *homogeneous* polynomials. Geometers prefer projective varieties because of their nice properties, explained in some of the results we present, such as Theorem 2.22. But, for starters, readers are invited to peruse the pictures shown in this chapter.

2.1. Affine Varieties

Algebraic varieties represent solutions to systems of polynomial equations. Fix a field K and consider polynomials $f_1, \ldots, f_k$ in $K[\mathbf{x}] = K[x_1, \ldots, x_n]$. The *variety* defined by these polynomials is the set of their common zeros:

$$\mathcal{V}(f_1, \ldots, f_k) \ := \ \{\, \mathbf{p} = (p_1, \ldots, p_n) \in K^n : f_1(\mathbf{p}) = \cdots = f_k(\mathbf{p}) = 0 \,\}.$$

Different sets of polynomials can define the same variety. For instance,

$$\mathcal{V}(f_1, f_2) \ = \ \mathcal{V}(f_1^2, f_2^5) \ = \ \mathcal{V}(f_1, f_1 + f_2). \tag{2.1}$$

Instead of thinking about the polynomials themselves, we consider the ideal they generate, $I = \langle f_1, \ldots, f_k \rangle$, and we define $\mathcal{V}(I) := \mathcal{V}(f_1, \ldots, f_k)$. A subset of K^n is a *variety* if it has the form $\mathcal{V}(I)$ for some ideal $I \subset K[\mathbf{x}]$.

Given any ideal $I \subset K[\mathbf{x}]$, by Hilbert's Basis Theorem (Corollary 1.14), we can always find a finite set of generators. By Exercise 1, the definition of $\mathcal{V}(I)$ does not depend on the choice of generators of I.

Remark 2.1. Two distinct ideals may define the same variety. Consider two nonconstant homogeneous polynomials f_1 and f_2. Then the ideal $\langle f_1^2, f_2^5 \rangle$ is strictly contained in the ideal $\langle f_1, f_2 \rangle = \langle f_1, f_1 + f_2 \rangle$. But these two distinct ideals define the same variety in (2.1). Chapter 6 on the Nullstellensätze deals with this issue for fields K that are either algebraically closed, like the complex numbers $K = \mathbb{C}$, or real closed, like the real numbers $K = \mathbb{R}$. In the latter case, our variety is unchanged if we pass to the principal ideal $\langle f_1^2 + f_2^2 \rangle$.

Algebraic geometry is the study of the geometry of varieties. As in many branches of mathematics, one considers the basic, irreducible building blocks for the objects of study. A variety $\mathcal{V}(I)$ is called *irreducible* if it cannot be written as a finite union of proper subvarieties in K^n. In symbols, $\mathcal{V}(I)$ is irreducible if and only if for any ideals J and J' in $K[\mathbf{x}]$ we have

$$\mathcal{V}(I) = \mathcal{V}(J) \cup \mathcal{V}(J') \implies \mathcal{V}(I) = \mathcal{V}(J) \text{ or } \mathcal{V}(I) = \mathcal{V}(J').$$

Any variety can be decomposed into irreducible varieties. The relevant algebraic tool is primary decomposition. This is the topic of the next chapter.

Example 2.2. Consider the ideal $I = \langle xy \rangle \subset \mathbb{R}[x, y]$. Its variety $\mathcal{V}(I) = \mathcal{V}(x) \cup \mathcal{V}(y)$ is a union of two lines in the plane $\mathbb{R}^2$. Hence, this is a reducible variety. Algebraically, I is the intersection of two larger ideals $\langle x \rangle$ and $\langle y \rangle$. Their respective varieties $\mathcal{V}(x)$ and $\mathcal{V}(y)$ are irreducible. This follows from Proposition 2.3 because $\langle x \rangle$ and $\langle y \rangle$ are prime ideals.

For any field K, we can turn K^n into a topological space, using the *Zariski topology*. In this topology, the closed sets are the varieties in K^n. In this setting, the definition of an irreducible variety coincides with the definition of an irreducible topological space. If $K = \mathbb{R}$ or $K = \mathbb{C}$, then we also have the classical Euclidean topology on K^n. The Euclidean topology is much finer than the Zariski topology because it has many more open sets.

Our aim is to relate geometric properties of the variety $\mathcal{V}(I)$ to algebraic properties of the ideal I. Consider a maximal ideal of the form $M := \langle x_1 - p_1, \ldots, x_n - p_n \rangle$ in $K[\mathbf{x}]$. The point $(p_1, \ldots, p_n)$ lies in $\mathcal{V}(I)$ if and only if $I \subseteq M$. Given any subset $W \subset K^n$, we consider the set of all polynomials that vanish on W. This set is a radical ideal, denoted by

$$\mathcal{I}(W) \; := \; \{ f \in K[\mathbf{x}] \, : \, f(\mathbf{p}) = 0 \text{ for all } \mathbf{p} \in W \}.$$

The set W is a variety if and only if $W = \mathcal{V}(\mathcal{I}(W))$. Furthermore, given any two varieties V and W in K^n, we have $V \subseteq W$ if and only if $\mathcal{I}(W) \subseteq \mathcal{I}(V)$.

Proposition 2.3. *A variety $W \subset K^n$ is irreducible if and only if its ideal $\mathcal{I}(W)$ is prime.*

Proof. Suppose $\mathcal{I}(W)$ is prime and $W = \mathcal{V}(J) \cup \mathcal{V}(J')$. If $W \neq \mathcal{V}(J)$ then there exists $f \in J$ and $v \in W$ such that $f(v) \neq 0$. Therefore, $f \notin \mathcal{I}(W)$. For any $g \in J'$ we know that fg vanishes on $\mathcal{V}(J)$ and $\mathcal{V}(J')$, hence on W. Thus $fg \in \mathcal{I}(W)$. As $\mathcal{I}(W)$ is prime, we have $g \in \mathcal{I}(W)$. We conclude that $J' \subseteq \mathcal{I}(W)$. By Exercise 2, this implies $W = \mathcal{V}(\mathcal{I}(W)) \subseteq \mathcal{V}(J')$.

For the converse, suppose that W is irreducible and $fg \in \mathcal{I}(W)$. Hence

$$W \;=\; W \cap \mathcal{V}(fg) \;=\; W \cap (\mathcal{V}(f) \cup \mathcal{V}(g)) \;=\; (W \cap \mathcal{V}(f)) \cup (W \cap \mathcal{V}(g)).$$

Without loss of generality, $W = W \cap \mathcal{V}(f)$. This means that $W \subseteq \mathcal{V}(f)$ and hence $f \in \mathcal{I}(W)$. This argument proves that $\mathcal{I}(W)$ is a prime ideal. □

Remark 2.4. Proposition 2.3 relates geometry and number theory. Prime ideals in a polynomial ring $K[\mathbf{x}]$ correspond to irreducible varieties, while prime ideals in the ring of integers $\mathbb{Z}$ correspond to prime numbers (or zero). Hence, irreducible varieties are to varieties what primes are to integers.

Prime ideals appear in applications as the constraints satisfied by a *generative model.* Such models are common in statistics. One considers a vector θ of real parameters and expresses probabilities (or moments of densities) as functions of θ. These functions are often polynomials or rational functions in θ, and one is interested in finding all valid polynomial constraints among the probabilities in question. Geometrically, this corresponds to computing the closure (in the Zariski topology) of the image of a polynomial map. This closure is an irreducible variety, so its ideal is prime by Proposition 2.3. That prime ideal represents the image and hence the generative model. It is computed as the kernel of the ring map dual to the polynomial map.

Example 2.5. We give an illustration for the most basic generative model, namely the *independence model* for two random variables X and Y, each with state space $\{1, \ldots, m\}$. Possible probability distributions of X (resp. Y) are represented by vectors $(p_1, \ldots, p_m)$ (resp. $(q_1, \ldots, q_m)$) in $\mathbb{R}^m$ with nonnegative entries that sum to 1. The probability that X (resp. Y) is in state i is p_i (resp. q_i). The joint random variable (X, Y) has m^2 states. Under the assumption that X and Y are independent, all possible probability distributions of (X, Y) belong to a variety in $\mathbb{R}^{m^2}$.

Consider the map that takes a distribution of X and a distribution of Y to the joint distribution of (X, Y). This map extends to a polynomial map

$$\begin{array}{ccc} \mathbb{R}^m \times \mathbb{R}^m & \rightarrow & \mathbb{R}^{m^2}, \\ (p_1, \ldots, p_m, q_1, \ldots, q_m) & \mapsto & (p_1q_1, p_1q_2, \ldots, p_1q_m, p_2q_1, \ldots, p_mq_m). \end{array} \tag{2.2}$$

In statistics one incorporates the requirement $\sum p_i = \sum q_i = 1$. We do so by restricting the domain. We write the resulting map explicitly for $m = 3$:

$$(p_1, p_2, q_1, q_2) \quad \mapsto \quad \big(p_1q_1,\, p_1q_2,\, p_1(1-q_1-q_2),\, p_2q_1,\, p_2q_2,\\ p_2(1-q_1-q_2), (1-p_1-p_2)q_1, (1-p_1-p_2)q_2, (1-p_1-p_2)(1-q_1-q_2)\big). \tag{2.3}$$

In Exercise 9 we ask for the variety and ideal given by the image of this map.

The algebraic study of the independence model was a point of departure for the development of *algebraic statistics*. In that subject one employs prime ideals to represent statistical models. This allows the use of algebraic invariants (such as dimension and degree) and algebraic methods (such as Gröbner bases) for data analysis and inference. Readers wishing to learn more about algebraic statistics should consult the textbooks [**19, 43, 57**].

We have argued that prime ideals are basic building blocks in algebraic geometry and its applications. This motivates the following definition on the algebra side. We now take R to be any commutative ring with unity.

Definition 2.6. The *spectrum* of the ring R is the set of proper prime ideals:

$$\mathrm{Spec}(R) := \big\{ p \subsetneq R \,:\, p \text{ is a prime ideal} \big\}.$$

This set is a topological space with the Zariski topology. Its closed sets are the *varieties* $\mathcal{V}(I) = \big\{ p \in \mathrm{Spec}(R) : I \subseteq p \big\}$ where I is any ideal in R.

The spectrum of the ring remembers a lot of information: all prime ideals and how they are related geometrically. Our most basic example of a ring R is the polynomial ring $K[\mathbf{x}]$ in n variables over a field K. Its spectrum is a topological space with a very rich structure. Among the points of the spectrum are the usual ones, written $(p_1, \ldots, p_n) \in K^n$. These correspond to maximal ideals of the form $\langle x_1 - p_1, \ldots, x_n - p_n \rangle$. However, the spectrum $\mathrm{Spec}(K[\mathbf{x}])$ has points corresponding to *all* irreducible subvarieties of K^n, not just those of dimension 0. In this manner, K^n is a proper subset of $\mathrm{Spec}(K[\mathbf{x}])$. Exercise 4 asks you to prove that the Zariski topology on K^n is the one induced from the Zariski topology on $\mathrm{Spec}\, K[\mathbf{x}]$. An example of a point in $\mathrm{Spec}\, \mathbb{R}[x]$ that is not of the form $\langle x - a \rangle$ for $a \in \mathbb{R}$ is the maximal ideal $\langle x^2 + 1 \rangle$. Indeed, $\mathbb{R}[x]/\langle x^2 + 1 \rangle \simeq \mathbb{C}$ by identifying x with i.

Our primary example is the *coordinate ring* $R = K[W]$ of a subvariety $W \subset K^n$. By definition, this is the quotient ring $R = K[\mathbf{x}]/J$ where $J = \mathcal{I}(W)$ is the radical ideal in the polynomial ring $K[\mathbf{x}]$ that encodes the variety. We interpret elements of $K[W]$ as polynomial functions on W. Indeed, as elements of J vanish on W, the evaluation of $f \in K[W]$ at a point of W does not depend on the choice of the representative polynomial. The prime ideals in $K[W]$ are in natural bijection with the prime ideals in $K[\mathbf{x}]$ that contain J. Geometrically, these correspond to irreducible subvarieties of W. Among these are the points $(p_1, \ldots, p_n) \in W$, which correspond to maximal ideals $\langle x_1 - p_1, \ldots, x_n - p_n \rangle$ in $K[W]$, just like before. The Zariski topologies on W and $\mathrm{Spec}(K[W])$ are compatible, in the sense of Exercise 4.

Example 2.7. A paraboloid in $\mathbb{R}^3$ is defined by the equation $z = x^2 + y^2$. Its ideal is $J = \langle z - x^2 - y^2 \rangle$. The coordinate ring of the paraboloid equals $\mathbb{R}[x, y, z]/J$. Elements of this ring represent polynomial functions on the paraboloid. What are the Gröbner bases of J and what are the standard monomials? What about the dimension and the degree?

We briefly discuss the points in $\mathrm{Spec}\,\mathbb{R}[x, y, z]/J$. Recall that this is a topological space with the Zariski topology. First, there are the classical real points on the surface. Second, there are pairs of complex conjugate points satisfying the equation $z = x^2 + y^2$. Next, there are all irreducible curves lying on the surface, one for each nonmaximal prime ideal of $\mathbb{R}[x, y, z]$ that strictly contains J. This includes curves that lie on the surface but have no real points, such as that for $\langle z^2 + 1, x^2 + y^2 - z \rangle$. Finally, there is the *generic point* corresponding to the zero ideal in $\mathbb{R}[x, y, z]/J$ or equivalently to the ideal J in $\mathbb{R}[x, y, z]$. Note that the closure of the generic point is the whole space $\mathrm{Spec}\,\mathbb{R}[x, y, z]/J$.

Remark 2.8. We continue the analogy from Remark 2.4 in order to provide geometric intuition for the *Chinese Remainder Theorem*. Fix $n_1, \ldots, n_k \in \mathbb{Z}$ that are pairwise coprime. In the language of varieties, the fact that $\langle n_i \rangle + \langle n_j \rangle = \mathbb{Z}$ is equivalent to the fact that the associated varieties do not intersect—recall that the ideal of the intersection is the sum of ideals. For each n_i we are given a residue class $a_i \in \mathbb{Z}/n_i\mathbb{Z}$, i.e. a function on the variety associated to n_i. As these varieties do not intersect, we expect to obtain a unique function on their union that restricts to the given functions on each piece.

The union of varieties is given by the intersection of the ideals, which corresponds to the product $N = \prod_{i=1}^{k} n_i$. This is precisely the Chinese Remainder Theorem: There exists a unique $a \in \mathbb{Z}/N\mathbb{Z}$ such that $a = a_i \bmod n_i$. We can push the analogy further. If the varieties intersect (i.e. the numbers are not pairwise coprime), then we expect the global function to exist if and only if the functions associated to the varieties agree on intersections.

Consider two varieties W_1 and W_2 and a map $f : W_1 \to W_2$ between them. Given a function on the target variety, say $g : W_2 \to K$, we define its pull-back to be the function $f^*(g) := g \circ f$ from W_1 to K. Of course, here we are interested in polynomial functions, so g is an element of the ring $K[W_2]$. Likewise, we want the pullback $f^*(g)$ to be an element of the ring $K[W_1]$.

Hence, given any polynomial map $f : W_1 \to W_2$ between varieties, we would like the map $f^* : K[W_2] \to K[W_1]$ to be a well-defined ring morphism. In Exercise 5 you will show that any ring morphism $K[W_2] \to K[W_1]$ induces a continuous map of topological spaces $\operatorname{Spec} K[W_1] \to \operatorname{Spec} K[W_2]$. Hence, we may think of maps between varieties as homomorphisms between their rings of functions in the opposite direction. In the language of *category theory*, our star $*$ is a *contravariant functor* from varieties to rings. It furnishes an *equivalence of categories* between irreducible affine varieties (over K) and finitely generated K-algebras that are integral domains.

Example 2.9. The following ring homomorphism is an isomorphism:

$$f^* : \mathbb{R}[x, y]/\langle y - x^2 \rangle \ \to \ \mathbb{R}[z], \quad x \mapsto z,\, y \mapsto z^2.$$

It arises from a map of varieties that takes a line to a parabola in the plane:

$$f : \mathbb{R} \to \mathcal{V}(y - x^2) \subset \mathbb{R}^2, \quad \lambda \mapsto (\lambda, \lambda^2).$$

Under this parametrization of the parabola, the coordinate functions x and y on $\mathbb{R}^2$ pull back to the functions z and z^2 on the line $\mathbb{R}$. We use the letter λ in the parametrization for extra clarity. It can get confusing when you pass to the map of spectra, a continuous map in the Zariski topologies.

Remark 2.10. Textbooks in algebraic geometry usually define affine varieties to be $\operatorname{Spec} R$, with its Zariski topology, for any (commutative, with unity) ring R. Here R need not be a finitely generated algebra over a field K. For us, in this book, affine varieties are zero sets of polynomials in $K[\mathbf{x}]$.

The dependence on the field is crucial for geometric properties of maps. Consider the squaring map $K^1 \to K^1, \lambda \mapsto \lambda^2$, from the affine line to itself.

- If $K = \mathbb{C}$ then the squaring map is surjective.
- If $K = \mathbb{R}$ then its image is the set of nonnegative real numbers. In both cases, the Zariski closure of the image is the whole line.
- If $K = \mathbb{F}_p$ and $p \neq 2$, then the image is a proper subset of K^1. It is Zariski closed. What if we replace K by its algebraic closure?
- In each case, is the map $\operatorname{Spec}(K[x]) \to \operatorname{Spec}(K[x])$ surjective?

From the perspective of spectra, it is instructive to study the ideal $I = \langle x^2 + 1 \rangle$ in $K[x]$. Exercise 6 asks the reader to give a description of $\mathcal{V}(I)$.

Example 2.11. Consider the three ideals $I_1 = \langle x^2 - y^2 \rangle$, $I_2 = \langle x^2 - 2y^2 \rangle$ and $I_3 = \langle x^2 + y^2 \rangle$ in $K[x, y]$. The first one is not prime for any K. The second one is not prime for $K = \mathbb{R}$ or $\mathbb{C}$, but it is a prime ideal when $K = \mathbb{Q}$. The ideal I_3 is not prime for $K = \mathbb{C}$, but it is a prime ideal for $K = \mathbb{Q}$ or $\mathbb{R}$.

We prove the last statement. Suppose $fg \in I_3 \subset \mathbb{R}[x,y]$. This means that $fg = (x^2+y^2)\cdot h$, where $f, g, h \in \mathbb{R}[x,y]$. By the Fundamental Theorem of Algebra, every homogeneous polynomial p in two variables has a unique (up to multiplication by constants) representation as a product of linear forms with complex coefficients $p = \prod l_j$. If p has real coefficients, then the decomposition is stable under complex conjugation: For every j, either l_j has real coefficients or $\overline{l_j}$ must also appear in the decomposition. We have $x^2+y^2 = (x+iy)(x-iy)$. In the ring $\mathbb{C}[x,y]$, without loss of generality, we may assume $(x+iy)|f$. But then, by the above argument, also $(x-iy)|f$. So $f = (x+iy)(x-iy)\prod_i \tilde{l}_i$ for $\tilde{l}_i \in \mathbb{C}[x,y]$. However, $\prod_i \tilde{l}_i$ is stable under conjugation, i.e. defines a real polynomial. So x^2+y^2 divides f in $\mathbb{R}[x,y]$.

Many models arise in applications as images of polynomial maps. It is important to note that the image need not be closed if $K = \mathbb{C}$. Also, it need not be dense in its Zariski closure if $K = \mathbb{R}$. This will be discussed in detail in Chapter 4. The following definition plays an important role.

Definition 2.12. A subset $A \subset K^n$ is *constructible* if it can be described as a finite union of differences of varieties. Over the real numbers, a subset $B \subset \mathbb{R}^n$ is *semialgebraic* if it can be described as the set of solutions of a finite system of polynomial inequalities (which may involve both $\geq$ and $>$) or a finite union of such.

Remark 2.13. Every constructible subset of $\mathbb{R}^n$ is semialgebraic, but the converse is not true. See below. The complement of a constructible set is constructible, and the complement of a semialgebraic set is semialgebraic.

Example 2.14. Take $n = 2$ and $K = \mathbb{R}$. The singleton $\mathcal{V}(x,y) = \{(0,0)\}$ is constructible and hence so is $\mathbb{R}^2 \backslash \{(0,0)\} = \mathcal{V}(0) \backslash \mathcal{V}(x,y)$. The orthant $\mathbb{R}^2_{\geq 0} = \{(u,v) \in \mathbb{R}^2 : u \geq 0 \text{ and } v \geq 0\}$ is semialgebraic. But it is not constructible, because the Euclidean closure of a constructible set is a variety. Its complement $B = \{(u,v) \in \mathbb{R}^2 : u < 0 \text{ or } v < 0\}$ is also semialgebraic. Can you write B as the set of solutions to a finite list of polynomial inequalities?

The two most important invariants of a variety V in K^n are its *dimension* and its *degree*. We defined these in Section 1.3, via the ideal $\mathcal{I}(V) \subset K[\mathbf{x}]$.

Example 2.15. Let V be a linear subspace of K^n. The dimension of V as a variety equals its dimension as a linear space. The degree of V is 1. Indeed, we may assume $\mathcal{I}(V) = \langle x_1, \ldots, x_s \rangle$ where $s = n - \dim(V)$. The result follows from Example 1.22 because $K[\mathbf{x}]/\langle x_1, \ldots, x_s \rangle \simeq K[x_{s+1}, \ldots, x_n]$.

Here is an important fact: If $V_1 \subsetneq V_2$ then $\dim(V_1) \leq \dim(V_2)$. The inequality is strict if V_2 is irreducible. The latter is not easy to prove from the definition we gave. It helps to consult a textbook in commutative algebra for alternative (but equivalent) definitions of dimension, e.g. [**3**, Chapter 11].

Here is a method for computing the dimension of a variety $\mathcal{V}(I)$. Note that $\mathcal{V}(\mathrm{in}_\prec(I))$ is a union of linear spaces in K^n, for any monomial order $\prec$.

(1) Compute a Gröbner basis of I and hence the monomial ideal $\mathrm{in}_\prec(I)$.
(2) Let $m_1, \ldots, m_k$ be monomials that generate $\mathrm{in}_\prec(I)$. Find the smallest (with respect to cardinality) set of variables $S = \{x_{i_1}, \ldots, x_{i_d}\}$ such that every generator m_j is divisible by some variable in S.
(3) The difference $n - d$ is the dimension of both $\mathcal{V}(\mathrm{in}_\prec(I))$ and $\mathcal{V}(I)$.

The second most important invariant of a variety $V = \mathcal{V}(I) \subset K^n$ is the degree. We now provide its geometric interpretation. A general subspace $L \subset K^n$ with $\dim(L) + \dim(V) = n$ intersects V in finitely many points. For K algebraically closed, their number is the degree of V. Indeed, this follows inductively from the fact that a general linear polynomial is not a zero divisor in $K[V]$. Adding it to I changes the Hilbert function in such a way that the dimension drops by 1 and the degree remains the same. If $\dim(I) = 0$ and I is radical, then the degree is the cardinality of $V = \mathcal{V}(I)$.

The following theorem is a generalization of the above considerations to polynomials of arbitrary degree. Its first appearance, for two polynomials in two variables, goes back to Isaac Newton.

Theorem 2.16 (Bézout's Theorem). *Let $f_1, \ldots, f_k$ be general polynomials in n variables of degrees $d_1, \ldots, d_k > 0$. For $I = \langle f_1, \ldots, f_k \rangle$ we have $\dim(I) = n - k$ and $\mathrm{degree}(I) = d_1 \cdots d_k$.*

As the reader may have guessed, the crucial point is that each f_i is not a zero divisor in the ring $K[\mathbf{x}]/\langle f_1, \ldots, f_{i-1}\rangle$. For the complete proof we refer to [**47**, §IV.2]. Note that we always have $\dim\langle f_1, \ldots, f_k\rangle \geq n - k$. This follows from *Krull's Principal Ideal Theorem*, a result in commutative algebra. If equality holds, then the ideal I is called a *complete intersection*.

Some points on a variety are singular, such as the four nodes of the cubic surface in Figure 1.1. Our aim is now to discuss singularities in general. We start with the case of a hypersurface defined by one polynomial $f \in K[\mathbf{x}]$. A point $p \in \mathcal{V}(f)$ is *singular* if all partial derivatives vanish, i.e. $\frac{\partial f}{\partial x_i}(p) = 0$ for $i = 1, \ldots, n$. Thus the *singular locus* of f is the variety of the ideal $\langle f, \frac{\partial f}{\partial x_1}, \ldots, \frac{\partial f}{\partial x_n}\rangle$. If this ideal has no zeros, we say that the hypersurface $\mathcal{V}(f)$ is *smooth*. Smoothness is a very important condition. It tells us that our variety can be locally approximated by a linear space: the *tangent space*.

Let $I = \langle f_1, f_2, \ldots, f_k\rangle \subset K[\mathbf{x}]$ be a prime ideal defining an irreducible variety $Y = \mathcal{V}(I)$ in K^n of dimension d. A point $p \in Y$ is *singular* if and only if the rank of the Jacobian matrix at p is smaller than the codimension:

$$\operatorname{rank}\begin{pmatrix} \frac{\partial f_1}{\partial x_1} & \cdots & \frac{\partial f_1}{\partial x_n} \\ \frac{\partial f_2}{\partial x_1} & \cdots & \frac{\partial f_2}{\partial x_n} \\ \vdots & \ddots & \vdots \\ \frac{\partial f_k}{\partial x_1} & \cdots & \frac{\partial f_k}{\partial x_n} \end{pmatrix}(p) \;<\; n-d.$$

This definition does not depend on the choice of ideal generators (Exercise 14). A point that is not singular is called *smooth*. For a smooth point the inequality above turns into an equality. The singular locus $\operatorname{Sing}(Y)$ is a variety in K^n. It may be defined by the ideal that is the sum of I and the ideal generated by $(n-d)\times(n-d)$ minors of the Jacobian matrix. The kernel of this matrix evaluated at the point p is, by definition, the vector space parallel to the tangent space to Y at p. The definition of smooth point assures that the tangent space and the variety have the same dimension.

If a variety $X \subset K^n$ is reducible and p lies in more than one irreducible component of X, then p is singular in X. If p belongs to a unique irreducible component Y, then p is singular in X if and only if it is singular in Y.

By Bézout's Theorem we know the dimension and degree of an ideal I that is generated by general polynomials. *Bertini's Theorem* tells us that such an ideal I is prime and its variety $\mathcal{V}(I)$ is smooth, provided $\dim(I) > 0$.

2.2. Projective Varieties

The geometric objects we have encountered so far are subsets of K^n. We called them *varieties*, but more precisely we should refer to them as *affine varieties*. We now change our perspective by focusing on *projective varieties*.

We start by recalling the construction of the projective space $\mathbb{P}(V)$ over a vector space V of dimension $n+1$. The points of $\mathbb{P}(V)$ are the lines through the origin in V. The symbol $[a_0 : \cdots : a_n] \in \mathbb{P}(V)$ denotes the line that also goes through the point $(a_0,\ldots,a_n) \in V$. Here not all the a_i are zero. Formally, $\mathbb{P}(V)$ is the set of equivalence classes $[v]$, for $v \in V\backslash\{0\}$, modulo the relation $v_1 \sim v_2$ if and only if $v_1 = \lambda v_2$ for some $\lambda \in K^* = K\backslash\{0\}$. For the topological construction over $\mathbb{R}$ or $\mathbb{C}$, we note that each line through the origin in V intersects the unit sphere at precisely two points. Thus $\mathbb{P}(V)$ may be regarded as a quotient of the sphere, identifying antipodal points. In particular, $\mathbb{P}(V)$ is compact in the classical topology. On the subset $S_i = \{a_i \neq 0\}$ of $\mathbb{P}(V)$ we rescale to get $a_i = 1$. We thus identify S_i with K^n. The affine spaces $S_i = K^n$ cover $\mathbb{P}^n := \mathbb{P}(V)$, because every point has some nonzero coordinate. We obtain $\mathbb{P}^n$ by glueing these $n+1$ charts.

As before, we are interested in functions on $\mathbb{P}^n$. The first problem is that it does not make sense to evaluate a polynomial f on $[a_0 : \cdots : a_n]$, as the result depends on the choice of representative. It may even happen that f vanishes for some representatives but not for others. Thus, we focus on *homogeneous* polynomials, i.e. linear combinations of monomials of fixed degree. If f is a homogeneous polynomial of degree d in $n+1$ variables, then $f(ta_0, \ldots, ta_n) = t^d f(a_0, \ldots, a_n)$. In particular, f vanishes on some representative of $[a_0 : \cdots : a_n]$ if and only if it vanishes on any representative.

Let $f_1, \ldots, f_k$ be homogeneous polynomials in $K[\mathbf{x}]$. They are allowed to have distinct degrees. We define the associated *projective variety*

$$\mathcal{V}(f_1, \ldots, f_k) = \big\{[a_0 : \cdots : a_n] \in \mathbb{P}(V) : f_i(a_0, \ldots, a_n) = 0 \text{ for } i = 1, \ldots, k\big\}.$$

An ideal I in $K[\mathbf{x}]$ is *homogeneous* if it is generated by homogeneous polynomials $f_1, \ldots, f_k$. Just like in the affine case, we set $\mathcal{V}(I) := \mathcal{V}(f_1, \ldots, f_k)$.

Remark 2.17. Homogeneous ideals contain (many) nonhomogeneous polynomials. For instance, $\langle x + y^2, y\rangle$ is a homogeneous ideal. See Exercise 11.

For any projective variety $X \subset \mathbb{P}^n$ one defines the *affine cone* $\hat{X}$ over it, i.e. the variety defined by the same ideal but in $V = K^{n+1}$. The dimension and degree of a projective variety can be defined via its affine cone:

$$\dim(X) := \dim(\hat{X}) - 1 \quad \text{and} \quad \operatorname{degree}(X) := \operatorname{degree}(\hat{X}). \tag{2.4}$$

It is usually preferable to work with projective varieties. Algebraic geometry is simpler in $\mathbb{P}^n$ than in K^n. For instance, parallel lines in K^2 do not intersect, but any two lines in $\mathbb{P}^2$ intersect. If X is any projective variety of degree ≥ 2, then the affine cone $\hat{X}$ is always singular at the point $0 \in V$. However, if this is the only singular point of $\hat{X}$, then $X \subset \mathbb{P}^n$ is smooth.

If Y is any variety in K^n, then there is an associated projective variety $\bar{Y}$ in $\mathbb{P}^n$, called the *projective closure* of Y, which is defined via its ideal. If $I \subset K[x_1, \ldots, x_n]$ is the ideal of Y, then the ideal $\bar{I}$ of $\bar{Y}$ lives in $K[x_0, x_1, \ldots, x_n]$. It is generated by the following infinite set of homogeneous polynomials:

$$\left\{ x_0^{\deg(g)} \cdot g\Big(\frac{x_1}{x_0}, \ldots, \frac{x_n}{x_0}\Big) \ : \ g \in I \right\}. \tag{2.5}$$

Here is an algorithm for computing the ideal $\bar{I}$ of the projective closure $\bar{Y}$.

Proposition 2.18. *Let I be an ideal in $K[x_1, \ldots, x_n]$ and let $\mathcal{G}$ be its reduced Gröbner basis for a degree-compatible monomial ordering. Then $\bar{I}$ is generated by the homogeneous polynomials in* (2.5) *where g runs over $\mathcal{G}$.*

Proof. Let $f = f(x_0, x_1, \ldots, x_n)$ be any homogeneous polynomial in $\bar{I}$. Suppose $\mathcal{G} = \{g_1, g_2, \ldots, g_s\}$. The dehomogenization $f(1, x_1, \ldots, x_n)$ lies in

I and hence its normal form modulo the Gröbner basis $\mathcal{G}$ is zero. This gives a representation

$$f(1, x_1, \ldots, x_n) \quad = \quad \sum_{i=1}^{s} h_i(x_1, \ldots, x_n) g_i(x_1, \ldots, x_n),$$

where $\deg(h_i g_i) \leq \deg(f)$ for all i. By homogenizing the summands in this identity we obtain

$$f(x_0, x_1, \ldots, x_n) \quad = \quad \sum_{i=1}^{s} x_0^{\deg(f)} \cdot h_i(\frac{x_1}{x_0}, \ldots, \frac{x_n}{x_0}) \cdot g_i(\frac{x_1}{x_0}, \ldots, \frac{x_n}{x_0}).$$

Hence, f lies in the ideal generated by the set (2.5) with I replaced by $\mathcal{G}$. □

Corollary 2.19. *We fix the degrevlex monomial ordering. The dimension and degree of an affine variety $Y \subset K^n$ are preserved when passing to its projective closure $\bar{Y} \subset \mathbb{P}^n$:*

$$\dim(\bar{Y}) \; = \; \dim(Y) \quad \text{and} \quad \text{degree}(\bar{Y}) \; = \; \text{degree}(Y).$$

Proof. The initial ideal of $\bar{I}$ and the initial ideal of I have the same generators. These monomials in $x_1, \ldots, x_n$ determine the dimension and degree. Note that the ideal $\bar{I}$ belongs to a ring with one more variable which increases the dimension by 1 and does not change the degree. But, by the definition in (2.4), the dimension of the projective variety is 1 less than that of the affine cone over it. This implies the corollary. □

Example 2.20. Let I be the ideal generated by $x_i - x_1^i$ for $i = 2, 3, \ldots, n$. Then $Y = \mathcal{V}(I)$ is a curve of degree n in K^n. For the degree reverse lexicographic monomial order $\prec$, the reduced Gröbner basis has $\binom{n}{2}$ elements, and $\text{in}_\prec(I) = \langle x_1, x_2, \ldots, x_{n-1} \rangle^2$. The ideal $\bar{I}$ is minimally generated by the 2×2 minors of the $2 \times n$ matrix

$$\begin{pmatrix} x_0 & x_1 & x_2 & \cdots & x_{n-1} \\ x_1 & x_2 & x_3 & \cdots & x_n \end{pmatrix}.$$

The initial monomials of the $\binom{n}{2}$ minors are the antidiagonal products. The projective variety $\bar{Y} = \mathcal{V}(\bar{I})$ is the *rational normal curve* of degree n in $\mathbb{P}^n$.

We now return to discussing desirable properties of projective varieties.

Remark 2.21. If $K = \mathbb{C}$ or $K = \mathbb{R}$, then every projective variety is compact in the classical topology. Indeed, the projective space $\mathbb{P}^n$ is compact, and every subvariety X is closed in the classical topology. Hence X is compact. If X is also smooth of dimension d, then X is a compact real manifold, of dimension d if $K = \mathbb{R}$ and of dimension $2d$ if $K = \mathbb{C}$. Many interesting manifolds arise in this manner. Well-known examples are the 1-dimensional complex varieties, which give rise to Riemann surfaces. The case of elliptic curves corresponding to tori is discussed in Section 2.3.

One nice property of projective varieties is that their images under polynomial maps are closed. This is known as Chevalley's Theorem and we will see the precise statement in Theorem 4.22. Another nice fact about projective varieties is the following result about their intersections.

Theorem 2.22 ([**47**, Theorem 6 in Section 6.2]). *Fix an algebraically closed field K. Let X and Y be two projective varieties in the n-dimensional ambient space $\mathbb{P}^n$, where $d_1 = \dim(X)$ and $d_2 = \dim(Y)$. Then their intersection $X \cap Y$ has dimension at least $d_1 + d_2 - n$. In particular, if $d_1 + d_2 \geq n$ then $X \cap Y$ is always nonempty.*

The hypotheses in this theorem are necessary. Consider the intersection of two surfaces in $\mathbb{P}^3$, where $n = 3$ and $d_1 = d_2 = 2$. Its dimension is at least $d_1 + d_2 - n = 1$. The statement fails in the affine space $\mathbb{C}^3$, where we can take two parallel planes. It also fails in $\mathbb{P}^3$ if the field is $K = \mathbb{R}$.

Example 2.23. Consider the two surfaces $X = \mathcal{V}(x_0^2 + x_1^2 - x_2^2 + x_3^2)$ and $Y = \mathcal{V}(x_0^2 + x_1^2 + x_2^2 - x_3^2)$ in $\mathbb{P}^3$. Over $\mathbb{C}$, their intersection is the union of four lines, so $\dim(X \cap Y) = 1$ as expected. However, over $\mathbb{R}$, the intersection consists of two points, so $\dim(X \cap Y) = 0$, which would violate Theorem 2.22.

Many models in the sciences and engineering are given by homogeneous polynomial equations. Typically, these constraints arise from a construction familiar from linear algebra. Whenever one encounters such a model, it makes much sense to regard it as a projective variety. We close this section with two examples.

Example 2.24 (Nilpotent matrices). An $n \times n$ matrix $A = (a_{ij})$ is a point in a projective space $\mathbb{P}^{n^2-1}$. The set of nilpotent matrices A is an irreducible projective variety $X \subset \mathbb{P}^{n^2-1}$. We have $\dim(X) = n^2 - n - 1$ and $\operatorname{degree}(X) = n!$. Indeed, X is a complete intersection. Its prime ideal $\mathcal{I}(X)$ is generated by the coefficients of the characteristic polynomial of A, except for the leading coefficient, which is a unit. For instance, if $n = 2$ then $\mathcal{I}(X) = \langle \operatorname{trace}(A), \det(A) \rangle$.

Example 2.25 (Kalman varieties). In control theory, one is interested in the set of $n \times n$ matrices A that have an eigenvector in a given linear subspace of K^n. This set is a projective variety in $\mathbb{P}^{n^2-1}$. For instance, let $n = 4$ and consider 4×4 matrices that have an eigenvector with the last two coordinates equal to zero. This Kalman variety has dimension 13 and degree 4 in $\mathbb{P}^{15}$. It is defined by the 2×2 minors of

$$\begin{pmatrix} a_{31} & a_{41} & a_{11}a_{31}+a_{21}a_{32}+a_{31}a_{33}+a_{34}a_{41} & a_{11}a_{41}+a_{21}a_{42}+a_{31}a_{43}+a_{41}a_{44} \\ a_{32} & a_{42} & a_{12}a_{31}+a_{22}a_{32}+a_{32}a_{33}+a_{34}a_{42} & a_{12}a_{41}+a_{22}a_{42}+a_{32}a_{43}+a_{42}a_{44} \end{pmatrix}.$$

2.3. Geometry in Low Dimensions

Smooth projective varieties in low dimensions give rise to interesting manifolds. Studying the geometry and topology of these manifolds leads to insights that are useful also for understanding higher-dimensional scenarios.

We work in projective spaces over the real numbers $\mathbb{R}$ and over the complex numbers $\mathbb{C}$. To distinguish these spaces, we denote them by $\mathbb{P}^n_{\mathbb{R}}$ and $\mathbb{P}^n_{\mathbb{C}}$. We regard both of them as compact real manifolds, of dimensions n and $2n$ respectively. Students of topology are encouraged to review the homology groups of the manifolds $\mathbb{P}^n_{\mathbb{R}}$ and $\mathbb{P}^n_{\mathbb{C}}$.

We start with $n = 1$. The real projective line $\mathbb{P}^1_{\mathbb{R}}$ is a circle. The complex projective line $\mathbb{P}^1_{\mathbb{C}}$ is a sphere, known as the *Riemann sphere*. Every proper subvariety of $\mathbb{P}^1_{\mathbb{R}}$ or $\mathbb{P}^1_{\mathbb{C}}$ is a finite collection of points defined by a *binary form* $f(x, y)$, i.e. a homogeneous polynomial in two variables. For instance, let $f = x^{11}y - 11x^6y^6 - xy^{11}$. The variety $\mathcal{V}(f)$ has dimension 0 and degree 12 in $\mathbb{P}^1_{\mathbb{C}}$. These 12 points on the Riemann sphere are famous in the history of geometry and arithmetic. They are the vertices of the icosahedron in Felix Klein's *Lectures on the Icosahedron.* Out of these 12 complex solutions, four are real. The remaining eight come in four conjugate pairs.

We now move on to $n = 2$. The real projective plane $\mathbb{P}^2_{\mathbb{R}}$ is a surface. However, it cannot be embedded homeomorphically in $\mathbb{R}^3$ (only in $\mathbb{R}^4$), so it is impossible to make a good picture. The simplest curve in the projective plane $\mathbb{P}^2_K$ is a line L, defined by one linear form in three variables. Of course, L is a projective line, $L \simeq \mathbb{P}^1_K$, so the discussion in the previous paragraph applies. The complement $\mathbb{P}^2_K \backslash L$ is the affine plane K^2. In particular, this complement is connected when $K = \mathbb{R}$. The decomposition into L and K^2 may be used to give a schematic picture of $\mathbb{P}^2_{\mathbb{R}}$. We identify $\mathbb{R}^2$ with the interior of the square. The boundary of the square should represent the line L. However, we need to identify opposite points of the boundary; this identification is represented by directed arrows on the boundary as in Figure 2.1.

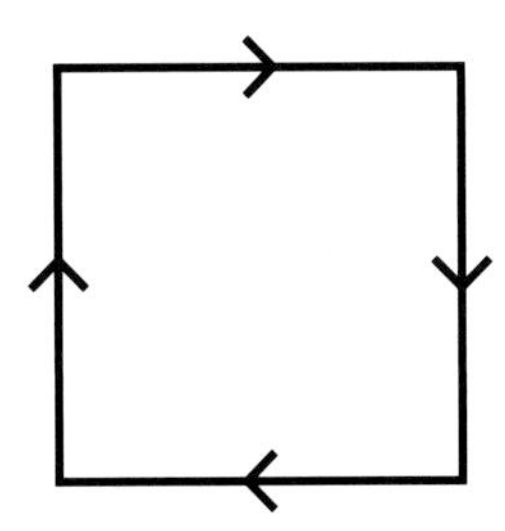

Figure 2.1. Schematic representation of $\mathbb{P}^2_{\mathbb{R}}$.

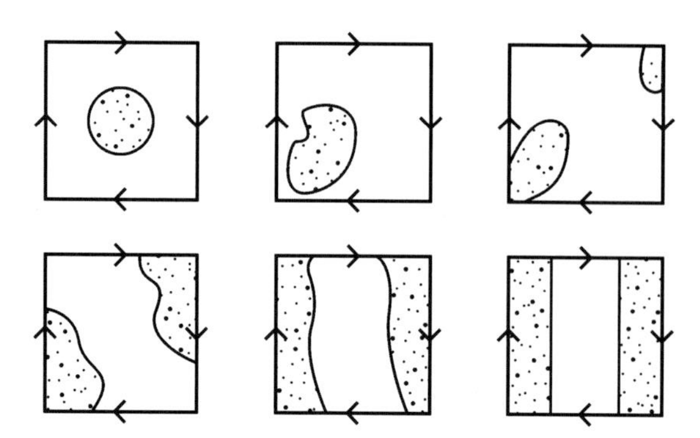

Figure 2.2. Six topological ovals in $\mathbb{P}^2_{\mathbb{R}}$ with shaded interiors. It is possible to cut out the interior of the lower right oval from the square and glue together the remaining antipodal points on the boundary. This shows that the complement of the interior of the oval is the Möbius strip.

For any curve C in $\mathbb{P}^2_{\mathbb{C}}$, the complement $\mathbb{P}^2_{\mathbb{C}} \backslash C$ is connected because C is a surface in the 4-dimensional manifold $\mathbb{P}^2_{\mathbb{C}}$. By contrast, consider a smooth conic, i.e. a curve of degree 2, $C \subset \mathbb{P}^2_{\mathbb{R}}$. Then $\mathbb{P}^2_{\mathbb{R}} \backslash C$ has two connected components. One is a disk and the other is a Möbius strip (see Figure 2.2). The former is the *inside* of C and the latter is the *outside* of C. A curve D in $\mathbb{P}^2_{\mathbb{R}}$ is called a *pseudoline* if $\mathbb{P}^2_{\mathbb{R}} \backslash D$ is connected and is called an *oval* otherwise. Every oval behaves like a conic in $\mathbb{P}^2_{\mathbb{R}}$; it has an inside and an outside.

Theorem 2.26. *Let C be a smooth curve of degree d in the projective plane $\mathbb{P}^2_K$. If $K = \mathbb{C}$ then C is an orientable surface of genus $g = \frac{(d-1)(d-2)}{2}$. If $K = \mathbb{R}$ then C is a curve with at most $g + 1$ connected components. If d is even then all components are ovals. If d is odd then one component is a pseudoline but all others are ovals.*

Proof. The first part of the theorem is standard and may be found in many classical books, including [**25**, I Ex. 7.2] and [**28**]. For the statements about real curves we refer the reader to [**6**, Chapter 12, §6]. □

Let us illustrate the above theorem for $d = 3$, i.e. $g = 1$. For example, consider a smooth cubic curve given in Weierstrass form, such as

$$f(x, y, z) = zy^2 - x^3 + xz^2.$$

Such curves are also known as *elliptic curves*. We decompose the projective plane $\mathbb{P}^2_{\mathbb{R}}$ into a line L given by the equation $z = 0$ and its complement, the affine plane $A = \mathbb{R}^2$. The cubic has two components: an oval and a pseudoline. We can see both of them by intersecting C with A, as depicted in Figures 2.3 and 2.4.

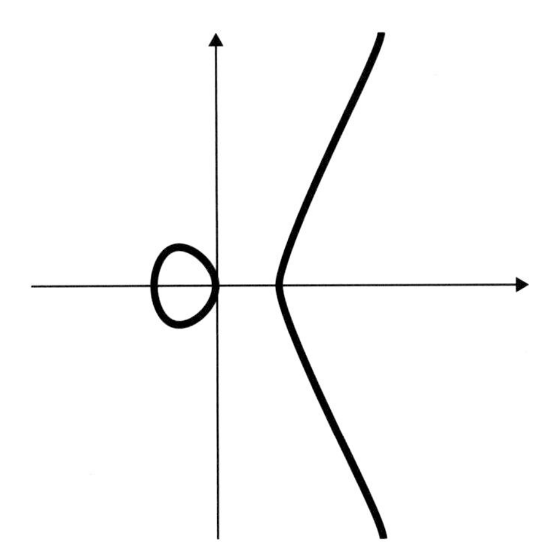

Figure 2.3. Real elliptic curve $y^2 = x^3 - x$. The connected component on the left is an oval. The connected component on the right is a pseudoline. For a more complete picture see Figures 2.4 and 2.5.

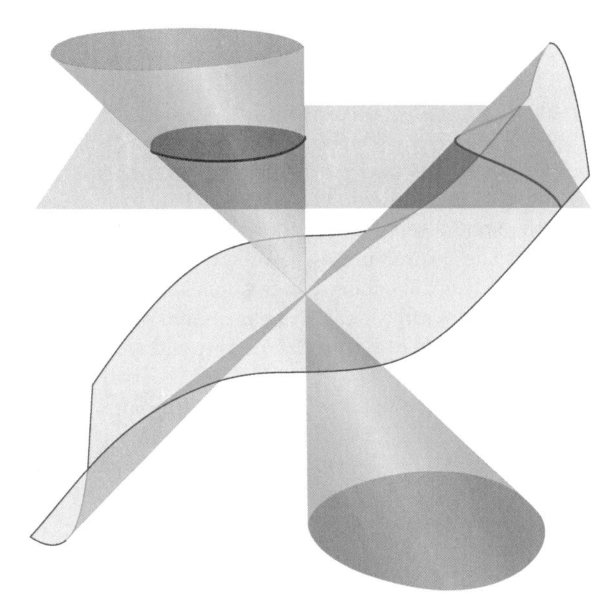

Figure 2.4. The cone $zy^2 - x^3 + xz^2 = 0$ over an elliptic curve. The *irreducible* variety is depicted in orange and blue, representing two connected components in the real projective space. We intersect the cone with the grey plane given by $z = 1$. This corresponds to the affine chart. The red and blue curve C we obtain is exactly the same as in Figure 2.3.

There is an additional point P of the curve that we do not see in Figure 2.3. The point P belongs to the line L. It is given by $z = x = 0$ and $y = 1$. We may consider the surface in $\mathbb{R}^3$ that is the affine cone over our curve. This surface is shown in Figure 2.4.

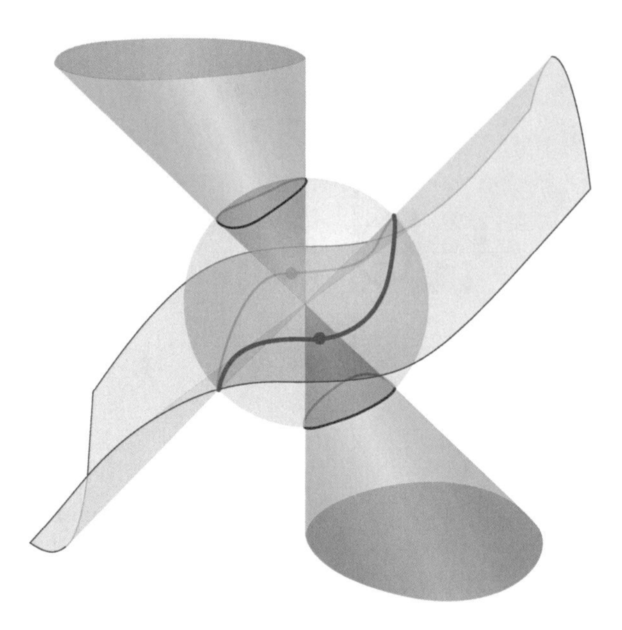

Figure 2.5. The cone $zy^2 - x^3 + xz^2 = 0$ over an elliptic curve, as in Figure 2.4. The grey sphere represents the projective space $\mathbb{P}^2_{\mathbb{R}}$, where we have to identify the antipodal points. The intersection of the surface with the sphere has three connected components. When we identify the antipodal points, the two red connected components become one. These red components correspond to the oval in $\mathbb{P}^2_{\mathbb{R}}$. Indeed, cutting them out of the sphere separates it into three pieces. After identifying antipodal points these three pieces become two. The other component, represented by the thick blue curve, corresponds to a pseudoline. It does not separate the sphere after identifying antipodal points. The points on the red and blue curve C in Figure 2.4 correspond to pairs of antipodal points on the red and blue curve in this picture, with one exception. The blue curve has one more pair of antipodal points, represented by the blue dots. Indeed, the line through these points is parallel to the grey plane in Figure 2.4. This pair of points corresponds to the unique point of the projective curve that does not belong to the affine chart given by $z = 1$. It is given by $z = 0$, $x = 0$ and $y = 1$.

How can we imagine the complex elliptic curve? This is not easy, as the correct picture of just the affine part would be in $\mathbb{C}^2 \simeq \mathbb{R}^4$. However, there exists a homeomorphism (but not a polynomial map!) of the complex curve C with the real topological torus, i.e. the product of two circles, $\mathbb{S}^1 \times \mathbb{S}^1$. It can be described as follows. We fix a point $p \in C$. For any point $q \in C$ consider a path γ from p to q. This is always possible since C is connected over the complex numbers. To a point q we associate the *complex* number $\int_\gamma \frac{dx}{y}$. Identifying the complex plane with $\mathbb{R}^2$, we obtain a map $f : C \to \mathbb{R}^2$. It turns out that $f(q)$ depends on our choice of γ. Indeed, let us choose p given by $z = 1$, $y = 0$ and $x = -1$. We may choose $q = p$ and γ equal to

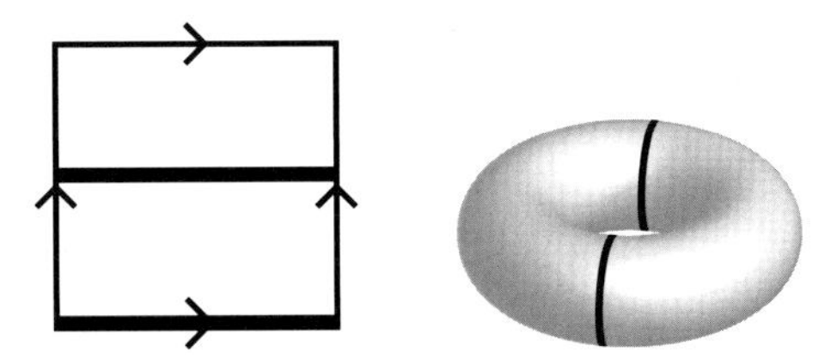

Figure 2.6. Two pictures showing a real torus that is homeomorphic to the elliptic curve. The left picture presents the torus as $\mathbb{R}^2/\mathbb{Z}^2$. The right picture is the familiar figure we know from topology. The two thick lines on both pictures correspond to the real part of the curve.

the oval depicted in Figure 2.3. The integral $\int_\gamma \frac{dx}{y}$ will be a nonzero real number λ. Thus $f(p)$ may be equal to any integer multiple of λ. Further, on the curve C there exists another loop γ' giving rise to the integral $\int_{\gamma'} \frac{dx}{y}$, which is a complex number τ. We consider the *lattice* M in $\mathbb{C} \simeq \mathbb{R}^2$ whose elements are the integer combinations $a\lambda + b\tau$ for $a, b \in \mathbb{Z}$. We know that $f(p)$ may be any point in M. Let $\pi : \mathbb{R}^2 \to \mathbb{R}^2/M$ be the natural projection. The map $\pi \circ f : C \to \mathbb{R}^2/M$ is now well-defined!

As $\mathbb{R}^2/M$ may be identified with the torus, we indeed obtain a homeomorphism $C \simeq \mathbb{R}^2/M$. The real part of the curve C is mapped to two disjoint circles, as shown in Figure 2.6. Indeed, both the oval and the pseudoline are circles; they are distinguished only by their embedding into $\mathbb{P}^2_\mathbb{R}$.

Remark 2.27. We contrast the topological torus mentioned here with the *algebraic* torus $(\mathbb{C}^*)^n$ playing a central role in Chapters 8 and 10. Indeed, a variety with a dense algebraic torus action will be called *toric*. The elliptic curve C is a basic example of a smooth projective variety that is *not* toric.

Remark 2.28. Elliptic curves made their first appearance in the third century. Diophantus of Alexandria asked for, in modern terms, a (positive) rational point on a specific elliptic curve $y(6 - y) = x^3 - x$. As we argued above, an elliptic curve has the structure of a group (torus). The geometric interpretation of this was already well known in the 19th century. Since the early 20th century, elliptic curves have played a central role in (modern) number theory (studied mainly over fields of finite characteristic or rational numbers). By the end of the 20th century, the group structure (over fields of finite characteristic!) had started to be used intensively in applied cryptography.

Example 2.29. Let us consider a *cuspidal curve* defined by $x^3 - y^2$. Over $\mathbb{R}$ it is presented in Figure 2.7 on the left. How can we draw it over $\mathbb{C}$? If we identify $\mathbb{C}$ with $\mathbb{R}^2$ we obtain a *surface* in $\mathbb{R}^4$. Indeed,

$$(x_1 + ix_2)^3 - (y_1 + iy_2)^2 = 0 \Leftrightarrow x_1^3 - 3x_1x_2^2 = y_1^2 - y_2^2 \text{ and } 3x_1^2x_2 - x_2^3 = 2y_1y_2.$$

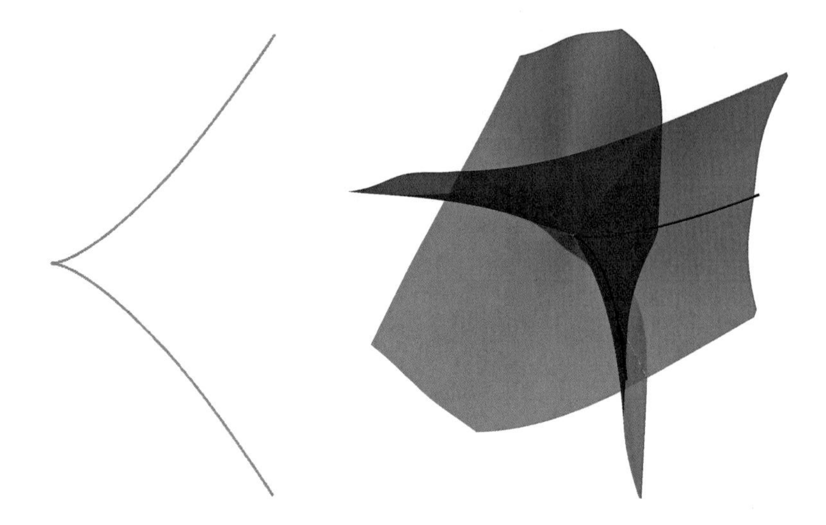

Figure 2.7. Real part of a cuspidal curve.

Hence, interpreted as a surface in $\mathbb{R}^4$, the variety is cut out by two polynomials. Although we cannot draw a picture in $\mathbb{R}^4$, we can project the given surface onto $\mathbb{R}^3$. The result is the surface presented in Figure 2.7 on the right, together with the black line, which is the real part.

The surface seems more singular; this is the result of projection. The original surface in $\mathbb{R}^4$ has just one singular point. In Chapter 4 methods for the computation of projections of algebraic varieties are presented.

Exercises

(1) Prove that the definition of $\mathcal{V}(I)$ does not depend on the choice of generators of I.

(2) (a) Show that $J \subseteq I$ implies $\mathcal{V}(I) \subseteq \mathcal{V}(J)$.
 (b) Show that for any subsets $A, B \subseteq K^n$, if $A \subseteq B$ then $\mathcal{I}(B) \subseteq \mathcal{I}(A)$.
 (c) Give counterexamples to both opposite implications.

(3) Prove that varieties (in K^n) satisfy the axioms of closed sets.

(4) By identifying the point $(p_1, \ldots, p_n) \in K^n$ with the prime ideal $\langle x_1 - p_1, \ldots, x_n - p_n \rangle$, we view K^n as a subset of $\operatorname{Spec} K[\mathbf{x}]$. Show that the Zariski topology induced from $\operatorname{Spec} K[\mathbf{x}]$ is the Zariski topology on K^n.

(5) Show that a morphism of rings $f\colon R_1 \to R_2$ gives a map $f^*\colon \operatorname{Spec} R_2 \to \operatorname{Spec} R_1$, by proving that the preimage of a prime ideal is prime. Show that the induced map is continuous with respect to the Zariski topology.

(6) Describe the variety $\mathcal{V}(I)$ in the affine line K^1 for $I = \langle x^2+1\rangle$ when $K = \mathbb{C}, \mathbb{R}, \mathbb{Q}$. Also, describe $\mathcal{V}(I) \subset \mathrm{Spec}(K[x])$ for each of these fields.

(7) Realize the set of $n \times n$ nilpotent matrices as an affine variety. What is its dimension? What is its degree?

(8) (a) Consider a polynomial $f \in K[\mathbf{x}]$. Let D be the (open) set $D_f = \{p \in K^n : f(p) \neq 0\}$. Construct an affine variety V and a polynomial map inducing a bijection $V \to D$.
  (b) Realize nondegenerate $n \times n$ matrices as an affine variety.

(9) (a) Use (or not) your favorite computer algebra system to determine the ideal of the image of the map given by formula (2.3). What is the meaning of the lowest-degree polynomial in this ideal?
  (b) Describe the ideal of the image of the map given by formula (2.2).
  (c) Generalize the previous point to more (independent) variables, possibly with different (but finite) numbers of states.

(10) Determine the set of all prime numbers p such that $I = \langle x^2 - 2y^2\rangle$ is a prime ideal in the polynomial ring $\mathbb{F}_p[x, y]$ over the finite field $\mathbb{F}_p$.

(11) Consider a polynomial $f = \sum_{\mathbf{a}} c_{\mathbf{a}} x^{\mathbf{a}}$. Its degree-$k$ part is the homogeneous polynomial $\sum_{\mathbf{a}:|\mathbf{a}|=k} c_{\mathbf{a}} x^{\mathbf{a}}$.
  (a) Provide an example of a homogeneous ideal generated by nonhomogeneous polynomials.
  (b) Prove that an ideal $I = \langle f_1, \dots, f_j\rangle$ is homogeneous if and only if for any i and k the degree-k part of f_i belongs to I.
  (c) Propose an algorithm that, given a set of generators of $I \subset K[\mathbf{x}]$, decides whether I is a homogeneous ideal.

(12) (a) Let $I \subset K[\mathbf{x}]$ be a monomial ideal. Prove that $\mathcal{V}(I)$ is a union of (some) vector subspaces of K^n spanned by standard basis vectors.
  (b) How do you characterize the sets of basis vectors that span a subspace belonging to $\mathcal{V}(I)$?

(13) Draw various pseudolines in $\mathbb{P}^2_{\mathbb{R}}$, in analogy to Figure 2.2. Topologically, what is the complement of a pseudoline?

(14) Prove that the rank of the Jacobian matrix does not depend on the choice of generators of the ideal.

(15) Can you solve the problem of Diophantus of Alexandria in Remark 2.28? Hint: Consider a tangent line to the elliptic curve at the point $(-1, 0)$.

(16) Extending Example 2.25, determine the Kalman variety of all 6×6 matrices that have an eigenvector whose last three coordinates are zero.

(17) Let $X \subset \mathbb{C}^4$ be the variety defined by two random polynomials of degree 3 in four variables. What do we learn about X from Bézout's Theorem, and what do we learn about X from Bertini's Theorem?

Chapter 3

Solving and Decomposing

"*Divide et impera*", Roman principle

Solving systems of polynomial equations is a key task in nonlinear algebra. But what does it mean to solve such a system? How should the solutions be presented? The answers to these questions depend on the dimension of the variety of solutions. If the variety is 0-dimensional, then it consists of finitely many points in K^n and we aim to list each point explicitly. If $K = \mathbb{R}$ or $K = \mathbb{C}$, then this is done in applications by displaying a floating point approximation to each of the n coordinates of a solution.

If the solution variety has positive dimension, then it has infinitely many points and we cannot list them all. In that case, the answer consists of a description of each irreducible component. Algebraically, this leads us to the topic of primary decomposition. If the given ideal is not radical, then its constituents are primary ideals, and we distinguish between minimal primes and embedded primes. To some readers, these objects may seem unnatural at first. However, they become quite natural when we interpret multivariate polynomials as linear partial differential equations with constant coefficients.

3.1. 0-Dimensional Ideals

Let K be a field and consider the polynomial ring $K[x]$ in one variable x. Every ideal in $K[x]$ is principal, so it has the form $I = \langle f \rangle$. The variety $\mathcal{V}(I)$ consists of the zeros of f and is 0-dimensional (unless $f = 0$). The polynomial f has a unique factorization $f = \prod_{i=1}^{k} g_i^{a_i}$, where each g_i is irreducible and the a_i are positive integers. The set of solutions decomposes as

$$\mathcal{V}(I) \;=\; \mathcal{V}(g_1) \cup \cdots \cup \mathcal{V}(g_k).$$

On the level of ideals we have the following decomposition as an intersection:

$$I \;=\; \langle g_1 \rangle^{a_1} \cap \cdots \cap \langle g_k \rangle^{a_k}.$$

This primary decomposition remembers the multiplicity a_i of each factor g_i, so it contains more information than the irreducible decomposition of $\mathcal{V}(I)$.

The decomposition depends on the field K. If K is algebraically closed, such as $K = \mathbb{C}$, then each factor g_i is a linear polynomial $g_i(x) = x - u_i$, where $u_1, \ldots, u_k$ are the zeros of f. If $K = \mathbb{R}$ then each g_i is either linear or quadratic. If $K = \mathbb{Q}$ then g_i can have arbitrarily high degree. In each case, the quotient ring $K[x]/\langle g_i \rangle$ is a field. It is an algebraic extension of K.

Example 3.1. The polynomial $f = x^3 - 2x^2 + x - 2 \in \mathbb{R}[x]$ satisfies

$$\langle f \rangle \;=\; \langle x - 2 \rangle \cap \langle x^2 + 1 \rangle.$$

The first ideal in the intersection corresponds to the real zero 2, while the second ideal corresponds to the pair of complex zeros i and $-i$. Such factorizations are easy to find using a computer algebra system when the given polynomial is not too large. What if we now replace the given f by $g = x^3 - 2x^2 + x - 1$? How does the ideal $\langle g \rangle$ decompose in $\mathbb{R}[x]$? And in $\mathbb{C}[x]$?

Polynomials of degree m in one variable can have up to m zeros. The number m can be large. Often we are not interested in all the zeros, but only in specific ones. For instance, we might only be interested in solutions that are real and positive. This restriction is very important for many applications, e.g. in statistics where the solutions represent probabilities.

Example 3.2. Let $I = \langle x^m - x - 1 \rangle$, where $m \geq 2$. The variety $\mathcal{V}(I)$ consists of m complex numbers, but only one of them is real and positive. Thus, $\mathcal{V}(I) \cap \mathbb{R}_{>0}$ is a singleton. This follows from *Descartes' rule of signs*, which states that the number of positive real solutions is bounded above by the number of sign alternations in the coefficient sequence. If m is even then there is also one negative solution.

In many applications one encounters polynomials whose coefficients depend on parameters. For instance, let ϵ be an unknown that represents a small positive real number. Let $\mathbb{Q}(\epsilon)$ be the field of rational functions in that unknown and $K = \overline{\mathbb{Q}(\epsilon)}$ its algebraic closure. Elements in K can be expressed as series in ϵ with rational exponents. These are known as *Puiseux series*. Expressing in Puiseux series is analogous to floating point expansion of numbers in $\mathbb{R}$.

Example 3.3. The polynomial $f = \epsilon^2 x^3 + x^2 + x - \epsilon$ is irreducible in $\mathbb{Q}(\epsilon)[x]$. It factors into three linear factors, $f = \epsilon^2(x-u_1)(x-u_2)(x-u_3)$ in $K[x]$, where

$$\begin{aligned} u_1 &= -\epsilon^{-2} + 1 + \epsilon^2 + \epsilon^3 + 2\epsilon^4 + 3\epsilon^5 + 5\epsilon^6 + 10\epsilon^7 + \cdots, \\ u_2 &= -1 - \epsilon - 3\epsilon^3 + 3\epsilon^4 - 16\epsilon^5 + 32\epsilon^6 - 121\epsilon^7 + \cdots, \\ u_3 &= \epsilon - \epsilon^2 + 2\epsilon^3 - 5\epsilon^4 + 13\epsilon^5 - 37\epsilon^6 + 111\epsilon^7 + \cdots. \end{aligned}$$

Each of these three roots is an algebraic number over $\mathbb{Q}(\epsilon)$. We have written them as Puiseux series. If we think of ϵ as a very small positive quantity, then $u_1 \sim -\epsilon^{-2}$, $u_2 \sim -\epsilon^0$ and $u_3 \sim \epsilon^1$. The exponents -2, 0 and 1 tell us the asymptotic behavior. These exponents are known as *tropical solutions*; cf. Chapter 7.

We have seen that *solving* a polynomial equation $f = 0$ amounts to *decomposing* the principal ideal $I = \langle f \rangle$, i.e. presenting it as an intersection of simpler ideals. The situation is analogous for systems of polynomials in $n \geq 2$ variables, i.e. ideals $I \subset K[\mathbf{x}]$ where $\mathbf{x} = (x_1, \ldots, x_n)$. Suppose now that K is algebraically closed and assume that $\mathcal{V}(I)$ is 0-dimensional. This means that the quotient ring $K[\mathbf{x}]/I$ is a finite-dimensional vector space over K. By Theorem 1.17, a basis is given by the standard monomials for a given monomial order. The number of standard monomials is an upper bound for the cardinality of $\mathcal{V}(I)$. Equality holds if and only if I is radical.

In the next section we will decompose our 0-dimensional ideal I as

$$I \;=\; \bigcap_{i=1}^{k} q_i,$$

where $\mathrm{rad}(q_i)$ is a prime ideal. Every prime ideal of dimension 0 in $K[\mathbf{x}]$ is a maximal ideal, so each $\mathrm{rad}(q_i)$ is a maximal ideal. Since K is algebraically closed, $\mathcal{V}(q_i)$ is a point in K^n. These points are the solutions to our system.

Example 3.4. Let $n = 2$ and $I = \langle xy, x^2 - x, y^2 - y \rangle$. This ideal is radical:

$$I \;=\; \langle x, y \rangle \cap \langle x - 1, y \rangle \cap \langle x, y - 1 \rangle.$$

The variety of this ideal consists of three points: $\mathcal{V}(I) = \{(0,0), (1,0), (0,1)\}$.

If the given ideal is not radical, then we cannot express it as an intersection of maximal ideals. This should not be surprising; already in the case of one variable, if a root has a multiplicity then we need a power of its ideal.

Example 3.5. Let $I = \langle xy, y^2 - y, x^2 y - x^2 \rangle$. We have the decomposition

$$I \;=\; \langle x, y - 1 \rangle \cap \langle x^2, y \rangle.$$

The varieties of both ideals are points: $(0,1)$ and $(0,0)$ respectively. However, the second ideal remembers additional data. It is not just $\langle x, y \rangle$ but also indicates a multiplicity of the solution $(0,0)$. We are now equipped to measure this multiplicity! The degree of I equals 3. The first ideal in the decomposition contributes with degree 1, while the second contributes with degree 2.

We now discuss an example that was seen in Exercise 10 of Chapter 1.

Example 3.6. Fix the field $K = \mathbb{Q}$ and $I = \langle x^3 - yz, y^3 - xz, z^3 - xy\rangle$ in $K[x,y,z]$. This ideal is a nonredundant intersection of 11 distinct ideals:

$$\begin{aligned} I = Q &\cap \langle x-1, y-1, z-1\rangle \cap \langle x+1, y+1, z-1\rangle \cap \langle x+y, y^2+1, z-1\rangle \\ &\cap \langle x+1, y-1, z+1\rangle \cap \langle x-1, y+1, z+1\rangle \cap \langle x-y, y^2+1, z+1\rangle \\ &\cap \langle x-1, y+z, z^2+1\rangle \cap \langle x+1, y-z, z^2+1\rangle \\ &\cap \langle y-1, x+z, z^2+1\rangle \cap \langle y+1, x-z, z^2+1\rangle. \end{aligned}$$

The first intersectand is a primary ideal with radical $\mathrm{rad}(Q) = \langle x,y,z\rangle$:

$$Q \;=\; \big\langle\, x^2y, x^2z, xy^2, xz^2, y^2z, yz^2, x^3 - yz, y^3 - xz, z^3 - xy \,\big\rangle.$$

Each of the other 10 intersectands is a prime ideal. If we replace K by the complex numbers $\mathbb{C}$, then six of the prime ideals decompose further:

$$\langle x-1, y+z, z^2+1\rangle \;=\; \langle x-1, y-i, z+i\rangle \cap \langle x-1, y+i, z-i\rangle.$$

We learn that $\mathcal{V}(I)$ consists of 17 complex points. Only five are real.

The idea of decomposing a mathematical object into simpler pieces is important. In the next section we present a theory of decomposing ideals. We shall express ideals as intersections of simpler ideals. Our point of departure is the following proposition. It shows how algebraic varieties may be decomposed. This proposition applies to varieties of any dimension.

Proposition 3.7. *Any variety in K^n can be uniquely represented as a finite union of irreducible varieties (pairwise not contained in each other).*

Proof. We start by proving the existence of such a decomposition. Any variety W is either irreducible or a union $W_1 \cup V_1$. In the latter case we apply the same reasoning to the pieces. For instance, we may write W_1 as a union $W_2 \cup V_2$, and so on. We obtain an ascending chain of ideals, $\mathcal{I}(W_1) \subseteq \mathcal{I}(W_2) \subseteq \cdots$. This chain stabilizes by Hilbert's Basis Theorem. Thus the decomposition finishes with finitely many irreducible varieties.

Suppose we have two irreducible decompositions of the same variety:

$$V_1 \cup \cdots \cup V_k \;=\; W_1 \cup \cdots \cup W_s.$$

Fix any index i_0 in $\{1,\ldots,s\}$. The identity implies $W_{i_0} = \bigcup_{j=1}^k (V_j \cap W_{i_0})$. Since W_{i_0} is irreducible, there exists an index j_0 such that $W_{i_0} = V_{j_0} \cap W_{i_0}$, and hence $W_{i_0} \subseteq V_{j_0}$. But we similarly find that $V_{j_0} \subseteq W_{i_1}$ for some $i_1 \in \{1,\ldots,k\}$. We cannot have $W_{i_0} \subsetneq W_{i_1}$, so $W_{i_0} = V_{j_0}$. Hence, for every component W_{i_0} on the right there is a unique component V_{j_0} on the left that is equal to W_{i_0}. The uniqueness of the decomposition follows. □

3.2. Primary Decomposition

This section develops a generalization of the following two basic facts:

(1) Every integer $n > 1$ can be uniquely decomposed as a product of powers of prime numbers:

$$n = p_1^{a_1} \cdots p_k^{a_k}.$$

(2) Any variety can be uniquely decomposed as a union of irreducible varieties. We saw this in Proposition 3.7.

The algebraic notion of an ideal connects the first (number-theoretic) fact and the second (geometric) fact. Indeed, any integer n can be identified with the ideal $\langle n \rangle$ in the ring $\mathbb{Z}$. The elements of $\langle n \rangle$ are the integer multiples of n. The ideal $\langle n \rangle$ is prime in $\mathbb{Z}$ if and only if n is a prime number. We can restate fact (1) in terms of intersections of powers of prime ideals as follows:

(1') Every nonzero ideal $I \subset \mathbb{Z}$ has a unique decomposition

$$I = (I_1)^{a_1} \cap \cdots \cap (I_k)^{a_k}$$

where the I_j are prime ideals.

Over an algebraically closed field, we have an identification of varieties with radical ideals (see Chapter 6). This yields the following restatement of (2):

(2') Every radical ideal $I \subset \mathbb{C}[\mathbf{x}]$ has a unique decomposition as an intersection of prime ideals that are pairwise not contained in each other:

$$I = p_1 \cap \cdots \cap p_k.$$

These examples suggest that our aim should be to decompose ideals I in a ring R. Here, a decomposition of I is a presentation as an intersection of other ideals. At this point, we need to answer the following questions:

(1) What kind of ideals should be allowed in the intersection?

(2) What restrictions should be put on the ring R?

(3) Can we expect the decomposition to be unique?

We start with the first question. The number-theoretic example suggests that all ideals are intersections of powers of prime ideals. But this is not true.

Example 3.8. The ideal $I = \langle x^2, y\rangle$ is not an intersection of powers of prime ideals in $\mathbb{C}[x,y]$. Indeed, suppose $I = \bigcap_i p_i^{a_i}$. For all i, we have $p_i \supset I$. Hence $p_i = \langle x, y\rangle$, as this is the only prime ideal containing I. The ideal $\bigcap_i \langle x, y\rangle^{a_i}$ would be a power of $\langle x, y\rangle$, whereas I is no such power.

The correct constituents are *primary* ideals. Recall that I is primary if and only if $ab \in I$ and $a \notin I$ implies $b^n \in I$ for some n, given any $a, b \in R$.

Next consider question (2): which rings R to take? Clearly, $\mathbb{Z}$ and $K[\mathbf{x}]$ share a lot of nice properties. But there is a larger class of rings that works.

Definition 3.9. A ring R is *Noetherian* if every ascending chain of ideals

$$I_1 \subseteq I_2 \subseteq I_3 \subseteq \cdots$$

stabilizes, i.e. there exists k such that $I_k = I_{k+1} = I_{k+2} = \cdots$.

Noetherian rings are named after the German algebraist Emmy Noether. A hint as to how important they are is given in Exercise 4. Note that $\mathbb{Z}$ and $K[\mathbf{x}]$ are Noetherian rings because their ideals are finitely generated. Before stating our main existence theorem, let us introduce a technical definition.

Definition 3.10. An ideal I in a ring R is *irreducible* if and only if whenever $I = J_1 \cap J_2$ for some ideals J_1 and J_2 in R, we have $I = J_1$ or $I = J_2$.

Theorem 3.11. *Let I be an ideal in a Noetherian ring R. Then there exist primary ideals $q_1, q_2, \ldots, q_k$ in R such that*

$$I = q_1 \cap q_2 \cap \cdots \cap q_k.$$

Proof. First we show that every ideal in R is a finite intersection of irreducible ideals. Suppose not, and let I_1 be an ideal that cannot be presented in this way. In particular, it is not irreducible. Thus, $I_1 = J_1 \cap J_2$ and each J_i strictly contains I_1. If J_1 and J_2 are finite intersections of irreducible ideals, then so is I_1. Hence, we may assume that J_1 cannot be presented as such a finite intersection. Let $I_2 := J_1$. We have $I_1 \subsetneq I_2$. We repeat the construction starting with I_2 and get an ideal I_3 with $I_1 \subsetneq I_2 \subsetneq I_3$, where I_3 is not a finite intersection of irreducible ideals. Continuing, we get a chain of strictly ascending ideals. However, this is not possible in a Noetherian ring.

We next prove that every irreducible ideal q is primary. By replacing the ring R with R/q, we may assume $q = \{0\}$. Suppose $ab = 0$ and $a \neq 0$. We must prove that b is nilpotent. Consider the following ascending chain:

$$\{x \in R : bx = 0\} =: \mathrm{Ann}(b) \subseteq \mathrm{Ann}(b^2) \subseteq \mathrm{Ann}(b^3) \subseteq \cdots.$$

Since R is Noetherian, ascending chains of ideals become stationary. Hence $\mathrm{Ann}(b^n) = \mathrm{Ann}(b^{n+1})$ for some n. We claim that $\langle a\rangle \cap \langle b^n\rangle = \{0\}$. Indeed, suppose $\lambda a = \mu b^n \in \langle a\rangle \cap \langle b^n\rangle$ for some $\lambda, \mu \in R$. Clearly,

$$0 = \lambda ab = \mu b^{n+1}.$$

Hence, $\mu \in \mathrm{Ann}(b^{n+1}) = \mathrm{Ann}(b^n)$. Thus, $\mu b^n = 0$. As $\{0\}$ was assumed to be irreducible and $\langle a \rangle \supsetneq \{0\}$, we have $b^n = 0$. This completes the proof. □

We now turn to the third question, concerning uniqueness. We need not assume that R is Noetherian, as long as the ideal I in question is an intersection of finitely many primary ideals: $I = \bigcap_{i=1}^k q_i$. Here it is assumed that each q_i is necessary, i.e. $\bigcap_{j \neq i_0} q_j \not\subset q_{i_0}$ for all $1 \leq i_0 \leq k$. The next two lemmas suggest grouping the primary ideals q_i by their radical.

Lemma 3.12. *The radical of a primary ideal q is the unique smallest prime ideal containing it.*

The proof is left as Exercise 5 for the reader. A primary ideal q whose radical equals a given prime ideal p is called *p-primary.* In many cases, the powers of a prime ideal p are primary. This is the case for polynomial rings in one variable; see Exercise 2. However, it is not true in general that the power of a prime ideal is a primary ideal, even in the polynomial ring $\mathbb{C}[\mathbf{x}]$.

Example 3.13. Let P be the ideal generated by the nine 2×2 minors of a 3×3 matrix $X = (x_{ij})$ of unknowns. This ideal is prime and contains none of the x_{ij}. We claim that the ideal P^2 is not primary. To see this, we verify (using Gröbner bases) that $x_{ij} \cdot \det(X)$ lies in P^2 for all $1 \leq i, j \leq 3$. However, P^2 is generated by quartics and contains no cubics. Thus, neither $\det(X)$ nor any power of x_{ij} is in P^2. We conclude that P^2 is not primary.

We next focus on p-primary ideals for a fixed prime ideal p.

Lemma 3.14. *If $q_1, \ldots, q_k$ are p-primary ideals, then so is $q_1 \cap \cdots \cap q_k$.*

Proof. The following shows that the radical of $I := \bigcap_{i=1}^k q_i$ equals p:

$$\begin{aligned} a \in \mathrm{rad}(I) &\iff \exists n : a^n \in I \iff \exists n \forall i : a^n \in q_i \\ &\iff \forall i : a \in \mathrm{rad}(q_i) = p \iff a \in p. \end{aligned}$$

To see that I is primary, we assume that $ab \in I$ and $a \notin I$. Then $a \notin q_{i_0}$ for some i_0. Since $ab \in q_{i_0}$ and q_{i_0} is primary, $b \in \mathrm{rad}(q_{i_0}) = p = \mathrm{rad}(I)$. Hence $b^n \in I$ for some n. □

Lemma 3.14 suggests that given any primary decomposition $I = \bigcap_{i=1}^k q_i$, we aggregate the q_i's with the same radical and replace them by their intersection. The result is still a primary decomposition of I. This motivates the following definition. A *minimal primary decomposition* is a representation

$$I = q_1 \cap q_2 \cap \cdots \cap q_k \tag{3.1}$$

where the q_i's are primary ideals that have pairwise distinct radicals whose intersection is nonredundant, meaning $\bigcap_{j \neq i_0} q_j \not\subset q_{i_0}$ for all $1 \leq i_0 \leq k$.

To sum up, we have proved the following result for a Noetherian ring R:

(1) every ideal has a (finite) primary decomposition, and

(2) every (finite) primary decomposition of an ideal can be changed to a minimal one (apply Lemma 3.14 and remove unnecessary ideals).

We next show that minimal primary decompositions may still not be unique.

Example 3.15. The following are two minimal primary decompositions:

$$\langle x^2, xy\rangle \;=\; \langle x\rangle \cap \langle x,y\rangle^2 \;=\; \langle x\rangle \cap \langle x^2, y\rangle \quad \subset \quad \mathbb{C}[x,y]. \tag{3.2}$$

It turns out that while the primary ideals q_i in the decomposition (3.1) need not be unique, their radicals are unique. Recall that the quotient of an ideal I by a ring element a is the ideal $(I : a) = \{b \in R : ab \in I\}$.

Theorem 3.16. *For any ideal I in a ring R with a minimal primary decomposition* (3.1), *the k prime ideals* $\mathrm{rad}(q_i)$ *do not depend on the choice of that decomposition. They are precisely the prime ideals that have the form* $\mathrm{rad}(I : a)$ *for some element a in R. If R is Noetherian then the last radical is not needed: They are precisely the prime ideals $(I : a)$ for some $a \in R$.*

Remark 3.17. The ideal $(I : a)$ is usually not prime. It is prime only for some very special elements $a \in R$. It is those special a we are interested in.

Proof. Fix a minimal primary decomposition $I = \bigcap_{i=1}^k q_i$. Intersection commutes with ideal quotients, so $(I : a) = \bigcap_{i=1}^k (q_i : a) = \bigcap_{a \notin q_j} (q_j : a)$. It also commutes with radicals: $\mathrm{rad}(I : a) = \bigcap_{a\notin q_j} \mathrm{rad}(q_j : a)$. We next argue that $a \notin q_i$ implies $\mathrm{rad}(q_i : a) = \mathrm{rad}(q_i)$. Suppose $b \in \mathrm{rad}(q_i : a)$, i.e. $b^n a \in q_i$. As q_i is primary and $a \notin q_i$, we have $(b^n)^m \in q_i$, i.e. $b \in \mathrm{rad}(q_i)$. Hence, $\mathrm{rad}(q_i : a) \subseteq \mathrm{rad}(q_i)$. The other inclusion is obvious. We conclude that $\mathrm{rad}(I : a)$ equals the intersection of the prime ideals $\mathrm{rad}(q_j)$ satisfying $a \notin q_j$.

By Exercise 7, if $\mathrm{rad}(I : a)$ is prime then it is equal to $\mathrm{rad}(q_j)$ for some j. Next, consider any $\mathrm{rad}(q_{i_0})$. As the primary decomposition is minimal, there exists $a \in \bigcap_{j\neq i_0} q_j \backslash q_{i_0}$. The conclusion above shows that $\mathrm{rad}(I : a) = \mathrm{rad}(q_{i_0})$.

It remains to prove the last assertion. If $(I : a)$ is prime, then it is equal to its radical. Thus, we must consider a prime ideal of the form $\mathrm{rad}(I : a)$ and show that it equals $(I : a')$ for some $a' \in R$. We already know that $\mathrm{rad}(I : a) = \mathrm{rad}(q_{i_0})$ for some i_0. By Exercise 8, $\mathrm{rad}(q_{i_0})^n \subset q_{i_0}$ for some positive integer n. Hence, there exists n such that $(\bigcap_{j\neq i_0} q_j)\cdot(\mathrm{rad}(q_{i_0}))^n \subseteq I$. We fix the smallest n with this property. Then we pick an element

$$a' \in \left((\bigcap_{j\neq i_0} q_j) \cdot (\mathrm{rad}(q_{i_0}))^{n-1} \right) \backslash I.$$

(Here, if $n = 1$, then $\text{rad}(q_{i_0})^{n-1}$ is the ring R.) By definition, $a' \cdot \text{rad}(q_{i_0}) \subseteq I$, and thus $\text{rad}(q_{i_0}) \subseteq (I : a')$. However, $a' \in (\bigcap_{j \neq i_0} q_j) \backslash I$, so $a' \notin q_{i_0}$. We have the inclusions $\text{rad}(q_{i_0}) \subseteq (I : a') \subseteq \text{rad}(I : a') = \text{rad}(q_{i_0})$, which are in fact equalities. The last equation follows from the previous paragraph. □

Definition 3.18. The *associated primes* of an ideal I are the radicals of the primary ideals appearing in a minimal primary decomposition. Equivalently, these are the prime ideals of the form $\text{rad}(I : a)$ for some element a of the ring. If the ring is Noetherian, these are the prime ideals of the form $(I : a)$.

Before going further, let us discuss the geometric meaning of the associated primes. If $I = \bigcap_{i=1}^{k} q_i$ then $\text{rad}(I) = \bigcap_{i=1}^{k} \text{rad}(q_i)$. Thus, every component in the irreducible decomposition of the variety $\mathcal{V}(I)$ corresponds to one of the associated primes of I. However, the converse is not true.

Example 3.19. Let $I = \langle x^2, xy \rangle$ as in Example 3.15. We have $\text{rad}(I) = \langle x \rangle$. The variety $\mathcal{V}(I)$ is irreducible. It is a line in a plane. However, the minimal primary decompositions (3.2) reveal that I has two associated primes: the expected prime $\langle x \rangle$ and the unexpected prime $\langle x, y \rangle$, which is a point on the line. Thus, the associated primes remember more information than just the variety. There is a point corresponding to the ideal $\langle x, y \rangle$ on the line defined by $\langle x \rangle$. This point is hidden, or embedded, inside the line.

The formal replacement of varieties (corresponding to radical ideals) by arbitrary ideals made possible a tremendous advance of 20th century algebraic geometry. One is now able to work with functions that are nonzero but have square zero, using basic well-understood algebra. This advance should be compared to the development of complex numbers in the 18th and 19th centuries. Basically in the same way, instead of answering the question "does there exist a square root of -1?" one introduces imaginary numbers and shows how to use them in an efficient way. Still, we should not forget the classical geometry we started from. The line from Example 3.19 is of a different nature than the point, and the two should be distinguished.

Definition 3.20. For an ideal I, let $\text{Ass}(I)$ be the set of associated primes. The minimal (with respect to inclusion) elements of $\text{Ass}(I)$ are the *minimal primes* of I. Associated primes that are not minimal are called *embedded*.

In what follows, our standing assumption is that R is a Noetherian ring. An embedded prime p of an ideal I must contain a minimal prime p'. This means that the irreducible component $\mathcal{V}(p')$ of $\mathcal{V}(I)$ strictly contains the irreducible variety $\mathcal{V}(p)$. We say that $\mathcal{V}(p)$ is *embedded* in $\mathcal{V}(p')$. It is not necessary to describe $\mathcal{V}(I)$ as a set. The minimal primes correspond exactly to irreducible components of $\mathcal{V}(I)$. They are the nonredundant intersectands in

$$\mathrm{rad}(I) \;=\; \bigcap_{i=1}^{k} \mathrm{rad}(q_i).$$

The next lemma offers another explanation for the name minimal primes.

Lemma 3.21. *A prime ideal is a minimal prime of I if and only if it is a minimal element (with respect to inclusion) among the primes that contain I.*

Proof. It is enough to prove that every prime p containing I also contains a prime in $\mathrm{Ass}(I)$. Then p also contains a minimal prime. They are equal if p is minimal with respect to inclusion. Thus, suppose p contains $I = \bigcap_{i=1}^{k} q_i$. By Exercise 7, $p \supseteq q_{i_0}$ for some some i_0. Hence, $p = \mathrm{rad}(p) \supset \mathrm{rad}(q_{i_0})$. □

The geometry that distinguishes embedded and minimal primes suggests an idea of how to get additional uniqueness properties in primary decompositions. Indeed, in Example 3.15 it is the ideal corresponding to the embedded component that changes, while the minimal prime remains the same.

Theorem 3.22. *Let $I = \bigcap_{i=1}^{k} q_i$ be a minimal primary decomposition. The primary ideals q_i corresponding to the minimal primes are determined by I.*

Proof. Let q_{i_0} be such that $\mathrm{rad}(q_{i_0})$ is a minimal prime. We claim that

$$q_{i_0} \;=\; \big\{a \,:\, ab \in I \text{ for some } b \notin \mathrm{rad}(q_{i_0})\big\}. \tag{3.3}$$

We already saw that the right-hand side does not depend on the decomposition of I. Thus the equation implies the theorem. We prove both inclusions.

Let $a \in q_{i_0}$. For every $i \neq i_0$ we have $q_i \not\subset \mathrm{rad}(q_{i_0})$. Otherwise, $\mathrm{rad}(q_i) \subset \mathrm{rad}(q_{i_0})$, which would contradict the hypothesis that $\mathrm{rad}(q_{i_0})$ is minimal. Hence, there exists $b_i \in q_i \backslash \mathrm{rad}(q_{i_0})$. We define $b := \prod_{j \neq i_0} b_j$. As $\mathrm{rad}(q_{i_0})$ is prime, we have $b \notin \mathrm{rad}(q_{i_0})$. However, $ab \in q_j$ for $j \neq i_0$, as $b \in q_j$. Furthermore, $ab \in q_{i_0}$, as $a \in q_{i_0}$. This implies $ab \in I = \bigcap_{i=1}^{k} q_i$, which means that a is contained in the right-hand side of (3.3).

Now we pick a and $b \notin \mathrm{rad}(q_{i_0})$ such that $ab \in I$. In particular, $ab \in q_{i_0}$. If $a \notin q_{i_0}$ then we get a contradiction to the fact that q_{i_0} is primary. This shows that the right-hand side is contained in the left-hand side. □

Let I and J be ideals in a Noetherian ring R. Using the definition in Proposition 1.3, we consider the following chain of ideal quotients in R:

$$(I : J) \subseteq (I : J^2) \subseteq (I : J^3) \subseteq \cdots.$$

This chain stabilizes and we set $(I : J^\infty) = (I : J^m)$ for $m \in \mathbb{N}$ sufficiently large. The ideal $(I : J^\infty)$ is called the *saturation* of I with respect to J. If $J = \langle f \rangle$ is a principal ideal, then the notation $(I : f^\infty)$ is also used for this.

Corollary 3.23. *Let I be the ideal in Theorem 3.22. The primary ideal q_i corresponding to a minimal prime $p_i = \mathrm{rad}(q_i)$ of I can be computed as*

$$q_i \;=\; \big(I : (a_1 \cdots a_{i-1} a_{i+1} \cdots a_k)^\infty\big),$$

where a_j is an element of $p_j \backslash p_i$ for $j \in \{1, \ldots, k\} \backslash \{i\}$.

Proof. This is equivalent to (3.3) when $i = i_0$, since $p_i = \mathrm{rad}(q_i)$. By definition, the saturation on the right-hand side consists of all ring elements b such that $b(a_1 \cdots a_{i-1} a_{i+1} \cdots a_k)^m \in I$ for some positive integer m. □

Primary decomposition for monomial ideals is easier than for general polynomial ideals. The associated primes are generated by subsets of the variables and can be characterized combinatorially. We here just show this for one example. For more information we refer to the textbook [**41**].

Example 3.24. Let $n = 3$ and $I = \langle xy^2z^3, x^2yz^3, xy^3z^2, x^3yz^2, x^2y^3z, x^3y^2z \rangle$. This has seven associated primes. A minimal primary decomposition is

$$I \;=\; \langle x \rangle \cap \langle y \rangle \cap \langle z \rangle \cap \langle x^2, y^2 \rangle \cap \langle x^2, z^2 \rangle \cap \langle y^2, z^2 \rangle \cap \langle x^3, y^3, z^3 \rangle.$$

This example generalizes to $n \geq 4$ as follows. The ideal I is generated by the $n!$ monomials $\prod_{i=1}^n x_i^{\pi_i}$, indexed by permutations $\pi \in S_n$, and $\mathrm{Ass}(I)$ consists of all $2^n - 1$ ideals generated by nonempty subsets of $\{x_1, \ldots, x_n\}$.

There are many algorithms and implementations for computing primary decompositions. The input is an ideal I in a polynomial ring $K[\mathbf{x}]$, and the output is the set $\mathrm{Ass}(I)$ and primary ideals $q_1, \ldots, q_k$ satisfying (3.1). Traditionally, these are symbolic methods built upon Gröbner bases. In recent years, numerical tools for decomposing ideals and varieties have received much attention. Solving polynomial systems means running such software.

3.3. Linear PDEs with Constant Coefficients

In this section, we offer an alternative perspective on the problem of solving systems of polynomial equations. This highlights the role of embedded primes and primary ideals in a context of practical importance.

Every polynomial with real or complex coefficients can be interpreted as a linear differential operator with constant coefficients. This operator is obtained by replacing x_i with the differential operator $\frac{\partial}{\partial x_i}$. Every ideal I in $\mathbb{R}[x_1, x_2, \ldots, x_n]$ can thus be interpreted as a system of linear partial differential equations (PDEs) with constant coefficients. Suppose we are interested in the solutions to these PDEs within some nice class of functions, such as polynomial functions, real analytic functions $\mathbb{R}^n \to \mathbb{R}$, or complex holomorphic functions $\mathbb{C}^n \to \mathbb{C}$. Then the set of solutions to our PDEs is a linear space over $\mathbb{R}$ or $\mathbb{C}$. We are interested in computing a basis of that solution space. This computation rests on the primary decomposition of the ideal I. Both minimal primes and embedded primes will play a role, and all primary components will contribute to our basis of the solution space. However, first of all, let us interpret the usual points of $\mathcal{V}(I)$ in terms of PDEs.

Lemma 3.25. *Let I be an ideal in $\mathbb{C}[\mathbf{x}]$. A point $(a_1, \ldots, a_n) \in \mathbb{C}^n$ lies in the variety $\mathcal{V}(I)$ if and only if the exponential function $\exp(a_1x_1+\cdots+a_nx_n)$ is a solution of the system of partial differential equations given by I.*

Proof. Let $f(\mathbf{x}) = \exp(a_1x_1 + \cdots + a_nx_n)$. Then $\frac{\partial f}{\partial x_i} = a_i \cdot f$ for all i. Let g be any polynomial in n variables and $g(\frac{\partial}{\partial \mathbf{x}})$ the corresponding differential operator. By induction on the degree of g, with degree 1 as the base case, we find that applying the operator $g(\frac{\partial}{\partial \mathbf{x}})$ to the function $f(\mathbf{x})$ yields $g(a_1, \ldots, a_n)$ times $f(\mathbf{x})$. This is zero for all $g \in I$ if and only if $(a_1, \ldots, a_n) \in \mathcal{V}(I)$. □

Lemma 3.25 embeds the classical solutions of a polynomial system into the solution space of the associated linear PDEs. But if the ideal is not radical, then it has more solutions, which are governed by the primary decomposition. We shall explain this for an ideal that was already encountered twice.

Example 3.26. We revisit Example 3.6 and Exercise 10 of Chapter 1. Let $n = 3$ and $I = \langle x^3 - yz, y^3 - xz, z^3 - xy \rangle$. The corresponding system of linear PDEs asks for all functions $f = f(x, y, z)$ that satisfy

$$\frac{\partial^3 f}{\partial x^3} = \frac{\partial^2 f}{\partial y \partial z}, \quad \frac{\partial^3 f}{\partial y^3} = \frac{\partial^2 f}{\partial x \partial z} \quad \text{and} \quad \frac{\partial^3 f}{\partial z^3} = \frac{\partial^2 f}{\partial x \partial y}. \tag{3.4}$$

To make this problem precise, we must specify the class of functions f that are allowed. For instance, we might take all holomorphic functions $f : \mathbb{C}^3 \to \mathbb{C}$. Alternatively, we might seek real analytic solutions $f : \mathbb{R}^3 \to \mathbb{R}$ or, among these, all polynomial solutions. Let's leave this unspecified for now.

The degree of our ideal I is $27 = 3 \times 3 \times 3$, which comes from the degrees of the three generators of I. The number 27 is also the dimension of the space of holomorphic solutions f to (3.4). A basis of that solution space consists of

$$\begin{gathered} 1\,, x\,, y\,, z\,, x^2\,, y^2\,, z^2\,, x^3{+}6yz\,, y^3{+}6xz,\; z^3{+}6xy,\; x^4{+}y^4{+}z^4{+}24xyz, \\ \exp(x-y-z)\,,\ \exp(x+y+z)\,,\ \exp(-x-y+z)\,,\ \exp(-x+y-z)\,, \\ \exp(x{-}iy{+}iz)\,, \exp(x{+}iy{-}iz)\,, \exp(-x{-}iy{-}iz)\,, \exp(-x{+}iy{+}iz)\,, \\ \exp(ix-y+iz)\,, \exp(ix+y-iz)\,, \exp(ix-iy+z)\,, \exp(ix+iy-z)\,, \\ \exp(-ix{-}y{-}iz)\,, \exp(-ix{+}y{+}iz)\,, \exp(-ix{-}iy{-}z)\,, \exp(-ix{+}iy{+}z). \end{gathered} \tag{3.5}$$

The subspace of polynomial solutions has dimension 11 and is spanned by the first row in (3.5). The larger subspace of real analytic solutions has dimension 15 and is spanned by the first two rows. All other basis functions are exponentials of linear forms that have $i = \sqrt{-1}$ among their coefficients. The 16 basis solutions in the last four rows of (3.5), along with the solution $1 = \exp(0x + 0y + 0z)$, are explained by Lemma 3.25. They are the exponential functions corresponding to the 17 distinct points in $\mathcal{V}(I) \subset \mathbb{C}^3$.

The basis in (3.5) was derived from the minimal primary decomposition

$$(3.6) \qquad I \quad = \quad Q \;\cap \bigcap_{\substack{a+b+c\,\equiv\,0 \\ \text{mod } 4}} \langle\, x - i^a,\, y - i^b,\, z - i^c \,\rangle \qquad \text{in } \mathbb{C}[x, y, z].$$

This decomposition is obtained by refining the primary decomposition over the rational numbers shown in Example 3.6. The 16 ideals in the intersection on the right-hand side of (3.6) are maximal and hence prime. They correspond to the 16 exponential solutions in (3.5). The ideal Q is primary to the maximal ideal $\mathrm{rad}(Q) = \langle x, y, z\rangle$. Since all associated primes are minimal, by Theorem 3.22 this primary ideal is uniquely determined by I:

$$Q \;=\; \big\langle\, x^2y, x^2z, xy^2, xz^2, y^2z, yz^2, x^3 - yz, y^3 - xz, z^3 - xy \,\big\rangle.$$

This 0-dimensional primary ideal has degree 11. It contributes the 11 polynomial solutions to the three partial differential equations in (3.4).

Below is a general result explaining our observations from Example 3.26.

Theorem 3.27. *Let I be a 0-dimensional ideal in $\mathbb{C}[x_1, \ldots, x_n]$, here interpreted as a system of linear PDEs. The space of holomorphic solutions has dimension equal to the degree of I. There exist nonzero polynomial solutions if and only if the maximal ideal $M = \langle x_1, \ldots, x_n\rangle$ is an associated prime of I. In that case, the polynomial solutions are precisely the solutions to the system of PDEs given by the M-primary component $(I : (I : M^\infty))$.*

Proof. Fix a degree-compatible monomial order and let $\mathrm{in}(I)$ be the initial ideal of I for that order. The set $\mathcal{S}$ of standard monomials is finite. For each $\mathbf{x}^\mathbf{u} \in \mathcal{S}$ we will construct explicitly a power series solution to the PDE given by I. We will also show that these solutions form a basis for the space of holomorphic solutions. These are the solutions represented by power series.

Regarding I as a $\mathbb{C}$-vector space, it has a basis consisting of elements of the form $\mathbf{x}^\mathbf{v} + \sum_{\mathbf{x}^\mathbf{u} \in \mathcal{S}} \lambda_\mathbf{u} \mathbf{x}^\mathbf{u}$, where $\mathbf{x}^\mathbf{v} \notin \mathcal{S}$. Consider a polynomial $\tilde{p}$ that is a $\mathbb{C}$-linear combination of monomials in $\mathcal{S}$. We claim that $\tilde{p}$ can be uniquely extended to a power series p that is a solution to the associated PDEs. Indeed, the above basis operators uniquely determine the coefficients of all other monomials, so p is unique. Further, p has the property that when differentiated with any operator from I, the constant term in the result is zero. Thus, all operators in I annihilate p. Hence, the dimension of the solution space equals $|\mathcal{S}| = \text{degree}(I)$. The basis of this space is given by

$$(3.7) \quad p_{\mathbf{u}}(x_1, \ldots, x_n) \;=\; \mathbf{x}^{\mathbf{u}} \;+\; \text{higher-order terms}, \;\; \text{where } \mathbf{x}^{\mathbf{u}} \text{ runs over } \mathcal{S}.$$

The series (3.7) is a polynomial if and only if it is annihilated by $(\partial/\partial x_i)^d$ for some d and $i = 1, 2, \ldots, n$. This is always the case when I is M-primary.

Suppose now that I is primary in $\mathbb{C}[\mathbf{x}]$. Since I is 0-dimensional, its radical is the maximal ideal $\langle x_1 - a_1, \ldots, x_n - a_n \rangle$, where $\mathcal{V}(I) = \{(a_1, \ldots, a_n)\}$ in $\mathbb{C}^n$. By translating $(a_1, \ldots, a_n)$ to the origin $(0, \ldots, 0)$, we can apply the analysis in the previous paragraph. From this and Lemma 3.25, we obtain degree(I) many polynomials $p_{\mathbf{u}}$ with $\mathbf{x}^{\mathbf{u}} \in \mathcal{S}$ as in (3.7) such that

$$(3.8) \qquad p_{\mathbf{u}}(x_1, \ldots, x_n) \cdot \exp(a_1 x_1 + \cdots + a_n x_n)$$

solves the PDEs given by I. These functions form a basis of the holomorphic solutions to I. None of them is a polynomial unless $(a_1, \ldots, a_n) = (0, \ldots, 0)$.

Next, let I be an arbitrary 0-dimensional ideal. Its minimal primary decomposition (3.1) is unique, by Theorem 3.22. The solution space to I, regarded as a system of linear PDEs, contains the solution spaces of its primary components $q_1, q_2, \ldots, q_k$. For each of these primary ideals, we construct a basis of holomorphic solutions (3.8). The union of these bases is a basis of the solution space of I, and its cardinality equals degree(I).

Finally, we argue that if $M \in \text{Ass}(I)$, then the M-primary component of I is the double quotient $(I : (I : M^\infty))$. Actually, this is a special case of Corollary 3.23 since every maximal ideal is prime. In the primary decomposition (3.1), suppose that q_1 is M-primary. Then $(I : M^\infty) = q_2 \cap \cdots \cap q_k$. Taking the ideal quotient of I by $q_2 \cap \cdots \cap q_k$ recovers the ideal q_1. Hence $q_1 = (I : (I : M^\infty))$ as desired. This completes the proof. $\square$

In the preceding discussion, we studied the solutions to 0-dimensional polynomial systems in the guise of linear PDEs with constant coefficients. We saw that the solution space of such an ideal I is a vector space of dimension equal to the degree of I. This is different from the situation of solving polynomial equations. The variety $\mathcal{V}(I)$ of classical solutions in $\mathbb{C}^n$ changes its cardinality depending on whether I is radical or not. The solution space to the PDEs, on the other hand, always has the expected dimension degree(I), independently of whether the ideal I is radical or not.

The solution spaces to our PDEs vary gracefully with parameter changes. This underscores the utility of primary decompositions in the context of solving equations. We demonstrate this perspective in a simple example.

Example 3.28 ($n = 2$). Consider the ideal $I = \langle x^2 - \delta^2, y^2 - \epsilon^2 \rangle \subset \mathbb{R}[x, y]$, where δ and ϵ are small real parameters. We view I as a PDE system:

$$\frac{\partial^2 f}{\partial x^2} = \delta^2 f \quad \text{and} \quad \frac{\partial^2 f}{\partial y^2} = \epsilon^2 f.$$

For $\delta, \epsilon \neq 0$, the solution space is spanned by the four exponential functions

$$f_{ij} \; := \; \exp\big((-1)^i \delta x + (-1)^j \epsilon y\big) \quad \text{where } i, j \in \{0, 1\}.$$

However, these four functions become linearly dependent when $\delta\epsilon = 0$. We therefore change the basis of our 4-dimensional solution space as follows:

$$\begin{array}{rclcl}
g_{00} & = & \frac{1}{4}(f_{00} + f_{01} + f_{10} + f_{11}) & = & 1 + \frac{\delta^2}{2}x^2 + \frac{\epsilon^2}{2}y^2 + \cdots, \\
g_{01} & = & \frac{1}{4\epsilon}(f_{00} - f_{01} + f_{10} - f_{11}) & = & y + \frac{\delta^2}{2}x^2 y + \frac{\epsilon^2}{6}y^3 + \cdots, \\
g_{10} & = & \frac{1}{4\delta}(f_{00} + f_{01} - f_{10} - f_{11}) & = & x + \frac{\delta^2}{6}x^3 + \frac{\epsilon^2}{2}xy^2 + \cdots, \\
g_{11} & = & \frac{1}{4\epsilon\delta}(f_{00} - f_{01} - f_{10} + f_{11}) & = & xy + \frac{\delta^2}{6}x^3 y + \frac{\epsilon^2}{6}xy^3 + \cdots.
\end{array}$$

This family remains linearly independent for all values of δ and ϵ. In particular, for $\delta = \epsilon = 0$, we obtain the standard basis $\mathcal{S} = \{1, x, y, xy\}$ modulo the ideal $\mathrm{in}(I) = \langle x^2, y^2 \rangle$. This is a basis of the solutions to $\frac{\partial^2 f}{\partial x^2} = \frac{\partial^2 f}{\partial y^2} = 0$.

We next briefly discuss the PDEs arising from polynomial ideals I that are not 0-dimensional. It is still true that the primary decomposition of I reveals the solution space of these PDEs. The precise statement is an important result in analysis known as *Ehrenpreis's Fundamental Principle* or the *Palamodov-Ehrenpreis Theorem*. The details of this theorem are outside the scope of this book. For the statement see [**53**, §10.5] and the references given therein.

We here illustrate the role of primary decomposition in one example. The key observation is that embedded primes reveal spurious solution spaces.

Example 3.29. Let $n = 4$ and consider the ideal

$$J \; = \; \langle\, xw,\; xz + yw,\; yz \rangle.$$

Somewhat surprisingly, this is not radical. Its radical is the monomial ideal

$$\sqrt{J} \; = \; \langle x, y \rangle \cap \langle z, w \rangle \; = \; \langle xw, xz, yw, yz \rangle.$$

The given ideal J has three associated primes. The primes $\langle x, y \rangle$ and $\langle z, w \rangle$ are minimal primes, and the maximal ideal $\langle x, y, z, w \rangle$ is an embedded prime. A minimal primary decomposition of the given ideal is

$$J \; = \; \langle x, y \rangle \;\cap\; \langle z, w \rangle \;\cap\; (J + \langle x, y, z, w \rangle^3).$$

The third primary ideal is not unique. If we replace the third power of the maximal ideal by any higher power, then the intersection remains the same.

As before, we interpret the generators of J as a system of linear PDEs:

$$\frac{\partial^2 f}{\partial x \partial w} = \frac{\partial^2 f}{\partial x \partial z} + \frac{\partial^2 f}{\partial y \partial w} = \frac{\partial^2 f}{\partial y \partial z} = 0.$$

The linear space of solutions $f(x,y,z,w)$ is infinite-dimensional. It is spanned by all functions of the form $g(w,z)$ and $h(x,y)$, together with the one special function $xz - yw$. The former correspond to the two minimal primes. The latter spurious solution arises from the embedded primary component.

Whenever one encounters a system of polynomial equations with special structure and one is curious about the variety of solutions, it pays to explore the primary decomposition and ponder the solutions to the associated PDEs. Students who struggle with *schemes* in an algebraic geometry class may find our PDE interpretation a useful way to understand their structure.

Given a system of polynomial equations, the primary decomposition often reveals interesting structures. Most importantly, it tells us how to break up the solutions into meaningful pieces. As an illustration, we examine the following question from linear algebra: *Let A, B and C be 2×2 matrices. In which ways is it possible for the triple product ABC to be the zero matrix?*

We approach this problem as follows. We set $n = 12$ and fix the polynomial ring $\mathbb{R}[a_{ij}, b_{ij}, c_{ij}]$ whose variables are the 12 entries of the matrices A, B and C. Let I be the ideal in $\mathbb{R}[a_{ij}, b_{ij}, c_{ij}]$ that is generated by the four entries of the matrix product ABC. For example, one of the four generators of I is the upper left entry of ABC. This is the trilinear form

$$a_{11}b_{11}c_{11} + a_{12}b_{21}c_{11} + a_{11}b_{12}c_{21} + a_{12}b_{22}c_{21}. \tag{3.9}$$

In the back of our minds, we think of this as a partial differential equation

$$\frac{\partial^3 f}{\partial a_{11}\partial b_{11}\partial c_{11}} + \frac{\partial^3 f}{\partial a_{12}\partial b_{21}\partial c_{11}} + \frac{\partial^3 f}{\partial a_{11}\partial b_{12}\partial c_{21}} + \frac{\partial^3 f}{\partial a_{12}\partial b_{22}\partial c_{21}} = 0. \tag{3.10}$$

The scheme-theoretic version of our linear algebra question is this: *Which functions on matrix triples satisfy these four partial differential equations?*

A computation with a computer algebra system reveals that the ideal I is radical. It is the intersection of six prime ideals. Three of them are the ideals respectively generated by the entries of A, B and C. The next two associated primes are generated by the 2×2 minors of the matrices

$$\begin{pmatrix} a_{11} & a_{21} & -b_{21} & -b_{22} \\ a_{12} & a_{22} & b_{11} & b_{12} \end{pmatrix} \quad \text{and} \quad \begin{pmatrix} b_{11} & b_{21} & -c_{21} & -c_{22} \\ b_{12} & b_{22} & c_{11} & c_{12} \end{pmatrix}.$$

Finally, the last associated prime of I is the ideal $I + \langle \det(A), \det(C) \rangle$. Thus $\mathrm{Ass}(I)$ consists of six primes, and all are minimal. Using computer algebra, we check that I is indeed equal to the intersection of these six prime ideals.

Geometrically, we are studying a variety $\mathcal{V}(I)$ in the affine space $\mathbb{C}^{12}$. It is the solution set of four cubic equations. We found that $\mathcal{V}(I)$ is the union of six irreducible components. Three of them are linear spaces of dimension 8. The other three irreducible components have dimension 9 and are not linear spaces. Their degrees are 4, 4 and 8. In response to the original linear algebra question, the six irreducible components of $\mathcal{V}(I)$ correspond to the following six scenarios for a triple of 2×2 matrices:

$$\begin{gathered}\operatorname{rank}(A) = 0 \quad \text{or} \quad \operatorname{rank}(B) = 0 \quad \text{or} \quad \operatorname{rank}(C) = 0 \quad \text{or} \\ \operatorname{rank}(A) = \operatorname{rank}(B) = 1 \quad \text{or} \quad \operatorname{rank}(B) = \operatorname{rank}(C) = 1 \\ \text{or} \quad \operatorname{rank}(A) = \operatorname{rank}(C) = 1.\end{gathered}$$

Each of the six irreducible components $\mathcal{V}(I)$ admits a parametrization with polynomials. In particular, it is a rational variety. Using Lemma 3.25, we can then write down all exponential solutions to the four partial differential equations, like (3.10), that are given by I. The solutions come in six families.

The solutions contributed by the first irreducible component, $\{\operatorname{rank}(A) = 0\}$, are the functions $f(B, C)$ that do not depend on the matrix A. The solutions contributed by the last irreducible component have the form

$$\begin{aligned} f(A,B,C) = \exp\big[& r_1 s_1 a_{11} + r_1 s_2 a_{12} + r_2 s_1 a_{21} + r_2 s_2 a_{22} + (t_{11}u_2 - s_2 t_{12}) b_{11} \\ & + (s_2 t_{21} - t_{11} u_1) b_{12} + (s_1 t_{12} - t_{22} u_2) b_{21} + (t_{22} u_1 - s_1 t_{21}) b_{22} \\ & + u_1 v_1 c_{11} + u_1 v_2 c_{12} + u_2 v_1 c_{21} + u_2 v_2 c_{22} \big], \end{aligned}$$

where r_i, s_j, t_{ij}, u_i and v_j are arbitrary complex numbers. The functions f above satisfy the PDEs because the coefficients of $a_{11}, a_{12}, \ldots, c_{22}$ furnish a parametrization of the irreducible variety $\{ABC = 0, \operatorname{rank}(A) = \operatorname{rank}(C) = 1\}$.

Here is our conclusion for this example, valid for the entire book: Taking a fresh look at linear algebra offers a point of entry to nonlinear algebra.

Exercises

(1) Let $R = \mathbb{C}[x, y]/\langle x^2, xy, y^2 \rangle$. Is $\{0\}$ an irreducible ideal? Is it primary?

(2) Let $n = 1$. Prove that an ideal I in the univariate polynomial ring $\mathbb{Q}[x]$ is a power of a prime ideal if and only if I is primary.

(3) Prove that a ring is Noetherian if and only if every ideal is finitely generated.

(4) (a) Prove that if R is Noetherian, then so is R/I for any ideal I.
(b) Prove Hilbert's Basis Theorem: If R is Noetherian, then so is $R[x]$.

(5) Prove Lemma 3.12.

(6) Check that Example 3.15 provides two distinct minimal primary decompositions. What are all primary decompositions of this ideal?

(7) (a) Prove that a prime ideal p cannot be equal to an intersection of (finitely many, more than one, incomparable) ideals.
   (b) More generally, prove that if a prime ideal contains an intersection of finitely many ideals, then it contains one of them.

(8) Prove that in a Noetherian ring every ideal contains a power of its radical. Give a counterexample in the case of a non-Noetherian ring.

(9) Find three polynomials in three unknowns, each having degree precisely 5, whose variety in $\mathbb{C}^3$ consists of precisely 37 complex solutions.

(10) Find all solutions (x, y) of the two equations $x^2 + y = \epsilon$ and $y^2 + x = \epsilon$ over the algebraic closure of the field $\mathbb{Q}(\epsilon)$. Write down series solutions.

(11) Let $n = 10$ and $K = \mathbb{R}$, and consider the ideal generated by the 10 polynomials $x_i y_j + x_j y_i$ where $1 \leq i < j \leq 5$. These are the 2×2 *subpermanents* of a 2×5 matrix of unknowns. Find a minimal primary decomposition of I. Interpret your result in terms of solving partial differential equations.

(12) Which 2×3 matrices A and B satisfy $AB^T = BA^T$? What about 3×2 matrices?

(13) Let $K = \mathbb{F}_2$ be the field with two elements. Find an ideal I in $K[x, y]$ that has precisely 10 associated primes, of which five are embedded.

(14) Consider the ideal $I = \langle x + y + z, xy + xz + yz, xyz \rangle$ generated by the three elementary symmetric polynomials in $\mathbb{Q}[x, y, z]$. Interpret these as linear PDEs and determine the solution space of the PDEs.

(15) Interpret the ideal P^2 in Example 3.13 as a system of linear PDEs and determine the solution space. How do we see that P^2 is not primary?

(16) Let A, B and C be 3×3 matrices. How is it possible for the triple product ABC to be the zero matrix? In other words, write down the nine trilinear equations like (3.9) and describe their variety of solutions.

(17) What is numerical primary decomposition?

Chapter 4

Mapping and Projecting

"*A technique is a trick that works*", Gian-Carlo Rota

A frequently encountered challenge is to compute the image of a polynomial map. Such an image need not be an algebraic variety. However, a natural outer approximation of the image is given by its Zariski closure. The Zariski closure of the image is a variety, described by the polynomials that vanish on it. In this chapter we show how this variety can be found by eliminating variables. Gröbner bases and resultants serve as our primary tools. Further, we provide theorems that enable us to understand the difference between the image and its closure. The answer we obtain depends heavily on the setting, whether we work over the complex numbers $\mathbb{C}$ or over the real numbers $\mathbb{R}$, and whether the given polynomials are homogeneous or nonhomogeneous.

4.1. Elimination

In this section we introduce elimination of variables for polynomial ideals. This is our main tool for computing the closure of the image of a polynomial map. We show how to carry it out in practice using Gröbner bases.

We fix an algebraically closed field K and the polynomial ring $K[\mathbf{x}] = K[x_1, \ldots, x_n]$. Every ideal $I \subset K[\mathbf{x}]$ has an associated affine variety

$$\mathcal{V}(I) = \{ \mathbf{p} \in K^n \,:\, f(\mathbf{p}) = 0 \text{ for all } f \in I \}.$$

We consider the projection from K^n onto the linear subspace K^m that is given by the first m coordinates:

$$\pi : K^n \to K^m, \quad (p_1, \ldots, p_m, p_{m+1}, \ldots, p_n) \mapsto (p_1, \ldots, p_m).$$

If V is a variety in K^n, its image $\pi(V)$ need not be a variety.

Example 4.1 ($n = 2, m = 1$). The image of the hyperbola $V = \mathcal{V}(xy - 1)$ under the projection $K^2 \to K^1$ from the plane to the x-axis is $\pi(V) = K^1 \backslash \{0\}$. This is not a variety in K^1. Note that the image becomes closed if, prior to projecting, we first perform a change of coordinates. For instance, if we replace V by the hyperbola $V' = \mathcal{V}\big((x+y)(x-y) - 1\big)$, then $\pi(V') = K^1$.

The Zariski closure $\overline{\pi(V)}$ of the image $\pi(V)$ is a variety in K^m. It is the smallest variety containing $\pi(V)$. We call the variety $\overline{\pi(V)}$ the *closed image* of V under the map π. A more general case, where π is an arbitrary polynomial map instead of a coordinate projection, is discussed in the next section. The following theorem characterizes the ideal of the closed image $\overline{\pi(V)}$.

Theorem 4.2. *Let $I \subset K[\mathbf{x}]$ be an ideal and $V = \mathcal{V}(I)$ its variety in K^n, where K is an algebraically closed field. Then the closed image of V in K^m is the variety $\overline{\pi(V)} = \mathcal{V}(J)$ that is defined by the* elimination ideal

$$J \; = \; I \cap K[x_1, \ldots, x_m]. \tag{4.1}$$

If I is radical or prime, then the elimination ideal J has the same property.

Proof. If J is not a prime ideal, then there exist polynomials f and g in $K[x_1, \ldots, x_m]$ such that $fg \in J$ but $f, g \notin J$. The same polynomials show that I is not prime. Similarly, if J is not radical, then there exist f in $K[x_1, \ldots, x_m]$ and $r \geq 2$ such that $f^r \in J$ but $f \notin J$. The same f shows that I is not radical. Similar reasoning shows that all ideals $I \subset K[\mathbf{x}]$ satisfy

$$\mathrm{Rad}(I) \cap K[x_1, \ldots, x_m] \quad = \quad \mathrm{Rad}\big(I \cap K[x_1, \ldots, x_m]\big).$$

Since passing to the radical does not change the variety of a given ideal, we may assume that I and J are radical ideals. We shall now make a forward reference and use the Nullstellensatz (Chapter 6). A polynomial belongs to I if and only if it vanishes on $V = \mathcal{V}(I)$. This holds, in particular, for polynomials f in the subring $K[x_1, \ldots, x_m]$. Such an f belongs to J if and only if it vanishes on $\pi(V)$, which is the case if and only if it vanishes on $\overline{\pi(V)}$. The latter condition means that f lies in the radical ideal of $\overline{\pi(V)}$. We conclude that the radical ideal of the closed image $\overline{\pi(V)}$ is precisely the elimination ideal J. For further details we refer to [**10**, §4.4, Theorem 3]. □

Theorem 4.2 says that the algebraic operation of elimination corresponds to the geometric operation of projection. This holds in many settings, not just in algebraic geometry. For instance, Gaussian elimination in linear algebra corresponds to projection of linear subspaces, and Fourier-Motzkin elimination in convex geometry corresponds to projection of polyhedra. Alternatively, from the perspective of logic, we can think of our projection as quantifier elimination. We are eliminating the $n-m$ existentially quantified variables from the first-order logic statement $\exists\, x_{m+1}, \ldots, x_n \,:\, \mathbf{x} \in V$.

Elimination and projection are fundamental operations in many applications. One good example is the problem of matrix completion or tensor completion, which arises frequently in data science. Here is an illustration.

Example 4.3 (Matrix completion)**.** Fix $n = 15$ and let V be the irreducible variety of symmetric 5×5 matrices $X = (x_{ij})$ of rank ≤ 2. Its prime ideal $I = \mathcal{I}(V)$ is minimally generated by 50 homogeneous cubic polynomials, namely the 3×3 minors of the matrix X. In fact, these 50 cubics form a Gröbner basis for the degree reverse lexicographic order.

We now order the 15 variables x_{ij} in the given polynomial ring so that the five diagonal entries $x_{11}, x_{22}, x_{33}, x_{44}, x_{55}$ come last. We wish to eliminate these five variables from the prime ideal I. So, in the notation above, we have $m = 10$. A computation reveals that the elimination ideal is principal:

$$\begin{aligned} J \,=\, \langle\, & x_{14}x_{15}x_{23}x_{25}x_{34} - x_{13}x_{15}x_{24}x_{25}x_{34} - x_{14}x_{15}x_{23}x_{24}x_{35} \\ & + x_{13}x_{14}x_{24}x_{25}x_{35} + x_{12}x_{15}x_{24}x_{34}x_{35} - x_{12}x_{14}x_{25}x_{34}x_{35} \\ & + x_{13}x_{15}x_{23}x_{24}x_{45} - x_{13}x_{14}x_{23}x_{25}x_{45} - x_{12}x_{15}x_{23}x_{34}x_{45} \\ & + x_{12}x_{13}x_{25}x_{34}x_{45} + x_{12}x_{14}x_{23}x_{35}x_{45} - x_{12}x_{13}x_{24}x_{35}x_{45} \,\rangle. \end{aligned}$$

The ideal generator of degree 5 is known as the *pentad* in algebraic statistics [**19**, Example 4.2.8]. Its 12 terms correspond to the 12 Hamiltonian cycles in the complete graph K_5. The hypersurface $\mathcal{V}(J)$ equals the closed image $\pi(V)$ of the determinantal variety V under the projection from K^{15} onto the subspace K^{10} whose coordinates are the off-diagonal entries.

Our result has the following interpretation in terms of matrix completion. If the 10 off-diagonal entries of a symmetric 5×5 matrix are given, then they can be completed to a matrix of rank ≤ 2 only if the pentad vanishes. This pentad constraint appears in the statistical theory of *factor analysis*.

Our next example shows how to find algebraic relations via elimination.

Example 4.4. The first four power sums in three variables x, y, z are the polynomials $x^i + y^i + z^i$ for $i = 1, 2, 3, 4$. These four polynomials must be algebraically dependent since they involve only three variables. But what is the algebraic relation satisfied by these four power sums?

We approach this question by setting $n = 7$ and $m = 4$, with the ideal

$$I \;=\; \langle\, x+y+z-p_1,\; x^2+y^2+z^2-p_2,\; x^3+y^3+z^3-p_3,\; x^4+y^4+z^4-p_4 \,\rangle.$$

This ideal lives in a polynomial ring in seven variables. We wish to eliminate the three original variables x, y, z. Thus, we ask for the elimination ideal

$$J = I \cap K[p_1, p_2, p_3, p_4].$$

This is a principal prime ideal. Its generator is a polynomial of degree 4:

$$J = \langle p_1^4 - 6p_1^2 p_2 + 3p_2^2 + 8p_1 p_3 - 6p_4 \rangle.$$

This is the desired relation, as you can check by plugging in the power sums.

The computations in Examples 4.3 and 4.4 were carried out using Gröbner bases. Here is how this works. We first fix the lexicographic monomial order $\prec$ on $K[\mathbf{x}]$ with $x_1 \prec x_2 \prec \cdots \prec x_n$. We then compute the reduced Gröbner basis for the ideal generated by the given polynomials. And, finally, we select those polynomials from the output that use only the first m variables. This method is justified by the following theorem.

Theorem 4.5. *If $\mathcal{G}$ is a lexicographic Gröbner basis for an ideal I in the polynomial ring $K[\mathbf{x}]$, then its elimination ideal J in* (4.1) *has the Gröbner basis $\mathcal{G}' = \mathcal{G} \cap K[x_1, \ldots, x_m]$. If $\mathcal{G}$ is the reduced Gröbner basis of I, then $\mathcal{G}'$ is the reduced Gröbner basis of J.*

Proof. Clearly, the set $\mathcal{G}'$ is contained in $J = I \cap K[x_1, \ldots, x_m]$. Consider any nonzero polynomial $f \in J$. The initial monomial $\text{in}_\prec(f)$ is divisible by $\text{in}_\prec(g)$ for some $g \in \mathcal{G}$. None of the variables $x_{m+1}, \ldots, x_n$ appears in the monomial $\text{in}_\prec(g)$. Every nonleading term of g is lexicographically smaller and so cannot use any of the last $n - m$ variables. Hence, g lies in $\mathcal{G}'$.

We have shown that some initial monomial from $\mathcal{G}'$ divides $\text{in}_\prec(f)$. Since the polynomial f was chosen arbitrarily from $J \backslash \{0\}$, this means that $\mathcal{G}'$ is a Gröbner basis for J. If the given Gröbner basis $\mathcal{G}$ is reduced, then $\mathcal{G}'$ also satisfies the two requirements for being a reduced Gröbner basis. □

This result shows that the lexicographic Gröbner basis $\mathcal{G}$ solves the elimination problem simultaneously for all m. Thus, computing $\mathcal{G}$ means triangularizing a given system of polynomial equations. We saw in Example 1.19 that it can be costly to compute a lexicographic Gröbner basis. One therefore often uses different strategies to carry out the elimination process. But Theorem 4.5 represents the main idea that underlies these strategies. Lexicographic elimination is a key tool for solving systems of polynomial equations. It is instructive to try this procedure for some 0-dimensional varieties.

Example 4.6. Here is a simple question: Can you find three real numbers x, y and z whose ith power sum equals i for $i = 1, 2, 3$? To answer this question, we compute the lexicographic Gröbner basis of the following ideal:

$$I = \langle x+y+z-1,\, x^2+y^2+z^2-2,\, x^3+y^3+z^3-3\,\rangle.$$

This Gröbner basis is

$$\mathcal{G} = \{\,6\underline{z^3}-6z^2-3z-1,\, 2\underline{y^2}+2yz-2y+2z^2-2z-1,\, \underline{x}+y+z-1\,\}.$$

Theorem 4.2 says that we can solve our equations by back-substitution. Indeed, the equations have six complex zeros. We first compute the three roots of the cubic in z, we substitute them into the second equation and solve for y, and then we set $x = 1-y-z$. The cubic has one real root and two complex conjugate roots:

$$z \in \big\{1.4308, -0.21542-0.26471i, -0.21542+0.26471i\big\}.$$

By symmetry, the zeros of I are precisely the six points in $\mathbb{C}^3$ whose coordinates are permutations of the three complex numbers above. Hence, the answer to our question is "no". The variety $\mathcal{V}(I)$ has no real points.

4.2. Implicitization

Implicitization is a special instance of elimination. Here, the problem is to compute the image of a polynomial map between two affine spaces. This can be done by forming the graph of the map and then projecting onto the image coordinates. To be precise, we consider a map of the form

$$f : K^n \to K^m, \quad \mathbf{p} = (p_1,\ldots,p_n) \mapsto \big(f_1(\mathbf{p}),\ldots,f_m(\mathbf{p})\big), \tag{4.2}$$

where $f_1,\ldots,f_m$ are polynomials in $K[z_1,\ldots,z_n]$ and K is an algebraically closed field. We write image(f) for the image of K^n under this map. This subset of K^m need not be a variety, as the following example shows:

Example 4.7. Let $n=2$ and $m=3$, and consider the map $f=(z_1, z_1z_2, z_1z_2^2)$ from the plane K^2 into 3-space K^3. The Zariski closure of the image is the surface $V = \mathcal{V}(x_1x_3 - x_2^2)$. The point $(0,0,1)$ is in the surface but not in image(f). For $K = \mathbb{C}$ we can approximate $(0,0,1)$ by a sequence of points that do lie in the image, e.g. by taking $z_1 = \epsilon^2$ and $z_2 = \epsilon^{-1}$ for $\epsilon \to 0$.

The *closed image* of the map $f : K^n \to K^m$ is the Zariski closure of the set image(f). The closed image is denoted by $\overline{\text{image}}(f) \subset K^m$.

Corollary 4.8. *Given the map f in (4.2), let I be the ideal in the polynomial ring $K[\mathbf{x},\mathbf{z}]$ in $n+m$ variables which is generated by $f_i(z_1,\ldots,z_n)-x_i$ for $i = 1,2,\ldots,m$. The closed image of $f : K^n \to K^m$ is the variety defined by the elimination ideal $J = I \cap K[\mathbf{x}]$. In symbols, $\overline{\text{image}}(f) = \mathcal{V}(J)$. Similarly, let $X \subset K^n$ be a variety and I_X its ideal. The closed image $\overline{f(X)}$ of X is the variety defined by the elimination ideal $J = (I+I_X) \cap K[\mathbf{x}]$.*

Proof. Allowing X to be the whole space, it is enough to prove the last statement. The graph of f restricted to X is Zariski closed in K^{n+m}, and $I + I_X$ is the ideal that defines it. The image of X is the projection of the graph into K^m. With this, the claim follows from Theorem 4.2. □

Example 4.9 (Plücker relations). What are the algebraic relations between the 2×2 minors of a 2×5 matrix? We answer this question by setting $m = n = 10$ and considering the map $f : K^{10} \to K^{10}$ that takes a matrix $\begin{pmatrix} z_{11} & z_{12} & z_{13} & z_{14} & z_{15} \\ z_{21} & z_{22} & z_{23} & z_{24} & z_{25} \end{pmatrix}$ to the vector $(x_{12}, x_{13}, \ldots, x_{45})$ where $x_{ij} = z_{1i}z_{2j} - z_{1j}z_{2i}$ for $1 \leq i < j \leq 5$. The graph of f is described by an ideal I in the polynomial ring $K[\mathbf{x}, \mathbf{z}]$ in 20 variables. Note that I is generated by 10 polynomials. The desired elimination ideal equals

$$\begin{aligned} I \cap K[\mathbf{x}] \;=\; \langle\, & x_{12}x_{34} - x_{13}x_{24} + x_{14}x_{23}\,,\; x_{12}x_{35} - x_{13}x_{25} + x_{15}x_{23}\,, \\ & x_{12}x_{45} - x_{14}x_{25} + x_{15}x_{24}\,,\; x_{13}x_{45} - x_{14}x_{35} + x_{15}x_{34}\,, \\ & x_{23}x_{45} - x_{24}x_{35} + x_{25}x_{34}\,\rangle. \end{aligned}$$

These five quadrics are the Plücker relations between the maximal minors. They play a key role in our study of Grassmannians in Chapter 5. The 10 variables in $K[\mathbf{x}]$ can be written as the entries of a skew-symmetric matrix

$$X \;=\; \begin{pmatrix} 0 & x_{12} & x_{13} & x_{14} & x_{15} \\ -x_{12} & 0 & x_{23} & x_{24} & x_{25} \\ -x_{13} & -x_{23} & 0 & x_{34} & x_{35} \\ -x_{14} & -x_{24} & -x_{34} & 0 & x_{45} \\ -x_{15} & -x_{25} & -x_{35} & -x_{45} & 0 \end{pmatrix}.$$

The Plücker relations are the *Pfaffians* of size 4×4, that is, the square roots of the principal 4×4 minors of X. Thus $\mathcal{V}(I \cap K[\mathbf{x}])$ is the variety of skew-symmetric 5×5 matrices of rank ≤ 2. We shall see in Chapter 5 that, as a projective variety in $\mathbb{P}^9$, this is the Grassmannian of lines in $\mathbb{P}^4$. Each such line is written in Plücker coordinates as the image of the rank-2 matrix X.

The notion of determinant is central to linear algebra. In nonlinear algebra, there is an analogous notion of *hyperdeterminant* for tensors.

Example 4.10 (Hyperdeterminant). Let $X = (x_{ijk})$ be a tensor of format $2 \times 2 \times 2$, where the $m = 8$ tensor entries are variables. Tensors will be discussed in Chapter 9. We here just view X as an element of an 8-dimensional linear space, with basis elements indexed by three numbers $0 \leq i, j, k \leq 1$.

The tensor represents a polynomial in three variables z_1, z_2 and z_3:

$$f = x_{000} + x_{100}z_1 + x_{010}z_2 + x_{001}z_3 + x_{110}z_1z_2 + x_{101}z_1z_3 + x_{011}z_2z_3 + x_{111}z_1z_2z_3.$$

For any fixed X, this polynomial defines a surface $\mathcal{V}(f)$ in K^3. We are interested in determining the condition under which this surface is singular.

It is singular at the point $\mathbf{z}$ if and only if the pair $(X, \mathbf{z}) \in K^{11}$ lies in the variety of

$$I = \left\langle f, \frac{\partial f}{\partial z_1}, \frac{\partial f}{\partial z_2}, \frac{\partial f}{\partial z_3} \right\rangle.$$

The elimination ideal $I \cap K[\mathbf{x}]$ is principal. We find that its generator is

$$\begin{aligned} x_{110}^2 x_{001}^2 + x_{100}^2 x_{011}^2 + x_{010}^2 x_{101}^2 + x_{000}^2 x_{111}^2 + 4x_{000}x_{110}x_{011}x_{101} + 4x_{010}x_{100}x_{001}x_{111} \\ - 2x_{100}x_{110}x_{001}x_{011} - 2x_{010}x_{110}x_{001}x_{101} - 2x_{010}x_{100}x_{011}x_{101} \\ - 2x_{000}x_{110}x_{001}x_{111} - 2x_{000}x_{100}x_{011}x_{111} - 2x_{000}x_{010}x_{101}x_{111}. \end{aligned}$$

This quartic is the $2 \times 2 \times 2$ *hyperdeterminant*. It vanishes when the surface $V(f)$ fails to be smooth in K^3. Hyperdeterminants exist for tensors of many larger formats. Their study is a fascinating topic in nonlinear algebra. A standard reference is the book by Gel'fand, Kapranov and Zelevinsky [**22**].

The most basic scenario in elimination arises when m variables are eliminated from a system of $m+1$ equations. One expects the result to be a single equation in the coefficients of that system. We saw this for $m = 3$ in Examples 4.4 and 4.10. The theory of *resultants* is custom-designed to predict the eliminant in such cases. We shall explain this in the remainder of the section. To set this up, we work over the field $\mathbb{Q}$ of rational numbers.

Let $i \in \{1, 2, \ldots, m+1\}$ and fix a general inhomogeneous polynomial f_i of degree d_i in $z_1, \ldots, z_m$. This polynomial has $\binom{d_i+m}{m}$ unknown coefficients $x_{i,\mathbf{u}}$, one for each monomial $\mathbf{z}^{\mathbf{u}}$ of degree $\leq d_i$. The total number of unknown coefficients equals $n = \sum_{i=1}^{m+1} \binom{d_i+m}{m}$. We write $\mathbb{Q}[\mathbf{x}, \mathbf{z}]$ for the resulting polynomial ring in $n+m$ variables. Inside this ring we consider the ideal

$$I = \langle f_1, f_2, \ldots, f_m, f_{m+1} \rangle \subset \mathbb{Q}[\mathbf{x}, \mathbf{z}].$$

We are interested in the ideal in $\mathbb{Q}[\mathbf{x}]$ found by eliminating the m variables z_i. This ideal describes the polynomial conditions on the coefficients of $m+1$ polynomials in m variables which represent the statement that these polynomials have a common root over the algebraic closure $\overline{\mathbb{Q}}$ of our field $\mathbb{Q}$. To be very precise, in the affine space $\overline{\mathbb{Q}}^m$, the condition of having a common root is not closed. Therefore, the geometric interpretation is best seen after projectivization.

Theorem 4.11. *The elimination ideal $I \cap \mathbb{Q}[\mathbf{x}]$ is principal. Its generator is an irreducible polynomial in the entries of the coefficient vector $\mathbf{x}$, denoted by* $\mathrm{Res}(f_1, \ldots, f_{m+1})$ *and called the* resultant. *The degree of the resultant in the coefficients of f_i equals $d_1 \cdots d_{i-1} d_{i+1} \cdots d_{m+1}$ for $i = 1, 2, \ldots, m+1$.*

Proof. We refer to [**11**, Chapter 3] for the proof. In that source, and in many others, the f_i are taken to be homogeneous polynomials in $m+1$ variables. We here prefer the inhomogeneous case, since it admits a simpler formulation as an elimination ideal. The two versions are equivalent.

Here, we just briefly explain where the formula for the degree comes from. To obtain the degree D_i of the resultant in the coefficients of f_i, let us fix general coefficients of $f_1, \ldots, f_{i-1}, f_{i+1}, \ldots, f_{m+1}$. The degree D_i is the degree of the polynomial P we obtain from the resultant after substituting all of these coefficients. By Bézout's Theorem 2.16, the m fixed polynomials define $D_i' := d_1 \cdots d_{i-1} d_{i+1} \cdots d_{m+1}$ many points in the affine space $\overline{\mathbb{Q}}^m$. The condition that f_i passes through one of these points is a linear condition in the coefficients of f_i. Hence, P vanishes if and only if its variables satisfy one of D_i' linear conditions. Thus the degree D_i of P equals D_i'. □

Example 4.12 (Determinants). Let $d_1 = \cdots = d_{m+1} = 1$. The $m+1$ polynomials f_i are affine-linear. They can be written as a matrix-vector product:

$$\begin{pmatrix} f_1 \\ f_2 \\ \vdots \\ f_m \\ f_{m+1} \end{pmatrix} = \begin{pmatrix} x_{1,1} & x_{1,2} & \cdots & x_{1,m} & x_{1,m+1} \\ x_{2,1} & x_{2,2} & \cdots & x_{2,m} & x_{2,m+1} \\ \vdots & \vdots & \ddots & \vdots & \vdots \\ x_{m,1} & x_{m,2} & \cdots & x_{m,m} & x_{m,m+1} \\ x_{m+1,1} & x_{m+1,2} & \cdots & x_{m+1,m} & x_{m+1,m+1} \end{pmatrix} \begin{pmatrix} z_1 \\ z_2 \\ \vdots \\ z_m \\ 1 \end{pmatrix}.$$

The resultant $\mathrm{Res}(f_1, \ldots, f_{m+1})$ is the determinant of the coefficient matrix $(x_{i,j})$. This determinant is a homogeneous polynomial of degree $m+1$ in $n = (m+1)^2$ unknowns with $(m+1)!$ terms. It has degree 1 in the coefficients of each f_i. Note that if we set all columns but the last to zero, then the determinant clearly vanishes. The polynomials in this case become constants and do not have a common zero. To better understand this, one could pass to the projective space, i.e. homogenize the affine-linear forms f_i.

Example 4.13 (Eliminating one variable from two quadratic polynomials). Let $m = 1$ and $d_1 = d_2 = 2$ and write simply z for z_1. Our system consists of two univariate polynomials of degree 2 with six unspecified coefficients:

$$f_1 = x_{11}z^2 + x_{12}z + x_{13} \quad \text{and} \quad f_2 = x_{21}z^2 + x_{22}z + x_{23}.$$

The generator of the elimination ideal $\langle f_1, f_2 \rangle \cap \mathbb{Q}[\mathbf{x}]$ is the *Sylvester resultant*

$$\mathrm{Res}(f_1, f_2) = \det \begin{pmatrix} x_{11} & x_{12} & x_{13} & 0 \\ 0 & x_{11} & x_{12} & x_{13} \\ x_{21} & x_{22} & x_{23} & 0 \\ 0 & x_{21} & x_{22} & x_{23} \end{pmatrix}. \tag{4.3}$$

This determinant is a polynomial of degree 4, and its expansion has seven terms. It vanishes if the two quadrics have a common zero. Note that the resultant $\mathrm{Res}(f_1, f_2)$ is homogeneous of degree 2 in the three coefficients of f_1, and it is also homogeneous of degree 2 in the three coefficients of f_2. We say that this resultant is a bihomogeneous polynomial of bidegree $(d_1, d_2) = (2, 2)$.

The formula (4.3) generalizes to two polynomials in z of arbitrary degrees d_1 and d_2. The resultant is the determinant of the *Sylvester matrix* of format $(d_1+d_2)\times(d_1+d_2)$. This matrix is denoted by Syl_{d_1,d_2} and is displayed below. The first d_2 rows are formed by the coefficients of the first polynomial f_1, suitably shifted and padded with zeros. The last d_1 rows are similarly formed by the coefficients of the second polynomial f_2. The matrix is as follows:

$$(4.4) \quad \mathrm{Syl}_{d_1,d_2} = \begin{pmatrix} x_{11} & x_{12} & \cdots & \cdots & x_{1,d_1+1} & 0 & \cdots & 0 \\ 0 & x_{11} & x_{12} & \ddots & \ddots & x_{1,d_1+1} & 0 & 0 \\ \vdots & \ddots & \ddots & \ddots & \cdots & \ddots & \ddots & \vdots \\ 0 & 0 & \cdots & x_{11} & x_{12} & \cdots & \cdots & x_{1,d_1+1} \\ x_{21} & x_{22} & \cdots & \cdots & x_{2,d_2+1} & 0 & \cdots & 0 \\ 0 & x_{21} & x_{22} & \ddots & \ddots & x_{2,d_2+1} & \ddots & 0 \\ \vdots & \ddots & \ddots & \ddots & \cdots & \ddots & \ddots & \vdots \\ 0 & \cdots & 0 & x_{21} & x_{22} & \cdots & \cdots & x_{2,d_2+1} \end{pmatrix}.$$

For $d_1 = d_2 = 2$ this is the 4×4 matrix seen in (4.3).

Theorem 4.14. *The determinant of the Sylvester matrix* Syl_{d_1,d_2} *is equal to the resultant* $\mathrm{Res}(f_1, f_2)$ *of the two univariate polynomials*

$$\begin{aligned} f_1(z) &= x_{11}z^{d_1} + \cdots + x_{1,d_1}z + x_{1,d_1+1} \\ \textit{and} \quad f_2(z) &= x_{21}z^{d_2} + \cdots + x_{2,d_2}z + x_{2,d_2+1}. \end{aligned}$$

Proof. We first note that $\det(\mathrm{Syl}_{d_1,d_2})$ is a nonzero polynomial. We can see this by taking $f_1 = z^{d_1}$ and $f_2 = 1$. Here the Sylvester matrix Syl_{d_1,d_2} specializes to the identity matrix, so its determinant is nonzero.

Let Z denote the column vector with entries $z^{d_1+d_2-1}, z^{d_1+d_2-2}, \ldots, z, 1$ and F the column vector with entries $z^{d_2-1}f_1, \ldots, zf_1, f_1, z^{d_1-1}f_2, \ldots, zf_2$, f_2. Both vectors have length $d_1 + d_2$. They are related by the Sylvester matrix:

$$\mathrm{Syl}_{d_1,d_2} \cdot Z = F.$$

Multiplying on the left by the adjugate of the Sylvester matrix, we obtain

$$\det(\mathrm{Syl}_{d_1,d_2}) \cdot Z = \mathrm{adj}(\mathrm{Syl}_{d_1,d_2}) \cdot F.$$

The last coordinate of the column vector Z equals 1. Thus, the last coordinate in this equation shows that $\det(\mathrm{Syl}_{d_1,d_2})$ is a polynomial linear combination of the entries of F, and hence it lies in the ideal $\langle f_1, f_2\rangle$.

The Sylvester determinant is a nonzero homogeneous polynomial of degree d_1+d_2 that lies in the ideal $\langle f_1, f_2\rangle \cap \mathbb{Q}[\mathbf{x}]$. We know from Theorem 4.11 that this ideal is principal, and its generator $\mathrm{Res}(f_1, f_2)$ has degree $d_1 + d_2$ as well. This implies that the resultant $\mathrm{Res}(f_1, f_2)$ is equal to the Sylvester determinant $\det(\mathrm{Syl}_{d_1,d_2})$, up to a nonzero multiplicative constant. □

Example 4.15. Let $f_1(z)$ and $f_2(z)$ be univariate polynomials of degrees d_1 and d_2 in $\mathbb{Q}[z]$. This defines a map $f : \mathbb{C} \to \mathbb{C}^2$ whose closed image is an algebraic curve in the plane $\mathbb{C}^2$ with coordinates (x_1, x_2). The implicit equation of this curve is the resultant $\mathrm{Res}_z\big(x_1 - f_1(z), x_2 - f_2(z)\big)$, taken with respect to the variable z. For a concrete example consider the plane cubic curve given parametrically by $f = (z^3 + 4z, z^2 - 3)$. Its equation is

$$\det \begin{pmatrix} -1 & 0 & -4 & x_1 & 0 \\ 0 & -1 & 0 & -4 & x_1 \\ -1 & 0 & x_2+3 & 0 & 0 \\ 0 & -1 & 0 & x_2+3 & 0 \\ 0 & 0 & -1 & 0 & x_2+3 \end{pmatrix} = x_2^3 - x_1^2 + 17x_2^2 + 91x_2 + 147.$$

If $m \geq 2$ then the resultant $\mathrm{Res}(f_1, f_2, \dots, f_{m+1})$ is more difficult to compute, and there does not always exist a formula as a determinant whose entries are linear expressions in the coefficients of $f_1, f_2, \dots, f_{m+1}$. In some cases, however, such formulas are available in the literature. For instance, Sylvester already gave such a formula for $m = 2$ and $d_1 = d_2 = d_3$. A considerable body of information on matrix formulas for resultants can be found in the excellent book by Gel'fand, Kapranov and Zelevinsky [**22**].

4.3. The Image of a Polynomial Map

We have discussed methods for computing the Zariski closure of the image of a polynomial map. Can we say something about the image itself? The answer is yes, but the situation very much depends on the field K and whether we are in the projective case or the affine case. In this section we discuss methods for computing such images. We begin by highlighting the difference between the real numbers and the complex numbers with regard to this problem.

We start with a brief discussion of the situation over the real numbers $\mathbb{R}$. Let X be an affine variety in $\mathbb{R}^n$. We would like to understand the image $f(X)$ of X under a polynomial map $f = (f_1, \dots, f_m) : \mathbb{R}^n \to \mathbb{R}^m$. Easy examples show that the Zariski closure of the image and the image itself can differ a lot. For instance, this happens for $n = m = 1$, $X = \mathbb{R}$ and $f(z) = z^2$. Then $f(X) = \mathbb{R}_{\geq 0}$ is the set of nonnegative real numbers.

Is there a chance in general of describing the image using polynomials? The following example confirms that inequalities are needed to describe the image. Its conclusion is familiar from the quadratic formula learned in high school.

Example 4.16. Let $n = 4$ and let X be the hypersurface defined by

$$ax^2 + bx + c = 0,$$

where (a, b, c, x) are coordinates on $\mathbb{R}^4$. We take $m = 3$ and define f to be the projection $\mathbb{R}^4 \to \mathbb{R}^3$ onto the coordinates (a, b, c). The image is

$$f(X) = \{ (a,b,c) \in \mathbb{R}^3 \, : \, b^2 - 4ac \geq 0 \} \backslash \{ (0,0,c) \, : \, c \neq 0 \}.$$

In particular, we see that the image $f(X)$ is not a closed subset of $\mathbb{R}^3$.

The following theorem provides an answer to our question. It refers to the concept of semialgebraic sets, which was introduced in Definition 2.12.

Theorem 4.17 (Tarski-Seidenberg). *Working over the field of real numbers, the image of a variety in $\mathbb{R}^n$ under the map f is a semialgebraic set in $\mathbb{R}^m$.*

Proof. See [**6**, §1.4]. □

Thus, to provide a description of the image over $\mathbb{R}$ we need two ingredients, polynomial equations and polynomial inequalities, suitably combined.

Example 4.18. Let $n = 6$ and $m = 9$, and let f be the map that multiplies a 3×2 matrix Z by its transpose to get a 3×3 matrix $X = (x_{ij})$:

$$Z = \begin{bmatrix} z_{11} & z_{12} \\ z_{21} & z_{22} \\ z_{31} & z_{32} \end{bmatrix} \mapsto X = \begin{bmatrix} z_{11}^2 + z_{12}^2 & z_{11}z_{21} + z_{12}z_{22} & z_{11}z_{31} + z_{12}z_{32} \\ z_{11}z_{21} + z_{12}z_{22} & z_{21}^2 + z_{22}^2 & z_{21}z_{31} + z_{22}z_{32} \\ z_{11}z_{31} + z_{12}z_{32} & z_{21}z_{31} + z_{22}z_{32} & z_{31}^2 + z_{32}^2 \end{bmatrix}.$$

The image of f is the set of positive semidefinite symmetric 3×3 matrices of rank ≤ 2. This is a 5-dimensional semialgebraic set in the space $\mathbb{R}^9$ of 3×3 matrices. Its polynomial description consists of the four equations

$$x_{12} = x_{21}\,,\ x_{13} = x_{31}\,,\ x_{23} = x_{32} \ \text{ and } \ \det(X) = 0 \tag{4.5}$$

and the six inequalities

$$x_{11} \geq 0\,,\ x_{22} \geq 0\,,\ x_{33} \geq 0\,,\ x_{11}x_{22} \geq x_{12}^2\,,\ x_{11}x_{33} \geq x_{13}^2\,,\ x_{22}x_{33} \geq x_{23}^2.$$

It is generally difficult to compute the semialgebraic set $f(X)$ when we are given a real variety X and a polynomial map f. One algorithm that accomplishes this is known as cylindrical algebraic decomposition (CAD). We refer to [**4**, Chapter 5] for a textbook introduction to CAD. A friendly case study (with soccer balls) can be found in [**31**]. See [**31**, Example 6] for an illustration of how to call CAD in the computer algebra system `Mathematica`.

We now return to the setting of an algebraically closed field K, such as the complex numbers $K = \mathbb{C}$. Here, the situation is a bit easier than in the Tarski-Seidenberg Theorem 4.17. For instance, the image of $f : \mathbb{C}^6 \to \mathbb{C}^9$ in Example 4.18 is closed; it is precisely the subvariety defined by (4.5). However, the image is generally not closed. We saw this in Examples 4.1 and 4.7. Recall from Definition 2.12 that a subset of K^n is *constructible* if it is a finite union of differences of varieties. We now present a counterpart of Theorem 4.17, where the field K is assumed to be algebraically closed.

Theorem 4.19 (Chevalley). *The image of a constructible set $X \subset K^n$ under a polynomial map $f\colon K^n \to K^m$ is a constructible set.*

Proof. As in Section 4.2, we realize $f(X)$ as the image under a projection $K^{m+n} \to K^m$ of the intersection of $X \times K^m$ and the graph of f. Hence, by induction on m, it is enough to prove that the image of a constructible set $X \subset K^{m+1}$ under the projection $f\colon K^{m+1} \to K^m$ is constructible. In our proof we focus on the most important case, where X is a variety in K^{m+1}. We proceed by induction on $\dim X$. The base case $\dim X = 0$ is trivial.

By decomposing X into irreducible components (Proposition 3.7), we may assume that X is irreducible. Let Y be the closed image of X. This is the irreducible variety whose prime ideal is $\mathcal{I}(Y) = \mathcal{I}(X) \cap K[x_1, \ldots, x_m]$.

We claim that there exists a nonempty Zariski open set $U \subset Y$ which is contained in $f(X)$. Before proving the claim, we note that it will allow us to finish the proof. Indeed, $Y \backslash U$ is a variety whose preimage is a proper subvariety $X' \subset X$. By induction, $f(X')$ is constructible and $f(X) = f(X') \cup U$.

To prove the claim, we consider a reduced lexicographic Gröbner basis $\mathcal{G}$ of the ideal $\mathcal{I}(X)$, as in Theorem 4.5. Each element g_i of $\mathcal{G}$ can be written as

$$g_i \;=\; h_i(x_1, \ldots, x_m) x_{m+1}^{b_i} + \text{terms of lower degree in } x_{m+1}.$$

If all of the b_i are zero, then $\mathcal{I}(X) = \mathcal{I}(Y)$ and hence $f(X) = Y$. So we may assume $b_1 > 0$. We set $U = Y \backslash \mathcal{V}(h_1)$. This is a Zariski open subset of Y.

We first note that U is nonempty. Otherwise, h_1 would vanish on Y and hence $h_1 \in \mathcal{I}(X)$. This is not possible, since $b_1 > 0$ and the Gröbner basis $\mathcal{G}$ is reduced. Our claim is that $U \subseteq f(X)$. Hence, for any $y \in U$, we must find $x \in K$ such that $(y, x) \in X \subset K^{m+1}$. Fix the ideal $I = \{q(y, x_{n+1}) \,:\, q \in \mathcal{I}(X)\}$ in $K[x_{m+1}]$. We have $I = \langle p \rangle$ for some $p \in K[x_{m+1}]$. If p has positive degree or equals zero, then there exists $x \in K$ such that $p(x) = 0$. We claim that $(y, x) \in X$. Indeed, if we pick any $q \in \mathcal{I}(X)$, then $q(y, x_{m+1}) = p(x_{m+1}) p'(x_{m+1})$ for some $p' \in K[x_{m+1}]$. Thus, $q(y, x) = p(x) p'(x) = 0$. Hence, $(y, x) \in X$.

It remains to exclude the case of $p \in K \backslash \{0\}$. Suppose for contradiction that $q(y, x_{m+1}) \in K \backslash \{0\}$ for some $q \in \mathcal{I}(X)$. We regard g_1 and q as polynomials in x_{m+1} and compute their resultant $R \in K[x_1, \ldots, x_m]$. In other words, we replace x_{ij} in the Sylvester matrix (4.4) with the coefficients of g_1 and q. Then R is the determinant. By Theorem 4.11, R belongs to the elimination ideal $\langle g_1, q \rangle \cap K[x_1, \ldots, x_m]$. This implies $R \in \mathcal{I}(X) \cap K[x_1, \ldots, x_m] = \mathcal{I}(Y)$. In particular, $R(y) = 0$. However, $R(y)$ is the determinant of the matrix that is obtained from (4.4) by evaluating the coefficients of g_1 and q at y. As $q(y, x_{m+1})$ is a nonzero constant, we see that each row in the lower part of the matrix has precisely one nonzero entry, corresponding to $x_{2,d_2+1} = q(y, x_{m+1})$. Hence, the Sylvester matrix is zero below the diagonal and its determinant equals $R(y) = h_1(y)^{\deg_{x_{m+1}} q} q(y, x_{m+1})^{b_1}$, which is a product of nonzero numbers. This is a contradiction, and it finishes the proof. □

More information about Chevalley's Theorem 4.19 and its proof can be found in [**10**, §§3.6 and 5.6] and [**60**, §§7.4.6–7.4.8].

Corollary 4.20. *In Theorem* 4.19, *if* $K = \mathbb{C}$ *and* $X \subset \mathbb{C}^n$ *is a variety, then the Zariski closure of the image* $f(X) \subset \mathbb{C}^m$ *equals the Euclidean closure.*

Proof. Using bars for Zariski closure, the set U in the proof above satisfies

$$U = \overline{f(X)} \backslash \mathcal{V}(h_1) \subseteq f(X) \subseteq \overline{f(X)} \subseteq \mathbb{C}^m.$$

However, the complement of a proper subvariety of a complex affine variety in $\mathbb{C}^m$ is dense in that variety with respect to the Euclidean topology. This shows that U is dense in $\overline{f(X)}$ for the familiar Euclidean topology on $\mathbb{C}^m$. □

Suppose we want to check whether a given point $y \in K^m$ lies in the image $f(X)$. This can be done by examining a system of polynomial equations

$$x \in X \quad \text{and} \quad f(x) = y. \tag{4.6}$$

These are equations in n unknowns $x_1, \ldots, x_n$, and we must decide whether a solution exists. According to the Nullstellensatz (Theorem 6.1), this amounts to deciding whether 1 lies in the ideal of $K[\mathbf{x}]$ specified by (4.6).

Computing the full image is a more difficult task. The desired output is a description as a constructible set. We approach this problem as follows:

- Compute the closed image X_0.
- Subtract from it a proper subvariety X_1.
- Add back X_2, a proper subvariety of X_1, etc.

This procedure must finish in a finite number of steps, by Hilbert's Basis Theorem. For a recent algorithm and its implementation we refer to [**24**].

Example 4.21. Let $m = n = 3$ and consider the map

$$f : \mathbb{C}^3 \to \mathbb{C}^3, \quad (z_1, z_2, z_3) \mapsto (z_1 z_2, z_1 z_3, z_2 z_3).$$

Its image is Zariski dense in $\mathbb{C}^3$. Here is a description as a constructible set:

$$\text{image}(f) \;=\; (\mathbb{C}^3 \backslash \mathcal{V}(x_1 x_2 x_2)) \cup \mathcal{V}(x_1 x_2, x_1 x_3, x_2 x_3).$$

Note the difference between Chavelley's Theorem and the Tarski-Seidenberg Theorem. Over the real numbers, we also need the inequality $x_1x_2x_3 \geq 0$.

Applied mathematicians tend to wonder why algebraic geometers restrict themselves to complex numbers and projective varieties. One explanation is that images of polynomial maps behave more nicely in that setting. The following result, which is known as the *Main Theorem of Elimination Theory*, makes this statement precise. It can be regarded as an algebraic analogue of the fact that images of compact sets under continuous maps are compact.

Theorem 4.22. *Let X be a projective variety over an algebraically closed field K. Then the image of X under a regular map f is (Zariski) closed.*

Here being *regular* means that the map is defined by homogeneous polynomials of the same degree and these polynomials never vanish simultaneously on the domain X. The nonvanishing condition ensures that each point has a well-defined image in projective space. The map f in Example 4.21 does not define a regular map from the projective plane $\mathbb{P}^2$ to itself. This map would not be defined at the points $(1:0:0)$, $(0:1:0)$ and $(0:0:1)$.

Proof. We refer to [**47**, §5.2]. If we work over $\mathbb{C}$, then the projective variety X is compact in the Euclidean topology. Therefore, the image of X is also compact and hence closed. However, by Corollary 4.20 to Chevalley's Theorem 4.19, the Zariski closure and the Euclidean closures of the image coincide. We conclude that the image is also Zariski closed. □

Theorem 4.22 is a powerful result. For instance, it implies the following. Consider a map $f = (f_1, \ldots, f_m)$ given by homogeneous polynomials of the same degree. Assume that the affine variety $\mathcal{V}(f_1, \ldots, f_m)$ equals $\{0\}$. Then the image of f is Zariski closed—we can compute it using elimination. For an application see Exercise 20. In general, the following theorem holds.

Theorem 4.23. *Consider a map $f = (f_1, \ldots, f_m) : \mathbb{C}^{n+1} \to \mathbb{C}^m$ given by homogeneous polynomials of the same degree. Let* $\dim \mathcal{V}(f_1, \ldots, f_m) = b+1$. *Suppose that the closed image of f has affine dimension $d+1$, i.e. the projective dimension is d. If $d + b < n$, then the image of f is closed.*

Proof. We regard f as a map from $\mathbb{P}^n \backslash \mathcal{V}(f_1, \ldots, f_m)$ to $\mathbb{P}^{m-1}$. Let $\mathbb{P}^d \subset \mathbb{P}^n$ be a general projective subspace of dimension $d < n - b$. It is disjoint from $\mathcal{V}(f_1, \ldots, f_m)$. Thus we may assume that f is well-defined on $\mathbb{P}^d$. The image of $\mathbb{P}^d$ under f is closed by Theorem 4.22. It is contained in the closed image of $\mathbb{P}^n$ under f. The two have the same dimension and are both irreducible. So the images coincide, and we conclude that the image of f is closed. □

Exercises

(1) Eliminate the variable z from the equations $x^3y^3z^3 - x - y - z = 1$ and $x^5 + y^5 + z^5 = 2$. Discuss the resulting curve in the (x, y)-plane.

(2) If an ideal I in a polynomial ring is prime, then so are its elimination ideals, and the same is true for being radical. Find examples which show that the converse does not hold. What is the geometric meaning of these facts?

(3) Compute the determinants of the Sylvester matrices $\mathrm{Syl}_{1,5}$, $\mathrm{Syl}_{2,4}$ and $\mathrm{Syl}_{3,3}$. Each is a polynomial of degree 6 in eight unknowns. Which of these three polynomials has the most terms?

(4) You are given a plane curve with parametrization $z \mapsto (f(z), g(z))$ where f and g are polynomials of degree 10. At most how many terms do you expect the implicit equation of that plane curve to have?

(5) Can you find an invertible 5×5 matrix that is skew-symmetric?

(6) You are given all entries of a skew-symmetric 5×5 matrix $X = (x_{ij})$ except for x_{12} and x_{45}. Under what condition on the eight visible entries can you complete the matrix with $\mathrm{rank}(X) \leq 2$?

(7) Let π be the linear map from $\mathbb{C}^3$ to $\mathbb{C}^2$ given by the matrix $\begin{pmatrix} 1 & 2 & 3 \\ 3 & 2 & 1 \end{pmatrix}$. Given an algebraic curve V in $\mathbb{C}^3$, explain how one can compute the plane curve $\overline{\pi(V)} \subset \mathbb{C}^2$. What happens if you replace $\mathbb{C}$ by $\mathbb{R}$?

(8) Consider the Fermat curve $V = \mathcal{V}(x^3 + y^3 + z^3)$ in $\mathbb{P}^2$. Compute the prime ideal that defines the closed image of V under the *Veronese map*

$$\mathbb{P}^2 \to \mathbb{P}^5, \quad (x : y : z) \mapsto (x^2 : xy : xz : y^2 : yz : z^2).$$

(9) You are given a complex 3×3 matrix and asked whether it is the square of a traceless 3×3 matrix. Write this question in the formulation (4.6). What does Chavelley's Theorem tell us about a general answer?

(10) Determine the prime ideal of relations between the 3×3 minors of a 3×6 matrix. Where in the next chapter does this prime ideal appear?

(11) Consider the map $f : \mathbb{R}^4 \to \mathbb{R}^3$ in Example 4.16, but now take X to be the hypersurface given by $ax^4 + bx + c = 0$. Verify that the image $f(X)$ is a semialgebraic set. Find an explicit polynomial description of $f(X)$.

(12) Let V_1 and V_2 be algebraic curves in $\mathbb{C}^3$ and let $V_1 + V_2$ be their pointwise sum. The Zariski closure $\overline{V_1 + V_2}$ is an algebraic variety in $\mathbb{C}^3$. Explain how one can compute the vanishing ideal of this variety.

(13) Following Example 4.10, how would you define the hyperdeterminant of a $2 \times 2 \times 3$ tensor (x_{ijk})? This hyperdeterminant is a polynomial in the 12 unknowns x_{ijk}. Can you compute it? How many terms does it have?

(14) Use the resultant method in Example 4.15 to compute the implicit equation of the plane cubic curve that has the parametrization

$$z \mapsto \left(2z^3 + 3z^2 + 5z + 7,\ 11z^3 + 13z^2 + 17z + 19\right).$$

(15) Let $m = 2$, $d_1 = 1$ and $d_2 = d_3 = 2$. The resultant $\mathrm{Res}(f_1, f_2, f_3)$ is a polynomial in $n = 15 = 3 + 6 + 6$ unknowns, one for each coefficient of f_1, f_2 and f_3. What is the degree of this polynomial? How many terms does it have?

(16) What constraints hold for off-diagonal entries of a rank-1 3×3 matrix?

(17) What constraints hold for the off-diagonal entries of a nilpotent 3×3 matrix? Answer this question for the field of complex numbers $\mathbb{C}$.

(18) What constraints hold for the off-diagonal entries of an orthogonal 3×3 matrix? Answer this question for the field of real numbers $\mathbb{R}$. What does the Tarski-Seidenberg Theorem tell us about the form of the answer?

(19) Let $m = 2$ and $d_1 = d_2 = d_3 = 2$. Then $\mathrm{Res}(f_1, f_2, f_3)$ is the resultant of three quadrics in the plane. This is a polynomial in $18 = 6 + 6 + 6$ variables of degree $12 = 4 + 4 + 4$. How many terms does it have? Find a good formula for $\mathrm{Res}(f_1, f_2, f_3)$. Hint: See [**11**, Chapter 3] or [**22**, §3.4.D].

(20) Let V be the space of homogeneous polynomials in n variables of degree d. The dth powers of linear forms form a subset of V. Is it Zariski closed for any n and d? What happens over the field of real numbers?

(21) Why is Theorem 4.22 called the Main Theorem of Elimination Theory? Where does elimination come in? What is the geometric interpretation?

Chapter 5

Linear Spaces and Grassmannians

"*Geometry is not true, it is advantageous*", Henri Poincaré

In previous chapters we saw the construction of projective space. We argued that projective varieties are preferable to affine varieties in many applications. Points in a projective space correspond to lines through the origin in the underlying vector space. In this chapter we replace lines with higher-dimensional linear subspaces. The role of the projective space is now played by a *Grassmannian.* This is a smooth projective variety whose points correspond to linear subspaces of a fixed dimension. For instance, the Grassmannian of lines in projective 3-space is a 4-dimensional variety. Its subvarieties represent families of lines. Counting lines that satisfy a certain property (e.g. lying on a cubic surface) leads us to *enumerative algebraic geometry,* a subject in which Grassmannians play a fundamental role.

5.1. Coordinates for Linear Spaces

Let V be a vector space of dimension n over a field K. In Chapter 2 we constructed the projective space $\mathbb{P}(V)$. Its points are the 1-dimensional subspaces of V. We note that $\mathbb{P}(V)$ is the key example of a compact algebraic variety when $K = \mathbb{C}$. Our aim is to generalize this construction from lines to subspaces of arbitrary dimension k with $0 < k \leq \dim V$. We will construct a projective variety $G(k, V)$ whose points correspond bijectively to k-dimensional subspaces of V. This variety is called the Grassmannian, after the 19th-century mathematician Hermann Grassmann. If $V = K^n$, then we use the notation $\mathbb{P}^{n-1}$ for the projective space $\mathbb{P}(V)$, and we write $G(k,n)$ for the Grassmannian $G(k,V)$.

We start with an explicit construction in coordinates, by fixing a basis $e_1,\ldots,e_n$ of V. Consider any k linearly independent vectors $v_1,\ldots,v_k \in V$. We represent them in the form of a $k \times n$ matrix M of rank k. To these vectors, or equivalently to a full-rank $k \times n$ matrix, we associate the linear subspace $W := K\{v_1,\ldots,v_k\}$ in V. This association is surjective but not injective, as we may replace the v_i's by linear combinations. In other words, the group $\mathrm{GL}(k)$ of invertible $k\times k$ matrices acts on the set of $k\times n$ matrices by left multiplication, and this does not change the linear span of the rows.

We know some polynomial functions that do not change (up to rescaling) upon taking linear combinations of the rows; these are the $k \times k$ minors of the $k \times n$ matrix. Suppose that W is a k-dimensional subspace of V. Pick any basis and express W as the row space of a $k \times n$ matrix. We then write $\mathfrak{i}(W)$ for the vector of all $k \times k$ minors of that matrix, up to scaling.

Example 5.1. Let $k = 2$ and $n = 5$, and let W be the linear subspace of $V = K^5$ that is spanned by $(1,1,1,1,1)$ and (a_1,a_2,a_3,a_4,a_5) for some scalars a_i that are not all identical. Then $\mathfrak{i}(W)$ is the point in $\mathbb{P}^9 = \mathbb{P}(K^{\binom{5}{2}})$ whose 10 homogeneous coordinates are $a_i - a_j$ for $1 \le i < j \le 5$.

Forming the vector of all $k\times k$ minors of the $k\times n$ matrix defines a map

$$\mathfrak{i} : \{k\text{-dimensional subspaces of } V\} \to \mathbb{P}(K^{\binom{n}{k}}).$$

This is well-defined since $\mathfrak{i}(W)$ does not depend on the chosen basis of W.

Lemma 5.2. *The map $\mathfrak{i}$ is injective.*

Proof. Consider two k-dimensional subspaces $W_1, W_2 \subset V$. Assume $\mathfrak{i}(W_1) = \mathfrak{i}(W_2)$. The matrices M_{W_1} and M_{W_2} that represent W_1 and W_2 have rank k. Without loss of generality we may assume that the first k columns are linearly independent. By performing linear operations on the rows of both matrices, we transform M_{W_i} to a matrix $\tilde{M}_{W_i}$ whose leftmost $k \times k$ submatrix is the identity. We observe that any entry of $\tilde{M}_{W_i}$ not in the first k columns is equal to some maximal minor or its negative. Thus, if $\mathfrak{i}(W_1) = \mathfrak{i}(W_2)$ then the two matrices $\tilde{M}_{W_1}$ and $\tilde{M}_{W_2}$ must be equal. This implies $W_1 = W_2$. □

Example 5.3. The reasoning in the proof above is illustrated by the matrix

$$\tilde{M}_W = \begin{pmatrix} 1 & 0 & 0 & p_{234} & p_{235} & p_{236} \\ 0 & 1 & 0 & -p_{134} & -p_{135} & -p_{136} \\ 0 & 0 & 1 & p_{124} & p_{125} & p_{126} \end{pmatrix}. \tag{5.1}$$

Let $k = 3$ and $n = 6$. Each entry p_{ijk} in the 3×3 block on the right equals the 3×3 minor of the matrix $\tilde{M}_W$ given by the column indices i, j, k.

The image of $\mathfrak{i}$ is the *Grassmannian* $G(k,n)$. Its inclusion in $\mathbb{P}(K^{\binom{n}{k}})$ is the *Plücker embedding*. For readers familiar with the exterior power of a vector space, here is a more invariant description of the Grassmannian:

$$G(k,n)=\{[v_1\wedge\cdots\wedge v_k] \in \mathbb{P}(\bigwedge^k V) : v_1,\ldots,v_k \in V \text{ are linearly independent}\}.$$

Indeed, first we may identify $\mathbb{P}(K^{\binom{n}{k}})$ with $\mathbb{P}(\bigwedge^k V)$ by fixing a basis of V and the induced basis of $\bigwedge^k V$. Expanding $v_1 \wedge \cdots \wedge v_k$ in that basis, we indeed obtain the $k \times k$ minors of the $n \times k$ matrix $[v_1, \ldots, v_k]$. The group $\mathrm{GL}(V)$ acts naturally on V, taking subspaces to subspaces. This induces an action on $\mathbb{P}(\bigwedge^k V)$ which restricts to the Grassmannian. A matrix $g \in \mathrm{GL}(V)$ transforms $v_1 \wedge \cdots \wedge v_k$ to $g(v_1) \wedge \cdots \wedge g(v_k)$. We note that this action is *transitive*: For every pair $p_1, p_2 \in G(k,V)$ there exists a (nonunique) automorphism $g \in \mathrm{GL}(V)$ such that $g(p_1) = p_2$. This holds because any set of k linearly independent vectors may be transformed by an invertible linear map to any other such set. Hence, $G(k,V)$ is an *orbit* under the action of $\mathrm{GL}(V)$ on $\mathbb{P}(\bigwedge^k V)$. In fact, $G(k,V)$ is the unique closed orbit in this space.

Projective varieties that are orbits of algebraic matrix groups are said to be *homogeneous*. The Grassmannians are prominent instances of homogeneous varieties. We will study algebraic groups and their representations as matrix groups in Chapter 10. Homogeneous varieties are always smooth. Indeed, any algebraic variety always contains a smooth point, and an action of a group must take a smooth point to a smooth point. A version of this statement is given in Exercise 2.

Our next aim is to show that the Grassmannian is a projective variety. Equivalently, we need to express the property that $\binom{n}{k}$ given numbers are the minors of a $k \times n$ matrix in terms of the vanishing of (homogeneous) polynomials.

Theorem 5.4. *The Grassmannian $G(k,n)$ is Zariski closed and irreducible.*

Proof. Lemma 5.2 gives us an idea of how to proceed. Namely, first let us assume that the matrix M_W representing W has the form seen in (5.1),

$$\left[\begin{array}{cccc|c} 1 & 0 & \cdots & 0 & \\ 0 & 1 & \cdots & 0 & \\ \vdots & \vdots & \ddots & \vdots & A \\ 0 & \cdots & 0 & 1 & \end{array}\right], \tag{5.2}$$

where A is a $k \times (n-k)$ matrix. Each maximal minor of M_W is now, up to sign, a minor of A of some size. Further, by Laplace expansion, a $q \times q$ minor of A for $q > 1$ may be expressed, as a quadratic polynomial, in terms of smaller minors. This gives $\sum_{q=2}^{\min(k,n-k)} \binom{k}{q}\binom{n-k}{q}$ inhomogeneous quadratic equations in the minors of the $k \times (n-k)$ matrix A. These define the part of the image of our map $\mathfrak{i}$ that lies in the affine open set $K^{\binom{n}{k}-1} \subset \mathbb{P}(K^{\binom{n}{k}})$ given by the nonvanishing of the first Plücker coordinate.

If $\mathfrak{i}(W)$ has first coordinate zero, then some other coordinate will be nonzero, as the matrix M_W must have some invertible $k \times k$ submatrix. If we multiply M_W on the left by the inverse of that matrix, then we obtain a matrix that looks like (5.2) but with its columns permuted. The same construction as before gives us a system of $\sum_{q=2}^{\min(k,n-k)} \binom{k}{q}\binom{n-k}{q}$ inhomogeneous quadratic equations in the $k(n-k)$ entries of the new matrix A.

Each of the quadratic equations in $k(n-k)$ variables obtained above can be written as a homogeneous quadric in the $\binom{n}{k}$ coordinates on $\mathbb{P}(K^{\binom{n}{k}})$. Namely, a minor of A is replaced by the corresponding maximal minor of M_W, and then the quadric is homogenized by the special minor that corresponds to the identity matrix in (5.2). Each of the homogenized polynomials vanishes on the Grassmannian. Indeed, each set of polynomials, by construction, vanishes on the part of the Grassmannian that corresponds to matrices with a fixed nonzero minor. However, any full-rank matrix is in the Zariski closure of the set of full-rank matrices with all minors nonvanishing.

We also know that our equations define the Grassmannian on each open affine chart of the projective space $\mathbb{P}(K^{\binom{n}{k}})$. Hence, the collection of all constructed quadrics gives a full polynomial description of $G(k,n)$ as a set.

The affine cone over the Grassmannian $G(k,n)$ is the image of a polynomial map $W \mapsto \mathfrak{i}(W)$ from the matrix space $K^{k\times n}$ to the affine space $K^{\binom{n}{k}}$. This map takes all maximal minors of a $k \times n$ matrix W. In particular, the image is irreducible and hence so is the Grassmannian $G(k,n)$. □

5.2. Plücker Relations

We have proved that the set $G(k,n)$ is cut out by quadratic equations. In fact, with slightly more effort one can show that $I(G(k,n))$ may be generated by quadratic polynomials. These are known as *Plücker relations* [**39**, Chapter 3]. Further below we will discuss the Plücker relations for $k=2$.

From a more algebraic perspective the equations vanishing on $G(k,n)$ are exactly the *polynomial relations between maximal minors*. We point out that finding the ideal of polynomial relations between the nonmaximal minors of a fixed size is an open problem in commutative algebra.

Our argument shows that the intersection $G(k,n) \cap K^{\binom{n}{k}-1}$ of the Grassmannian with the open affine set $K^{\binom{n}{k}-1}$ is the affine space $K^{k\times(n-k)}$ indicated in (5.2). Just like the usual projective space, the Grassmannian

$G(k,n)$ is glued together from affine spaces. In the same way, as a point in $\mathbb{P}^{n-1}$ has a nonunique representation as an n-tuple of numbers, a point in $G(k,n)$ may be represented by a $k \times n$ matrix. To make the representation unique, for the projective space one chooses $x_i \neq 0$ and sets $x_i = 1$. This corresponds to choosing a nonzero maximal minor in the matrix and taking the submatrix to be the identity. Thus, our constructions are generalizations of those known for the projective space $\mathbb{P}^{n-1} = G(1,n)$ to arbitrary Grassmannians.

Corollary 5.5. *The dimension of the Grassmannian $G(k,n)$ equals $k(n-k)$.*

Remark 5.6. The Grassmannian $G(k,n)$ parametrizes k-dimensional vector subspaces of an n-dimensional vector space or, equivalently, $(k-1)$-dimensional projective subspaces of an $(n-1)$-dimensional projective space.

Example 5.7 ($k = 2, n = 4$). The Grassmannian $G(2,4)$ is the image of the map

$$\begin{bmatrix} a & b & c & d \\ e & f & g & h \end{bmatrix} \mapsto (af-be : ag-ce : ah-de : bg-cf : bh-df : ch-dg) \in \mathbb{P}^5.$$

Alternatively, after fixing a basis (v_1, v_2, v_3, v_4) of the vector space $V \simeq K^4$, the following identity holds in its second exterior power $\bigwedge^2 V$:

$$\begin{aligned} &(av_1 + bv_2 + cv_3 + dv_4) \wedge (ev_1 + fv_2 + gv_3 + hv_4) \\ &\quad = (af - be)v_1 \wedge v_2 + (ag - ce)v_1 \wedge v_3 + (ah - de)v_1 \wedge v_4 \\ &\qquad + (bg - cf)v_2 \wedge v_3 + (bh - df)v_2 \wedge v_4 + (ch - dg)v_3 \wedge v_4. \end{aligned}$$

The Grassmannian $G(2,4)$ has dimension 4. It is a hypersurface in $\mathbb{P}^5$. We write the coordinates on $\mathbb{P}^5$ as $(p_{12} : p_{13} : p_{14} : p_{23} : p_{24} : p_{34})$. The indices refer to the minors of a 2×4 matrix. Following the proof of Theorem 5.4, we look at the matrices (5.2). Here, they take the form

$$\begin{bmatrix} 1 & 0 & c & d \\ 0 & 1 & g & h \end{bmatrix}.$$

The expansion of the rightmost 2×2 minor yields the inhomogeneous quadratic equation $p_{34} = ch - dg = (-p_{23})p_{14} - (-p_{24})p_{13}$. We homogenize this equation with the extra variable p_{12}. We conclude that the Grassmannian $G(2,4)$ is the hypersurface in $\mathbb{P}^5$ that is defined by the *Plücker quadric*

$$p_{23}p_{14} - p_{13}p_{24} + p_{12}p_{34}. \tag{5.3}$$

We now discuss the homogeneous prime ideal $I(G(k,n))$ of the Grassmannian $G(k,n)$. A complete description is known in terms of certain quadratic relations that form a Gröbner basis. These are known as *straightening relations*. For a derivation and explanation see e.g. [**51**, Chapter 3].

We here present the answer in the special case of $k = 2$. The corresponding Grassmannian $G(2, n)$ is the space of lines in $\mathbb{P}^{n-1}$. The ambient space in this case is $\mathbb{P}(\bigwedge^2 \mathbb{C}^n)$. The vector space $\bigwedge^2 \mathbb{C}^n$ is simply the space of skew-symmetric $n \times n$ matrices. Hence, it is convenient to write the $\binom{n}{2}$ Plücker coordinates as the entries of a skew-symmetric $n \times n$ matrix $P = (p_{ij})$. The (affine cone over the) Grassmannian $G(2, n)$ consists of special skew-symmetric $n \times n$ matrices. Below we prove that these are exactly matrices of rank 2. As a skew-symmetric matrix cannot have odd rank, we consider the submatrices of P of size 4×4. To bound the rank of a skew-symmetric matrix, it is enough to consider *principal* submatrices, i.e. those that have the same index set of rows and columns. One such submatrix is obtained by taking the first four rows and first four columns. The determinant of that matrix is the square of the Plücker quadric (5.3). One refers to the square root of the determinant of a skew-symmetric matrix of even order as its *Pfaffian*. Thus the 4×4 Pfaffians of our matrix P are the $\binom{n}{4}$ quadrics

$$\underline{p_{il}p_{jk}} - p_{ik}p_{jl} + p_{ij}p_{kl} \qquad \text{for } 1 \leq i < j < k < l \leq n. \tag{5.4}$$

Theorem 5.8. *The $\binom{n}{4}$ quadrics in* (5.4) *form the reduced Gröbner basis of the Plücker ideal $I(G(2, n))$, for any monomial ordering on the polynomial ring in the $\binom{n}{2}$ variables p_{ij} that selects the underlined leading terms.*

Proof. The argument in the proof of Theorem 5.4 shows that the quadrics (5.4) cut out $G(2, n)$ as a subset in $\mathbb{P}^{\binom{n}{2}-1}$. In other words, our Grassmannian is given as the set of skew-symmetric $n \times n$ matrices whose 4×4 Pfaffians vanish. These are the skew-symmetric matrices of rank 2.

By Hilbert's Nullstellensatz (proved in Chapter 6), we can conclude that the ideal $I(G(2, n))$ is the radical of the ideal generated by (5.4). We need to argue that the latter ideal is radical. However, this follows from the assertion that (5.4) is a Gröbner basis. Indeed, the leading monomials $p_{il}p_{jk}$ are square-free, so they generate a radical monomial ideal. However, if the initial ideal $\text{in}(J)$ of an ideal J is radical, then J itself is also radical. So all we need to do is verify the Gröbner basis property for our quadrics. That Gröbner basis is then automatically reduced, because neither of the two nonleading terms in (5.4) is a multiple of some other leading term.

To verify the Gröbner basis property, we reason as follows. For $n = 4$ the property is trivial because there is only one generator. For $n = 5, 6, 7$, verifying the property is a direct computation, e.g. using `Macaulay2`. One checks that the S-polynomial of any two quadrics in (5.4) reduces to zero. Suppose that $n \geq 8$ and consider two Plücker quadrics. These involve at most 8 distinct indices. If the number of distinct indices is 7 or less, then we are done by the aforementioned computation, which verified the claim for $n \leq 7$. Hence we may assume that all 8 indices occurring in the two Plücker quadrics are distinct. In that case, the two underlined leading monomials are relatively prime. Here, Buchberger's second criterion applies, and we can conclude that the S-polynomial automatically reduces to zero. In conclusion, all S-polynomials formed by pairs from (5.4) reduce to zero. This completes the proof. □

The Plücker relations for arbitrary Grassmannians are hardwired into the computer algebra system `Macaulay2`. One finds generators for the ideal $I(G(k,n))$ with the convenient command `Grassmannian(k-1,n-1)`. Here, the parameters k and n are decreased by 1 because `Macaulay2` refers to the projective geometry interpretation: Points in $G(k,n)$ correspond to projective spaces of dimension $k-1$ in an ambient projective space of dimension $n-1$. Another thing that is tricky about the command `Grassmannian` is the ordering of the Plücker coordinates in `Macaulay2`. Here is how it works.

Example 5.9 ($k=3, n=6$)**.** The following two command lines yield equations defining the Grassmannian of 3-dimensional vector subspaces in K^6.

```
R = QQ[p123,p124,p134,p234,p125,p135,p235,p145,p245,
        p345,p126,p136,p236,p146,p246,p346,p156,p256,p356,p456];
I = Grassmannian(2,5,R)
```

This produces 35 quadratic relations. Note that $G(3,6)$ is a smooth projective variety of dimension 9 and degree 42 in $\mathbb{P}^{19}$, as is seen by also typing

```
dim I, degree I, betti mingens I
```

Among the 35 minimal generators of $I(G(3,6))$, there are 30 trinomials, such as $p_{134}p_{125} - p_{124}p_{135} + p_{123}p_{145}$, plus five additional relations that involve all six indices, such as $p_{345}p_{126} - p_{125}p_{346} + p_{124}p_{356} - p_{123}p_{456}$. What is the combinatorial pattern? Can you generalize it to larger values of k and n? You will find answers and much more information in the books [**39**, **41**, **51**].

5.3. Schubert Calculus

We now show how Grassmannians can help us to answer enumerative questions. We would like to know how many lines or planes in space satisfy certain properties. This subject area is known as *enumerative geometry*, and the specific answers we provide are based on *Schubert calculus*. Schubert calculus furnishes an intersection theory for subvarieties of a Grassmannian.

We illustrate the concepts for the special case of $G(2,4)$. Here is the simplest question of Schubert calculus: *How many lines L intersect four general lines L_1, L_2, L_3, L_4 in $\mathbb{P}^3$?* The answer to this question is *two*. To see this, we represent the line L by its corresponding point $p = (p_{12} : p_{13} : p_{14} : p_{23} : p_{24} : p_{34})$ in $G(2,4) \subset \mathbb{P}^5$. As we will see, the condition that L

intersects a fixed L_i is a linear condition in p. We must solve four such linear equations, and these have general coefficients since L_1, L_2, L_3, L_4 are general lines. Hence, we are intersecting the Grassmannian $G(2,4)$ defined by the quadric $p_{12}p_{34} - p_{13}p_{24} + p_{14}p_{23} = 0$ with four general hyperplanes. Over an algebraically closed field, the number of points we obtain is the degree of the variety, which in our case equals 2. Each of the two points represents a line that is a solution to the starting problem. Note that this is also the bound from Bézout's Theorem applied to polynomials of degrees $1,1,1,1,2$.

To study such intersection problems more systematically, one introduces some special subvarieties of Grassmannians. We continue with the example of $G(2,4)$, the manifold of all lines in $\mathbb{P}^3$. We fix a complete flag in $\mathbb{P}^3$, consisting of a point in a line in a plane: $f_0 = \mathbb{P}^0 \subset f_1 = \mathbb{P}^1 \subset f_2 = \mathbb{P}^2 \subset \mathbb{P}^3$. Our aim is to group families of projective lines according to how they intersect that flag. These will be subvarieties X_i of dimension i in $G(2,4)$.

First, the flag distinguishes a point in $G(2,4)$, namely $X_0 := \{f_1\} \subset G(2,4)$. It also distinguishes a curve X_1 in the Grassmannian $G(2,4)$. The points of X_1 are the lines l such that $f_0 \in l \subset f_2$. The most interesting is the case of surfaces in our family. There are two types of those in $G(2,4)$:

(1) the surface X_2 consisting of all lines l such that $f_0 \in l$; and

(2) the surface $X_{2'}$ consisting of all lines l such that $l \subset f_2$.

Finally, there is also one 3-dimensional variety X_3, consisting of all lines that intersect the given line f_1. The varieties $X_1, X_2, X_{2'}$ and X_3 we described are called the *Schubert varieties* in $G(2,4)$. In Exercise 1 you will generalize the construction of Schubert varieties to larger Grassmannians.

Let us now fix a basis v_1, v_2, v_3, v_4 for K^4 and take f_i to be the linear subspace in $\mathbb{P}^3$ spanned by $v_1, \dots, v_{i+1}$. The point $f_1 \in G(2,4)$ is given in $\mathbb{P}^5$ by the vanishing of all coordinates p_{ij} apart from p_{12}. We have $f_0 \in l \subset f_2$ if and only if the line l is spanned by the rows of a matrix of the form

$$\begin{bmatrix} 1 & 0 & 0 & 0 \\ 0 & f & g & 0 \end{bmatrix}.$$

Hence, the curve X_1 is a line $\mathbb{P}^1$ in $G(2,4) \subset \mathbb{P}^5$, defined by the vanishing of all p_{ij} apart from p_{12} and p_{13}. Similarly, X_2 is the $\mathbb{P}^2$ with coordinates p_{12}, p_{13}, p_{14}, and $X_{2'}$ is a different $\mathbb{P}^2$ with coordinates p_{12}, p_{13}, p_{23}. The common span of these two planes is the 3-dimensional variety $X_3 = \mathcal{V}(p_{34})$.

The relationship between X_2, $X_{2'}$ and X_1 inside $G(2,4)$ can also be understood as follows. For any integer $k \geq 1$, let Q be a nonsingular quadratic hypersurface in $\mathbb{P}^{2k+1}$ and consider a linear subspace $L = \mathbb{P}^{k-1}$ that is contained in Q. Then there exist precisely two k-dimensional subspaces that

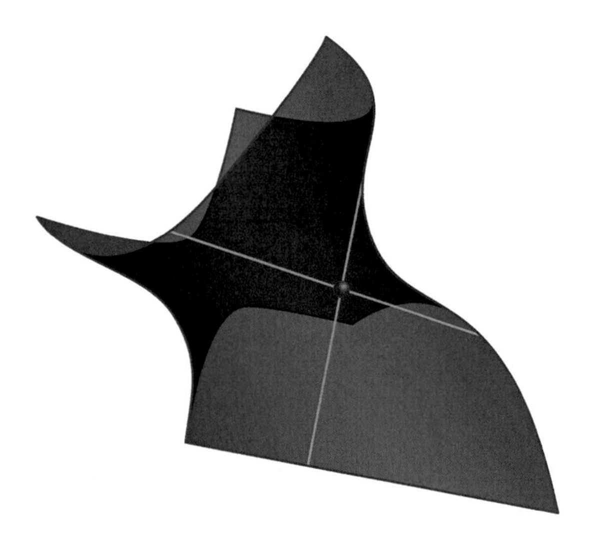

Figure 5.1. Every (red) point on a smooth quadratic surface lies on two (green) lines contained in the surface. This is a cartoon of the relationship between the Schubert subvarieties X_1, X_2 and $X_{2'}$ of $G(2,4)$.

contain L and are contained in Q. For $k = 2$ and $Q = G(2,4)$ containing the line $L = X_1$ in $\mathbb{P}^5$, the two subspaces are the planes X_2 and $X_{2'}$ in $\mathbb{P}^5$.

For $k = 1$, our statement is a classical fact from projective geometry. A quadratic surface in $\mathbb{P}^3$ is isomorphic to $\mathbb{P}^1 \times \mathbb{P}^1$. For point $p = L$ in the quadric, there are precisely two lines that contain p and lie on the quadric. In Figure 5.1, the blue quadric contains the red point p, and the two lines are green. This 3-dimensional picture is obtained by intersecting $G(2,4)$ with the subspace $H = \mathbb{P}^3$ that is defined by $p_{12} = p_{34} = 0$ inside $\mathbb{P}^5$. The red point p equals $H \cap X_1$. The quadric is simply $H \cap G(2,4) = \mathcal{V}(p_{23}p_{14} - p_{13}p_{24})$, and the two green lines are the intersections $X_2 \cap H$ and $X_{2'} \cap H$.

The Schubert subvarieties of a Grassmannian represent a basis for the cohomology ring of the Grassmannian when the underlying field is $K = \mathbb{C}$. One learns in *algebraic topology* that multiplication in a cohomology ring corresponds to intersection of submanifolds. This can then be used to answer enumerative questions, i.e. for counting the number of points in an intersection that turns out to be 0-dimensional. Schubert calculus is the art of making this precise when the ambient manifold is a Grassmannian.

When intersecting Schubert varieties X_i as cohomology classes $[X_i]$, one should think of them as coming from *distinct general* flags. For instance, consider the two surfaces X_2 and $X_{2'}$. They intersect in the line X_1. However, when thinking of the classes, the former represents all lines through some arbitrary point in $\mathbb{P}^3$ and the latter represents all lines contained in some entirely unrelated plane in $\mathbb{P}^3$. There are no lines satisfying both conditions, so the intersection of $[X_2]$ and $[X_{2'}]$ is the class of the empty set. We write this as $[X_2] \cdot [X_{2'}] = 0$. On the other hand, if we ask for lines going through two distinct points, or for lines contained in two distinct planes, then there is one solution. The self-intersections of the classes $[X_2]$ and $[X_{2'}]$ give one point. That one point is represented by the element 1 in the cohomology ring. Our discussion is summarized by the following relations that hold in the cohomology ring of the Grassmannian $G(2,4)$ of lines in $\mathbb{P}^3$:

$$[X_3][X_3] = [X_2] + [X_{2'}], \quad [X_2][X_2] = [X_{2'}][X_{2'}] = [X_3][X_1] = 1,$$
$$[X_2][X_{2'}] = 0.$$

Recall that multiplication represents intersection and sum represents union. The fact that the Schubert classes $[X_i]$ form a basis for the cohomology ring means that the class $[Z]$ of any subvariety Z is a linear combination of the $[X_i]$. Finding that linear combination for a given Z is similar to computing the degree of a subvariety in $\mathbb{P}^n$, as discussed at the end of Chapter 1.

We now have a conceptual framework for studying enumerative questions like *How many lines pass through four general lines in* $\mathbb{P}^3$*?* The set of such lines is a finite subset in $G(2,4)$. It is the intersection of four hypersurfaces, all of the form X_3. Since intersections are represented by multiplication in the cohomology ring, the following formal computation reveals the answer:

$$\begin{aligned}[X_3]^4 &= ([X_3][X_3])^2 = ([X_2] + [X_{2'}])^2 \\ &= [X_2]^2 + 2[X_2][X_{2'}] + [X_{2'}]^2 = 1 + 2 \cdot 0 + 1 = \mathbf{2}.\end{aligned}$$

Here is another question to be answered by Schubert calculus: *How many lines are simultaneously tangent to four general quadratic surfaces in* $\mathbb{P}^3$*?*

Consider a quadric Q in $\mathbb{P}^3$, represented as a symmetric 4×4 matrix. Let $\bigwedge^2 Q$ be the symmetric 6×6 matrix with entries given by the 2×2 minors of the matrix representing Q. This is known as the *second compound matrix* of Q. The condition for a line to be tangent to Q is expressed by the vanishing of the quadratic form $P(\bigwedge^2 Q)P^T$ in the Plücker coordinates $P = (p_{12}, p_{13}, \ldots, p_{34})$ of that line. This defines a threefold in $G(2,4)$. The cohomology class of that threefold is $2[X_3]$, since its equation is quadratic.

We conclude that the number of lines that are tangent to four given general quadratic surfaces in $\mathbb{P}^3$ is

$$(2[X_3])^4 = 16[X_3]^4 = 16 \cdot 2 = 32.$$

In order to actually compute the 32 lines over $\mathbb{C}$, given four concrete quadrics in $\mathbb{P}^3$, we would need to carry out some serious Gröbner basis computations.

Exercise 1 says that Schubert varieties in $G(k,n)$ correspond to Young diagrams contained in a $k\times(n-k)$ rectangle. Young diagrams play a fundamental role in representation theory. A formal definition appears in Definition 10.21. The classes of Schubert varieties form a $\mathbb{Z}$-basis for the cohomology ring of a Grassmannian. There is a purely combinatorial rule, named after Littlewood and Richardson, for multiplying (i.e. intersecting) Schubert varieties. This is expressed via Young diagrams [**21**, Appendix A.1].

The number of boxes in a Young diagram corresponds to the codimension of the Schubert variety. In particular, as there is only one Young diagram with one box, there is only one codimension-one class: the class of a hyperplane section under the Plücker embedding. The full $k\times(n-k)$ rectangle corresponds to the class of a point. We will not present the full Littlewood-Richardson rule here. Instead, we focus on the simpler *Pieri's rule.*

Let μ be any Young diagram and let λ be the Young diagram with s boxes contained in one column (resp. row). The product $\mu\cdot\lambda$ is the sum of all Young diagrams obtained by adding s boxes to μ, no more than one in a row (resp. column). Of course, if we work with the Grassmannian $G(k,n)$, then we disregard all diagrams that do not fit in a $k\times(n-k)$ rectangle.

Example 5.10. For $k=3$ and $n=6$, Pieri's rule gives the identity

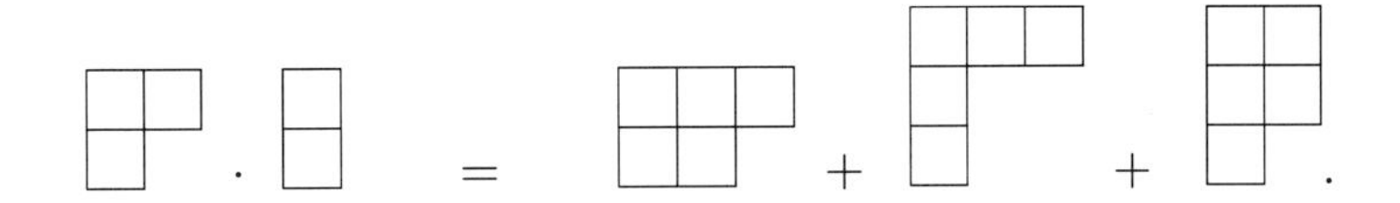

We close this section by deriving the degree of the Grassmannian $G(k,n)$ as a projective variety in $\mathbb{P}^{\binom{n}{k}-1}$. There are two different methods to do this.

The first method is geometric. Recall that the degree is the cardinality of the finite set that is obtained by intersecting $G(k,n)$ with $k(n-k)$ general hyperplanes in $\mathbb{P}^{\binom{n}{k}-1}$. The intersection of $G(k,n)$ with one hyperplane gives the hyperplane class in cohomology. The desired degree is found by multiplying the hyperplane class by itself $k(n-k)$ times. Each successive multiplication step is done using Pieri's rule. This leads us to Young tableaux.

Definition 5.11. A Young diagram with s boxes filled with the numbers $1,2,\ldots,s$ is a *standard Young tableau* if the entries are increasing in every row and column. These tableaux are ubiquitous in algebraic combinatorics.

Proposition 5.12. *The degree of the Grassmannian* $G(k,n)\subset\mathbb{P}^{\binom{n}{k}-1}$ *equals the number of standard Young tableaux of rectangular shape* $k\times(n-k)$.

Proof. The entries in such a standard Young tableau indicate where the box in each step is added when we apply Pieri's rule $k(n-k)$ times. $\square$

The second method is more algebraic, utilizing a Gröbner basis for the ideal $I(G(k,n))$. We explain this for $k=2$. By Theorem 5.8, the initial monomial ideal is $M = \langle p_{il}p_{jk} : 1 \leq i < j < k < l \leq n \rangle$. The variety $\mathcal{V}(M)$ is a union of coordinate subspaces of dimension $2n-3$; see e.g. Exercise 12 in Chapter 2. The number of these subspaces is the degree of M and hence the degree of $G(2,n)$. We find that this is the *Catalan number*:

$$\text{degree}(G(2,n)) \quad = \quad \frac{1}{n-1}\binom{2n-4}{n-2}. \tag{5.5}$$

By Proposition 5.12, this equals the number of standard Young tableaux of shape $2 \times (n-2)$. For more on Catalan numbers we refer to the entry A000108 in the *Online Encyclopedia of Integer Sequences.* In general, a closed formula for the degree of the Grassmannian may be derived from the hook-length formula for the number of standard Young tableaux [**21**, 4.12].

Theorem 5.13. *The degree of the Grassmannian $G(k,n) \subset \mathbb{P}^{\binom{n}{k}-1}$ equals*

$$\frac{(k(n-k))!}{\prod_{j=1}^{k} j \cdot (j+1) \cdots (j+n-k-1)}.$$

Remark 5.14. Grassmannians are named after Hermann Grassmann. However, it was Julius Plücker who first noted that lines in 3-space may be studied as a 4-dimensional object. Yet, Grassmann's earlier discoveries were fundamental; he was the one to realize that the algebraic setting of geometry allows us to consider objects not only in 3-dimensional space, but in any dimension. Can you imagine *data science* without Grassmann's insight?

Exercises

(1) Fix a complete flag in $\mathbb{P}^{n-1}$. Construct a bijection between
- subvarieties of $G(k,n)$ that consist of $l \in G(k,n)$ that intersect each element of the flag in at least the given dimension, and
- Young diagrams contained in a $k \times (n-k)$ rectangle.

Either the codimension or the dimension of the subvariety in $G(k,n)$ should equal the number of boxes in the corresponding Young diagram.

(2) Let G be a subgroup of the general linear group $\mathrm{GL}(V)$, and let $X \subset V$ be a variety such that the action of G on V restricts to X. Prove that if x is a smooth point of X and $g \in G$, then gx is also a smooth point.

(3) Consider the inclusion $G(2,4)\times G(2,4) \subset \mathbb{P}^5\times\mathbb{P}^5$. Using Plücker coordinates, describe the locus of pairs of lines $(l_1, l_2) \in G(2,4)\times G(2,4)$ such that l_1 intersects l_2 in $\mathbb{P}^3$. Hint: Represent both lines as 2×4 matrices. Note that two lines in $\mathbb{P}^3$ intersect if and only if they do not span the whole ambient space. Apply Laplace expansion of the determinant.

(4) Fix a variety $X \subset \mathbb{P}^n$. The spaces $\mathbb{P}^k \subset X$ form a subvariety of $G(k+1, n+1)$. This is known as the *Fano variety* of k-dimensional subspaces in X. Fix a nondegenerate quadric $Q \subset \mathbb{P}^3$. Describe the Fano variety of lines in it. Hint: One may solve this exercise either theoretically or by using algebra software. Also, Figure 5.1 gives a hint about the answer.

(5) How many *real* lines in 3-space can be simultaneously tangent to four given spheres? It is best to start by computing some explicit examples.

(6) The two lines incident to four given *real* lines in $\mathbb{P}^3$ can be either real or complex. In the latter case they form a complex conjugate pair. Write down a polynomial in the $24 = 4\cdot 6$ Plücker coordinates of four given lines whose sign distinguishes the two cases.

(7) How many lines in $\mathbb{P}^3$ are simultaneously incident to two given lines and tangent to two given quadratic surfaces?

(8) Consider the set of all lines in $\mathbb{P}^3$ that are tangent to the cubic Fermat surface $\{x_1^3+x_2^3+x_3^3+x_4^3=0\}$. This set is an irreducible hypersurface in the Grassmannian $G(2,4)$. Compute a polynomial in $p_{12}, p_{13}, \ldots, p_{34}$ that defines this hypersurface.

(9) Find a minimal generating set for the ideal of the Grassmannian $G(3,7)$.

(10) Prove that the determinant of a skew-symmetric $n\times n$ matrix is zero if n is odd, and is the square of a polynomial when n is even.

(11) Examine the monomial ideal that is generated by the underlined initial monomials in (5.4). Express this ideal as the intersection of prime ideals. How many primes occur? Use this result to derive the identity in (5.5).

(12) Fix six general planes $\mathbb{P}^2$ in $\mathbb{P}^4$. How many lines in $\mathbb{P}^4$ intersect all six planes? Describe Schubert calculus for the Grassmannian $G(2,5)$.

(13) Let $n = 2k$ and suppose that the $k\times k$ matrix $A = (a_{ij})$ in (5.2) is symmetric, i.e. its entries satisfy the equations $a_{ij} = a_{ji}$. Express these equations in terms of the $\binom{2k}{k}$ Plücker coordinates. The resulting subvariety of $G(k,2k)$ is known as the *Lagrangian Grassmannian.*

(14) Can you find four smooth quadratic surfaces in real 3-space such that 32 real lines are tangent to all four of the surfaces? Are these surfaces ellipsoids?

(15) The identity in Example 5.10 expresses a geometric statement about intersections inside the Grassmannian $G(3,6) \subset \mathbb{P}^{19}$. Explain this statement in your own words. Illustrate it by computing an explicit example.

Chapter 6

Nullstellensätze

"*This is not mathematics, it is theology!*", Paul Gordan

The German noun *Nullstellensatz* refers to a theorem that characterizes the existence of a zero (Nullstelle) for a system of polynomials. The classical version, due to Hilbert, works over algebraically closed fields. It says that the nonexistence of zeros is equivalent to the existence of a partition of unity for the given polynomials. A more general version furnishes a bijection between varieties and radical ideals. In this chapter we also discuss the analogous version over the field of real numbers. Here the main results are the real Nullstellensatz and the Positivstellensatz. These theorems give criteria for polynomial equations and inequalities to have no real solutions. This leads us to *real radical ideals* and their characterization via *sums of squares*.

6.1. Certificates for Infeasibility

In Chapter 3 we discussed how to find and represent solutions to a system of polynomial equations. But what if a solution does not exist? In this chapter we present methods of proving that a given system has no solution.

Throughout this section, we fix an algebraically closed field K, such as the complex numbers $K = \mathbb{C}$. We write $K[\mathbf{x}] = K[x_1, \ldots, x_n]$ for its polynomial ring in n variables. For an ideal $I \subset K[\mathbf{x}]$ we denote the associated variety in K^n by $\mathcal{V}(I)$, as in Chapter 2. We begin with the following version of the Nullstellensatz. This result appears as Theorem 1 in [**10**, §4.1].

Theorem 6.1. *If I is an ideal in $K[\mathbf{x}]$ that is proper, i.e. $1 \notin I$, then its variety $\mathcal{V}(I)$ in K^n is nonempty. Equivalently, if $\mathcal{V}(I) = \emptyset$ then $1 \in I$.*

Proof. Suppose $1 \notin I$. We use induction on n. For $n = 1$, the desired conclusion $\mathcal{V}(I) \neq \emptyset$ holds because $K[x_1]$ is a PID and every nonconstant polynomial in one variable has a zero in the algebraically closed field K.

Now let $n \geq 2$. For $a \in K$, we write $I_{x_n=a}$ for the ideal in $K[x_1, \ldots, x_{n-1}]$ obtained by setting $x_n = a$ in each element of I. One easily checks that this is indeed an ideal. We claim that there exists a scalar $a \in K$ such that $1 \notin I_{x_n=a}$. In such a case, by induction, there is a point $(a_1, \ldots, a_{n-1})$ in $\mathcal{V}(I_{x_n=a})$. This implies that $(a_1, \ldots, a_{n-1}, a)$ is a point in the variety $\mathcal{V}(I)$.

Consider the elimination ideal $I \cap K[x_n]$. To prove the claim, we distinguish two cases. First suppose that this ideal is not the zero ideal. Since $1 \notin I$, the principal ideal $I \cap K[x_n]$ is generated by a nonconstant polynomial

$$f(x_n) \;=\; \prod_{i=1}^{r}(x_n - b_i)^{m_i}.$$

Suppose that $1 \in I_{x_n=b_i}$ for $i = 1, 2, \ldots, r$. If this is not the case then we are done. Hence there exist $B_1, \ldots, B_r \in I$ such that $B_i(x_1, \ldots, x_{n-1}, b_i) = 1$ for all i. Note that B_i is congruent to 1 modulo $\langle x_n - b_i \rangle$ in $K[\mathbf{x}]$. This implies that the product $\prod_{i=1}^{r}(B_i - 1)^{m_i}$ belongs to the ideal $\langle f \rangle$. Since $f \in I$ and $B_i \in I$, the following identity holds modulo the ideal I:

$$0 \;=\; \prod_{i=1}^{r}(B_i - 1)^{m_i} \;=\; \prod_{i=1}^{r}(-1)^{m_i} \;=\; \pm 1, \qquad \text{i.e. } 1 \in I.$$

Next suppose $I \cap K[x_n] = \{0\}$. Let $\{g_1, \ldots, g_t\}$ be a Gröbner basis for I under the lexicographic order with $x_1 > \cdots > x_n$. Write $g_i = c_i(x_n)\mathbf{x}^{\alpha_i}$ + lower-order terms, where $\mathbf{x}^{\alpha_i}$ is a monomial in $x_1, \ldots, x_{n-1}$. The set $\{g_1, \ldots, g_t\}$ is also a Gröbner basis for the ideal generated by I in $K(x_n)[x_1, \ldots, x_{n-1}]$, a polynomial ring in $n-1$ variables over the field $K(x_n)$. By Buchberger's criterion, this means that every S-pair among the g_i reduces to zero. The number of S-pairs and reductions involved is finite. The set of coefficients appearing in this process is a finite subset of $K(x_n)$.

Since K is infinite, we can find a scalar $a \in K$ at which none of these rational functions vanishes. Consider $g_i' = g(x_1, \ldots, x_{n-1}, a) \in K[x_1, \ldots, x_{n-1}]$. The set $\{g_1', \ldots, g_t'\}$ generates the ideal $I_{x_n=a}$ in $K[x_1, \ldots, x_{n-1}]$. We claim that it is a Gröbner basis. Indeed, the monomials in $x_1, \ldots, x_{n-1}$ that appear in the reduction process for S-pairs among the g_i' over K are exactly the same as those for the g_i over $K(x_n)$. Note that the Gröbner basis element g_i' has the leading monomial $\mathbf{x}^{\alpha_i}$. None of the t monomials $\mathbf{x}^{\alpha_i}$ equals 1, since $I \cap K[x_n] = \{0\}$. This implies that 1 is not in the ideal $I_{x_n=a}$. $\square$

Theorem 6.1 gives a certificate for the nonexistence of solutions to a system of polynomial equations, namely the partition of unity we had promised.

Corollary 6.2. *Either a collection of polynomials* $f_1, \ldots, f_r \in K[\mathbf{x}]$ *has a common zero in* K^n *or there exists an identity* $g_1 f_1 + \cdots + g_r f_r = 1$ *with polynomial multipliers* $g_1, \ldots, g_r \in K[\mathbf{x}]$. *This is the desired certificate.*

Proof. Let $I = \langle f_1, \ldots, f_r \rangle$. Then either $\mathcal{V}(I) \neq \emptyset$ or $\mathcal{V}(I) = \emptyset$. In the latter case, $1 \in I$, and hence 1 is a polynomial linear combination of the f_i. $\square$

Example 6.3. Let $n = 2$ and consider the three polynomials

$$f_1 = (x+y-1)(x+y-2), \quad f_2 = (x-y+3)(x+2y-5), \quad f_3 = (2x-y)(3x+y-4).$$

These do not have a common zero. This is proved by a Nullstellensatz certificate $g_1 f_1 + g_2 f_2 + g_3 f_3 = 1$. One choice of multipliers is given by

$$g_1 = \frac{895}{756}x^2 - \frac{6263}{2160}x - \frac{2617}{2520}y + \frac{4327}{1008},$$

$$g_2 = \frac{5191}{3780}x^2 + \frac{358}{945}xy - \frac{6907}{3024}x - \frac{2123}{15120}y + \frac{3823}{7560},$$

$$g_3 = -\frac{179}{420}x^2 - \frac{716}{945}xy + \frac{1453}{1080}x - \frac{716}{945}y + \frac{13771}{7560}.$$

There are two possible methods for computing the multipliers $(g_1, \ldots, g_r)$ for the Nullstellensatz certificate, as in Corollary 6.2. The first method is to use the *extended Buchberger algorithm*. This is analogous to the extended Euclidean algorithm for the ring of integers or the ring of polynomials in one variable. For instance, given any collection of relatively prime integers, this method writes 1 as a $\mathbb{Z}$-linear combination of these integers.

In the extended Buchberger algorithm one keeps track of the polynomial multipliers that are used to generate new S-polynomials from current basis polynomials. In the end, each element in the final Gröbner basis is written explicitly as a polynomial linear combination of the input polynomials. If $\mathcal{V}(I) = \emptyset$, then that final reduced Gröbner basis is the singleton $\{1\}$.

The second method for computing Nullstellensatz certificates is to use degree bounds together with linear algebra. Let d be any integer that exceeds the degree of each f_i. Let g_i be a polynomial of degree $d \quad \deg(f_i)$ with coefficients that are unknowns, for $i = 1, 2, \ldots, r$. The desired identity $\sum_{i=1}^{r} g_i f_i = 1$ translates into a system of linear equations in all of these unknowns. We solve this system. If a solution is found, then this gives a certificate. If not, then there is no certificate in degree d, and we try a higher degree.

The two methods, in complete generality, can be very complicated to carry out in practice. The computation of Gröbner bases does not run in polynomial time. Worst-case complexity bounds for Gröbner bases are doubly exponential in the number of variables. Furthermore, the degrees of the multipliers g_i above are not polynomial in the input degrees either. Many mathematicians and computer scientists believe that there is no polynomial-time algorithm for deciding whether a given polynomial system has a complex solution. This is an active area of research. The situation is even worse if we want solutions with coordinates in $\mathbb{R}$, $\mathbb{Q}$ or $\mathbb{Z}$. In the last case, it is known that there exists no algorithm at all—irrespective of complexity—for deciding whether a system has an integral solution. This was Hilbert's 10th problem.

6.2. Hilbert's Nullstellensatz

Hilbert's Nullstellensatz provides a characterization of the set of all polynomials that vanish on a given variety. This classical result from 1890 is valid over any algebraically closed field K, such as the complex numbers $K = \mathbb{C}$. In this section we present this theorem and discuss some of its ramifications.

Recall that the *radical* of an ideal I in $K[\mathbf{x}]$ is the (possibly larger) ideal

$$\sqrt{I} \;=\; \{\, f \in K[\mathbf{x}] \;:\; f^m \in I \text{ for some } m \in \mathbb{N}\,\}.$$

This is a radical ideal, and hence it is an intersection of prime ideals.

Example 6.4. Consider the ideal $I = \langle\, x_1x_3,\ x_1x_4 + x_2x_3,\ x_2x_4 \,\rangle$ in the polynomial ring in four variables. It is not radical. To see this, note that the monomial $f = x_1x_4$ is not in I but f^2 is in I. The radical of I is

$$\sqrt{I} \;=\; \langle\, x_1x_3,\ x_1x_4,\ x_2x_3,\ x_2x_4 \,\rangle \;=\; \langle x_1, x_2\rangle \cap \langle x_3, x_4\rangle.$$

How many associated primes does the ideal I have? Do Gröbner bases of I give any hints? We refer to Example 3.29 for the answer.

We now show that $\sqrt{I}$ comprises all polynomials that vanish on $\mathcal{V}(I)$.

Theorem 6.5 (Hilbert's Nullstellensatz). *For any ideal I in the polynomial ring $K[\mathbf{x}]$ in n variables over an algebraically closed field K, we have*

$$\mathcal{I}(\,\mathcal{V}(I)\,) \;=\; \sqrt{I}. \tag{6.1}$$

Proof. The radical $\sqrt{I}$ is contained in the vanishing ideal $\mathcal{I}(\,\mathcal{V}(I)\,)$, because $f^m(\mathbf{a}) = 0$ implies $f(\mathbf{a}) = 0$ for all $\mathbf{a} \in K^n$. We must show that the left-hand side is a subset of the right-hand side in (6.1). Let $I = \langle f_1, \ldots, f_r\rangle$ and suppose that f is a polynomial which vanishes on $\mathcal{V}(I)$. Let y be a new variable and consider the ideal $J = \langle f_1, \ldots, f_r, yf - 1\rangle$ in the enlarged polynomial ring $K[\mathbf{x}, y] = K[x_1, \ldots, x_n, y]$. The variety $\mathcal{V}(J)$ in K^{n+1} is the empty set because $f = 0$ on every zero of $f_1, \ldots, f_r$ and $f \neq 0$ on every zero of $yf - 1$.

By Theorem 6.1, there exist multipliers $g_1, \ldots, g_r, h$ in $K[\mathbf{x}, y]$ such that

$$\sum_{i=1}^{r} g_i(\mathbf{x}, y) \cdot f_i(\mathbf{x}) \; + \; h(\mathbf{x}, y) \cdot (y f(\mathbf{x}) - 1) \;\; = \;\; 1.$$

We now substitute $y = 1/f(\mathbf{x})$ into this identity. This yields the following identity of rational functions in n variables:

$$\sum_{i=1}^{r} g_i\big(\mathbf{x}, \frac{1}{f(\mathbf{x})}\big) \cdot f_i(\mathbf{x}) \;\; = \;\; 1.$$

The common denominator is $f(\mathbf{x})^m$ for some $m \in \mathbb{N}$. Multiplying both sides by this common denominator, we obtain a polynomial identity of the form

$$\sum_{i=1}^{r} p_i(\mathbf{x}) \cdot f_i(\mathbf{x}) \;\; = \;\; f(\mathbf{x})^m.$$

This shows that f^m lies in I, and hence f lies in the radical $\sqrt{I}$. $\square$

Example 6.6. Which polynomial functions vanish on all nilpotent 3×3 matrices? We set $n = 9$ and take I to be the ideal generated by the entries of X^3, where $X = (x_{ij})$ is a 3×3 matrix with variables as entries. These are nine homogeneous cubic polynomials in nine unknowns x_{ij}. One of them is

$$x_{11}^3 + 2x_{11}x_{12}x_{21} + x_{12}x_{21}x_{22} + 2x_{11}x_{13}x_{31} + x_{12}x_{23}x_{31} + x_{13}x_{21}x_{32} + x_{13}x_{31}x_{33}.$$

But what are all the polynomials that vanish on nilpotent matrices? Can they be written as polynomial linear combinations of these nine cubics? The answer is no. The ideal $\mathcal{I}\big(\mathcal{V}(I)\big) = \sqrt{I}$ is larger than I. The radical of I is

$$\big\langle x_{11} + x_{22} + x_{33} \,,\, x_{11}x_{22} + x_{11}x_{33} - x_{12}x_{21} - x_{13}x_{31} + x_{22}x_{33} - x_{23}x_{32} \,,\, \det(X) \,\big\rangle.$$

In words, the ideal $\sqrt{I}$ is generated by the three coefficients of the nonleading terms in the characteristic polynomial of X. This reflects the fact that a square matrix is nilpotent if and only if it has no eigenvalues other than zero. Theorem 6.5 implies that every polynomial that vanishes on nilpotent 3×3 matrices is a polynomial linear combination of the three generators.

The Nullstellensatz implies a one-to-one correspondence between varieties in affine n-space and radical ideals in the polynomial ring in n variables.

Corollary 6.7. *The map $V \mapsto \mathcal{I}(V)$ defines a bijection between varieties in K^n and radical ideals in $K[\mathbf{x}]$. The inverse map that takes radical ideals to varieties is given by $I \mapsto \mathcal{V}(I)$.*

Proof. A variety V is Zariski closed and hence satisfies $V = \mathcal{V}(\mathcal{I}(V))$. The Nullstellensatz yields $I = \mathcal{I}(\mathcal{V}(I))$. These identities imply that both maps are one-to-one and onto, and that they are the inverses of each other. $\square$

Corollary 6.8. *The map $V \mapsto \mathcal{I}(V)$ defines a bijection between irreducible varieties in the affine space K^n and prime ideals in the polynomial ring $K[\mathbf{x}]$. As before, the inverse map is given by $I \mapsto \mathcal{V}(I)$.*

Proof. By Proposition 2.3, a variety V is irreducible if and only if its associated radical ideal $\mathcal{I}(V)$ is prime. □

Example 6.9 ($n = 2$)**.** There are only two kinds of proper irreducible subvarieties in the affine plane K^2. First, we have the points (a, b), corresponding to maximal ideals $I = \langle x - a, y - b\rangle$. Second, there are irreducible curves, one for each principal ideal $I = \langle f\rangle$ where f is an irreducible polynomial in $K[x, y]$. Arbitrary varieties are unions of these types. For instance, consider

$$J = \langle x^4 + 2x^2 + y^2 + 1\rangle \cap \langle y^3 - 4, 2x - y^2\rangle \quad \text{in } \mathbb{C}[x, y].$$

This ideal is radical. Its variety $\mathcal{V}(J)$ in $\mathbb{C}^2$ has five irreducible components, namely two quadratic curves and three points. Check that $\mathcal{I}(\mathcal{V}(J)) = J$.

In many applications one is interested in solving polynomial equations over the real numbers, and one cares less about nonreal complex solutions. This raises the following important question: Does there exist an analogue of Hilbert's Nullstellensatz over an ordered field, such as the real numbers $K = \mathbb{R}$? We shall see that the answer is affirmative. In the next section we discuss the real Nullstellensatz and the Positivstellensatz. These concern systems of polynomial equations and inequalities over the real numbers. They generalize linear programming duality for systems of linear equations and linear inequalities over $\mathbb{R}$. Moreover, as we shall see in Chapter 12, the Positivstellensatz plays an important role in nonlinear optimization.

The theorems above are false when $K = \mathbb{R}$ is the field of real numbers. To see this, let $n = 2$ and consider varieties in the plane $\mathbb{R}^2$. Theorem 6.1 fails for $I = \langle x^2 + y^2 + 1\rangle$. This is a proper ideal in $\mathbb{R}[x, y]$, but $\mathcal{V}_{\mathbb{R}}(I) = \emptyset$. Theorem 6.5 is also false for $I = \langle x^2 + y^2\rangle$. This is a radical ideal, but

$$\mathcal{I}\big(\mathcal{V}_{\mathbb{R}}(I)\big) = \langle x, y\rangle \quad \text{strictly contains} \quad \sqrt{I} = I.$$

We ask these two questions about ideals I in $\mathbb{R}[\mathbf{x}]$ and their varieties in $\mathbb{R}^n$:

- How can one best certify that the real variety $\mathcal{V}_{\mathbb{R}}(I)$ is empty?
- How can one compute the ideal $\mathcal{I}\big(\mathcal{V}_{\mathbb{R}}(I)\big)$ from generators of I?

The goal of the next section is to give answers to these questions.

6.3. Let's Get Real

We here present the *real Nullstellensatz*. Our point of departure is the fact that a polynomial f in $\mathbb{R}[\mathbf{x}]$ which is a sum of squares must be nonnegative, i.e. the inequality $f(\mathbf{u}) \geq 0$ holds for all $\mathbf{u} \in \mathbb{R}^n$. A natural question is whether the converse holds: Can every nonnegative polynomial be written as a sum of squares? The answer depends on what is being squared.

Hilbert showed in 1893 that the answer is no if one asks for squares of polynomials. However, it is yes if one allows squares of rational functions. This was the 17th problem in Hilbert's famous list from the International Congress of Mathematicians in 1900. It was solved by Emil Artin in 1927.

Theorem 6.10 (Artin's Theorem). *If $f \in \mathbb{R}[\mathbf{x}]$ is nonnegative on $\mathbb{R}^n$, then there exist polynomials $p_1, p_2, \ldots, p_r, q_1, q_2, \ldots, q_r \in \mathbb{R}[\mathbf{x}]$ such that*

$$f = \left(\frac{p_1}{q_1}\right)^2 + \left(\frac{p_2}{q_2}\right)^2 + \cdots + \left(\frac{p_r}{q_r}\right)^2.$$

Example 6.11 ($n = 2$). The *Motzkin polynomial* $M(x, y)$ is

$$x^4y^2 + x^2y^4 + 1 - 3x^2y^2 = \frac{(x^2+y^2+1) \cdot x^2y^2(x^2 + y^2 - 2)^2 + (x^2 - y^2)^2}{(x^2 + y^2)^2}.$$

Distributing the three terms of the factor $(x^2 + y^2 + 1)$, we see that the right-hand side is a sum of four squares of rational functions. This shows that $M(x, y)$ is nonnegative. However, it is not a sum of squares in $\mathbb{R}[x, y]$. Suppose it were equal to $\sum_i f_i^2$ where the f_i's are polynomials. None of the f_i's may contain a monomial x^d or y^d for $d > 0$; otherwise the largest such d would contribute positively to the coefficient of x^{2d} in $M(x, y)$. We have $f_i = \alpha_i + \beta_i xy + \tilde{f}_i$, where all terms of $\tilde{f}_i$ have degree ≥ 3 and $\alpha_i, \beta_i \in \mathbb{R}$. Further, $\tilde{f}_i$ cannot have terms of degree strictly greater than 3. Indeed, let us fix any degree-compatible monomial order. The leading term of $\sum f_i^2$ is the sum of squares of the top leading term in any of the f_i's that may contribute to it. In particular, the terms of top degree do not cancel. Hence, the coefficient -3 of x^2y^2 in $M(x, y)$ would then be equal to $\sum_i \beta_i^2 \geq 0$.

We shall derive Artin's Theorem 6.10 as a special case of the following more general statement. Theorem 6.12 is the real number analogue of the weak form of the Nullstellensatz which was established in Theorem 6.1.

Theorem 6.12. *Let I be an ideal in $\mathbb{R}[\mathbf{x}]$ whose variety $\mathcal{V}_{\mathbb{R}}(I)$ is empty. Then -1 is a sum of squares of polynomials modulo I, i.e. we have*

$$(6.2) \qquad 1 + p_1^2 + p_2^2 + \cdots + p_r^2 \in I \qquad \textit{for some } p_1, p_2, \ldots, p_r \in \mathbb{R}[\mathbf{x}].$$

For the proof of Theorem 6.12 see Murray Marshall's book [**40**, §2.3].

Proof of Theorem 6.10. Let y be a new variable. Consider the polynomial $g = f(\mathbf{x})y^2 + 1$ in $\mathbb{R}[\mathbf{x}, y]$. Since f is nonnegative, the real variety $\mathcal{V}_{\mathbb{R}}(g)$ is empty in $\mathbb{R}^{n+1}$. By Theorem 6.12, there exists a polynomial identity

$$(6.3) \qquad 1 + p_1(\mathbf{x}, y)^2 + p_2(\mathbf{x}, y)^2 + \cdots + p_r(\mathbf{x}, y)^2 + h(\mathbf{x}, y)g(\mathbf{x}, y) = 0.$$

We substitute $y = \pm\frac{1}{\sqrt{-f(\mathbf{x})}}$ into (6.3). This makes the last term cancel in both substitutions. Note that $p_i(\mathbf{x}, \frac{1}{\sqrt{-f(\mathbf{x})}}) = a_i(\mathbf{x}) + \frac{1}{\sqrt{-f(\mathbf{x})}} b_i(\mathbf{x})$ and $p_i(\mathbf{x}, -\frac{1}{\sqrt{-f(\mathbf{x})}}) = a_i(\mathbf{x}) - \frac{1}{\sqrt{-f(\mathbf{x})}} b_i(\mathbf{x})$ where a_i and b_i are rational functions.

We see that the two substitutions in (6.3) lead to the identities

$$1 + \sum_{i=1}^{r} a_i(\mathbf{x})^2 - \frac{1}{f(\mathbf{x})} \sum_{i=1}^{r} b_i(\mathbf{x})^2 \pm \frac{2}{\sqrt{-f(\mathbf{x})}} \sum_{i=1}^{r} a_i(\mathbf{x}) b_i(\mathbf{x}) = 0.$$

After adding these two expressions and dividing by 2, we obtain

$$1 + \sum_{i=1}^{r} a_i(\mathbf{x})^2 - \frac{1}{f(\mathbf{x})} \sum_{i=1}^{r} b_i(\mathbf{x})^2 = 0.$$

Clearing the denominator, we obtain

$$f = \frac{\sum_{i=1}^{r} b_i^2}{1 + \sum_{i=1}^{r} a_i^2} = \frac{(\sum_{i=1}^{r} b_i^2)(1 + \sum_{i=1}^{r} a_i^2)}{(1 + \sum_{i=1}^{r} a_i^2)^2}.$$

The right-hand side is a sum of squares of rational functions in n variables. This establishes Artin's Theorem 6.10. □

For systems consisting of both equations and inequalities, there is the *Positivstellensatz*. To motivate this, we review the corresponding result for linear polynomials, known as *Farkas' Lemma*. It lies at the heart of *linear programming duality*. Informally, Farkas' Lemma states that a system of linear equations and inequalities either has a solution in $\mathbb{R}^n$ or has a dual solution which certifies that the original system has no solution. Farkas' Lemma has many equivalent formulations. Here is one of them, selected to make the extension to higher-degree polynomials transparent.

Let $f_1, \ldots, f_r, g_1, \ldots, g_s$ be polynomials of degree 1 in $\mathbb{R}[\mathbf{x}]$, and consider

$$(6.4) \qquad f_1(\mathbf{u}) = 0, \ \ldots, \ f_r(\mathbf{u}) = 0, \quad g_1(\mathbf{u}) \geq 0, \ \ldots, \ g_s(\mathbf{u}) \geq 0.$$

In the dual problem, we seek numbers $a_1, \ldots, a_r, b_1, \ldots, b_s \in \mathbb{R}$ such that

$$(6.5) \quad a_1 \cdot f_1 + \cdots + a_r \cdot f_r + b_1^2 \cdot g_1 + \cdots + b_s^2 \cdot g_s = -1 \quad \text{in } \mathbb{R}[\mathbf{x}].$$

At most one of these two problems can have a solution. Indeed, since $b_1^2, \ldots, b_s^2 \geq 0$, the left-hand side of (6.5) is nonnegative for every vector $\mathbf{u}$ that solves (6.4).

Theorem 6.13 (Farkas' Lemma). *Given any linear polynomials $f_1, \ldots, f_r$ and $g_1, \ldots, g_s$ in $\mathbb{R}[\mathbf{x}]$, exactly one of the following two statements is true:*

(P) *There exists a point $\mathbf{u} \in \mathbb{R}^n$ such that (6.4) holds.*

(D) *There exist real numbers $a_1, \ldots, a_r, b_1, \ldots, b_s$ such that (6.5) holds.*

Consider the system (6.4) where the f_i and g_j are now arbitrary polynomials. In the dual problem, we seek polynomials a_i and $b_{j\nu}$ in $\mathbb{R}[\mathbf{x}]$ such that

$$(6.6) \qquad a_1 \cdot f_1 + \cdots + a_r \cdot f_r + \sum_{\nu \in \{0,1\}^s} \Big(\sum_j b_{j\nu}^2 \Big) \cdot g_1^{\nu_1} \cdots g_s^{\nu_s} = -1.$$

In the double sum we see linear combinations of square-free monomials in $g_1, \ldots, g_s$ whose coefficients are sums of squares. The set of polynomials that admit such a representation is the *quadratic module* generated by $g_1, \ldots, g_s$. Note that we allow the case of $s = 0$. In this case the sum of the g_i's disappears in (6.5), but the analogous sum in (6.6) is present. It represents the smallest quadratic module, i.e. the set of all sums of squares of polynomials. Quadratic modules associated with inequality constraints are fundamental in the study of semialgebraic sets [**40**, §2.1].

Theorem 6.14 (Positivstellensatz). *Given any polynomials $f_1, \ldots, f_r$ and $g_1, \ldots, g_s$ in $\mathbb{R}[\mathbf{x}]$, exactly one of the following two statements is true:*

(P) *There exists a point $\mathbf{u} \in \mathbb{R}^n$ such that* (6.4) *holds.*

(D) *There exist polynomials a_i and $b_{j\nu}$ in $\mathbb{R}[\mathbf{x}]$ such that* (6.6) *holds.*

Proof. See [**40**, §2.3]. □

The dual solution (D) in Theorem 6.14 is similar to the dual solution in Farkas' Lemma. One extra complication is that we now need products of the g_i. The result can be rephrased as follows: If a system of polynomial equations and inequalities is infeasible, then -1 lies in the sum of the ideal of equations and the quadratic module of inequalities. There is a more general version of the Positivstellensatz which also incorporates strict inequalities $h_1 > 0, \ldots, h_t > 0$. This is stated in [**53**, Theorem 7.5], proved in [**40**, §2.3].

The radical $\sqrt{I}$ of a polynomial ideal I was the main player in the strong form of Hilbert's Nullstellensatz (Theorem 6.5). It provides an algebraic representation for polynomials that vanish on a given complex variety. We now come to the analogous result for varieties over the real numbers.

Given an ideal I in $\mathbb{R}[\mathbf{x}]$, we define its *real radical* $\sqrt[\mathbb{R}]{I}$ to be the set

$$\big\{ f \in \mathbb{R}[\mathbf{x}] \, : \, f^{2m} + g_1^2 + \cdots + g_s^2 \in I \text{ for some } m \in \mathbb{N} \text{ and } g_1, \ldots, g_s \in \mathbb{R}[\mathbf{x}] \big\}.$$

One can check that $\sqrt[\mathbb{R}]{I}$ is an ideal in $\mathbb{R}[\mathbf{x}]$. Here is the analogue of Theorem 6.5:

Theorem 6.15 (Real Nullstellensatz). *For any ideal in $\mathbb{R}[\mathbf{x}]$, we have*

$$(6.7) \qquad \mathcal{I}\big(\mathcal{V}_{\mathbb{R}}(I)\big) = \sqrt[\mathbb{R}]{I}.$$

Proof. The argument is similar to the proof of Theorem 6.5. Clearly, $\sqrt[\mathbb{R}]{I}$ is contained in $\mathcal{I}(\mathcal{V}_{\mathbb{R}}(I))$. We need to show the reverse inclusion. Suppose that f vanishes on the real variety of $I = \langle f_1, \ldots, f_r \rangle \subset \mathbb{R}[\mathbf{x}]$. We introduce a new variable y and consider the ideal $J = \langle f_1, \ldots, f_r, yf - 1 \rangle$ in $\mathbb{R}[\mathbf{x}, y]$. It satisfies $\mathcal{V}_{\mathbb{R}}(J) = \emptyset$. By Theorem 6.12, there exists an identity of the form (6.2) for the ideal J. Substituting $y = 1/f(\mathbf{x})$ into that identity and clearing denominators, we find that some even power of f plus a sum of squares lies in I. This means that the polynomial f is in the real radical $\sqrt[\mathbb{R}]{I}$. □

Example 6.16. Consider the ideal generated by the Motzkin polynomial:

$$I \;=\; \langle\, M(x,y)\,\rangle \;=\; \langle\, x^4y^2 + x^2y^4 + 1 - 3x^2y^2\,\rangle.$$

Building on Example 6.11, we wish to compute the real radical $\sqrt[\mathbb{R}]{I}$. It must contain the numerators of the four summands in the rational sum of squares representation of M. This leads us to consider the ideal

$$J \;=\; \big\langle\, M, xy(x^2+y^2-2)\,,\, x^2 - y^2\,\big\rangle.$$

We find that the radical of J is the Jacobian ideal of the Motzkin polynomial:

$$\sqrt{J} \;=\; \Big\langle\, M, \frac{\partial M}{\partial x}\,,\,\frac{\partial M}{\partial y}\,\Big\rangle.$$

Furthermore, this radical ideal is precisely the real radical we are looking for:

$$\sqrt[\mathbb{R}]{I} \;=\sqrt{J} \;=\; \langle x-1, y-1\rangle \cap \langle x-1, y+1\rangle \cap \langle x+1, y-1\rangle \cap \langle x+1, y+1\rangle.$$

The real variety $\mathcal{V}_{\mathbb{R}}(M)$ defined by the Motzkin polynomial consists of the four points $(1,1)$, $(1,-1)$, $(-1,1)$ and $(-1,-1)$ in $\mathbb{R}^2$. Since M is nonnegative, these zeros are singular points of the complex curve $\mathcal{V}(M) \subset \mathbb{C}^2$.

Exercises

(1) Show that the real radical of an ideal I in $\mathbb{R}[\mathbf{x}]$ is a radical ideal and that it contains the radical of I.

(2) Find univariate polynomials g_1, g_2, g_3, g_4 in $\mathbb{Q}[x]$ such that

$$\begin{aligned} g_1(x-2)(x-3)(x-4) + g_2(x-1)(x-3)(x-4) \\ + g_3(x-1)(x-2)(x-4) + g_4(x-1)(x-2)(x-3) = 1. \end{aligned}$$

(3) An ideal I in $\mathbb{C}[\mathbf{x}]$ contains a monomial if and only if each point in its variety $\mathcal{V}(I)$ has at least one zero coordinate. Prove this fact, and describe an algorithm for testing whether an ideal contains a monomial.

(4) Let M be an ideal generated by monomials in $K[\mathbf{x}]$. How can we compute the radical $\sqrt{M}$? How can we compute the real radical $\sqrt[\mathbb{R}]{M}$? Hint: It could help to investigate the ideal in Example 3.24.

(5) Let I be the ideal generated by the two cubics $x_1^2x_2 - x_3^2x_4$ and $x_1x_2^2 - x_4^3$. Describe the projective variety $\mathcal{V}(I)$ in $\mathbb{P}^3$. Find the radical ideal $\sqrt{I}$. How many minimal generators does $\sqrt{I}$ have and what are their degrees?

(6) Let $V \subset \mathbb{R}^7$ be the variety of orthogonal Hankel matrices of format 4×4. Describe the ideal $\mathcal{I}(V)$. What are the irreducible components of V?

(7) Let I be the ideal generated by the two quartics $x_1^4 - x_1^2x_2^2$ and $x_2^4 - x_3^4$ in $\mathbb{R}[x_1, x_2, x_3]$. Determine the radical $\sqrt{I}$ and the real radical $\sqrt[\mathbb{R}]{I}$. Write each of these two radical ideals as an intersection of prime ideals.

(8) Let $f_1, \ldots, f_r$ and f be polynomials in $\mathbb{Q}[\mathbf{x}]$. Explain how Gröbner bases can be used to test whether f lies in the radical of $I = \langle f_1, \ldots, f_r \rangle$.

(9) Let $f \in \mathbb{R}[\mathbf{x}]$ be a nonnegative polynomial. Show that the real radical of the ideal generated by partial derivatives of f is contained in the real radical of $\langle f \rangle$. Find an example where the containment is strict.

(10) The circle given by $f = x^2 + y^2 - 4$ does not intersect the hyperbola given by $g = xy - 10$ in the plane $\mathbb{R}^2$. Find a real Nullstellensatz certificate for this, i.e. write -1 as a sum of squares modulo the ideal $\langle f, g \rangle$ in $\mathbb{R}[x, y]$.

(11) For an arbitrary $d \in \mathbb{N}$, exhibit an ideal I generated by quadratic polynomials and an additional polynomial f such that $f^d \notin I$ but $f^{d+1} \in I$.

(12) Let I be the ideal in $\mathbb{R}[x, y, z]$ generated by the *Robinson polynomial*

$$x^6 + y^6 + z^6 + 3x^2y^2z^2 - x^4y^2 - x^4z^2 - x^2y^4 - x^2z^4 - y^4z^2 - y^2z^4.$$

Determine the real radical $\sqrt[\mathbb{R}]{I}$ and the real variety $\mathcal{V}_{\mathbb{R}}(I)$ in $\mathbb{P}^2_{\mathbb{R}}$.

(13) Show that Theorem 6.15 implies Theorem 6.10.

(14) What is the effective Nullstellensatz?

(15) Find the radical and real radical of the ideal $I = \langle x^7 - y^7, x^8 - z^8 \rangle$ in $\mathbb{R}[x, y, z]$. Explain the difference between these two radical ideals.

(16) Prove Farkas' Lemma (Theorem 6.13).

(17) Consider the unit circles centered at the points $(0, 0)$, $(1, \frac{3}{2})$ and $(\frac{3}{2}, 1)$ in the plane $\mathbb{R}^2$. Do they intersect? Give a formulation like (6.4), and prove that the answer is no by exhibiting a Positivstellensatz certificate of the form (6.6).

Chapter 7

Tropical Algebra

"It is a theorem from algebraic geometry that all log-log plots look like straight lines", Maciej Zworski

The operations of addition and multiplication are familiar from elementary school. We here introduce tropical arithmetic. The new operations may at first seem unnatural to the reader, but we justify them with several applications, e.g. in the design of dynamic programming algorithms. A big part of our discussion centers on *tropical linear algebra.* The point is that the piecewise-linear structures of tropical mathematics provide yet another transition point between linear and nonlinear algebra. On the fully nonlinear side lies *tropical algebraic geometry.* We briefly touch on this subject with a discussion of tropical varieties and their geometric properties. This chapter is meant as an invitation to explore the textbook [**38**], whose setup we follow closely.

7.1. Arithmetic and Valuations

The *tropical semiring* $(\overline{\mathbb{R}}, \oplus, \odot)$ is the set $\overline{\mathbb{R}} = \mathbb{R} \cup \{\infty\}$, consisting of the set $\mathbb{R}$ of real numbers together with an extra element, ∞, that represents plus-infinity. It comes with two arithmetic operations, denoted by $\oplus$ and $\odot$.

We define addition and multiplication in the tropical semiring as follows:

$$x \oplus y \;:=\; \min(x,y) \qquad \text{and} \qquad x \odot y \;:=\; x+y.$$

The *tropical sum* of two real numbers is their minimum, and the *tropical product* of two real numbers is their usual sum. It takes some practice to carry out arithmetic in the tropical world. Here is an example with numbers:

$$4 \oplus 5 \;=\; 4 \qquad \text{and} \qquad 4 \odot 5 \;=\; 9.$$

Tropical addition and tropical multiplication are both commutative and associative. The distributive law holds, and the times operator $\odot$ takes precedence when plus $\oplus$ and times $\odot$ occur in the same expression. For example:

$$\begin{aligned} 4 \odot (5 \oplus 7) \;&=\; 4 \odot 5 \;=\; 9, \\ 4 \odot 5 \;\oplus\; 4 \odot 7 \;&=\; 9 \oplus 11 \;=\; 9. \end{aligned}$$

Both arithmetic operations have an identity element. Infinity is the *identity element* for addition, and zero is the *identity element* for multiplication:

$$x \oplus \infty \;=\; x \qquad \text{and} \qquad x \odot 0 \;=\; x.$$

We also note the following equations involving the two identity elements:

$$x \odot \infty = \infty \qquad \text{and} \qquad x \oplus 0 = \begin{cases} 0 & \text{if } x \geq 0, \\ x & \text{if } x < 0. \end{cases}$$

There is no subtraction in tropical arithmetic. There is no real number x that can be called "17 minus 8" because the equation $8 \oplus x = 17$ has no solution x. Tropical division is defined to be classical subtraction, so $(\mathbb{R} \cup \{\infty\}, \oplus, \odot)$ satisfies all ring axioms except for the existence of an additive inverse.

Such algebraic structures without additive inverse are called *semirings*, whence the name tropical semiring. It is essential to remember that "0" is the multiplicative identity element. If we write a term without an explicit coefficient, then that coefficient is zero. Thus, $x \oplus y$ means $0 \odot x \oplus 0 \odot y$.

Example 7.1 (Binomial Theorem)**.** We consider the third tropical power of a tropical sum. The following identities hold for all real numbers $x, y \in \mathbb{R}$:

$$\begin{aligned} (x \oplus y)^{\odot 3} &= (x \oplus y) \odot (x \oplus y) \odot (x \oplus y) \\ &= 0 \odot x^{\odot 3} \;\oplus\; 0 \odot x^{\odot 2} \odot y \;\oplus\; 0 \odot x \odot y^{\odot 2} \;\oplus\; 0 \odot y^{\odot 3}. \end{aligned}$$

Of course, the zero coefficients can be dropped here:

$$(x \oplus y)^{\odot 3} \;=\; x^{\odot 3} \oplus x^{\odot 2} \odot y \oplus x \odot y^{\odot 2} \oplus y^{\odot 3} \;=\; x^{\odot 3} \;\oplus\; y^{\odot 3}.$$

What is the relationship between classical arithmetic and tropical arithmetic? An informal answer is that the latter is the image of the former under taking logarithms. Indeed, if u and v are small positive real numbers, then $\log(u \cdot v)$ equals $\log(u) \odot \log(v)$, and $\log(u + v)$ is approximately the same as $\log(u) \oplus \log(v)$. Thus tropical geometry arises naturally when one draws a log-log plot of figures in $\mathbb{R}^2_{>0}$. This was the point of the quote at the beginning of this chapter. We refer to [**38**, Chapter 1] for further motivations.

The use of logarithms to simplify mathematical objects goes back to the 17th century. The original idea was really to reduce multiplication to addition. According to Laplace, the invention of logarithms "doubles the life of the astronomer, and spares him the errors and disgust inseparable from long calculations". For several centuries, scientists and engineers used a device called a *slide rule* to perform multiplications. These marvelous objects embody the logarithm. They were used until the late 1970s, also by school children, including the second author of this book. Eventually, slide rules were replaced by calculators. However, the idea of addition as a model for multiplication remains powerful, as it simplifies both algebra and geometry.

A more formal way of understanding the relationship between classical arithmetic and tropical arithmetic is to introduce fields with valuations.

Definition 7.2. A *valuation* on a field K is a function $\mathrm{val}: K \to \mathbb{R} \cup \{\infty\}$ that satisfies the following three axioms for all $a, b \in K$:

(1) $\mathrm{val}(ab) = \mathrm{val}(a) + \mathrm{val}(b)$,

(2) $\mathrm{val}(a+b) \geq \min\{\mathrm{val}(a), \mathrm{val}(b)\}$, and

(3) $\mathrm{val}(a) = \infty$ if and only if $a = 0$.

We often identify a valuation with its restriction to $K^* = K \backslash \{0\}$. With this, the image of val is an additive subgroup of $\mathbb{R}$, known as the *value group*.

A field K with valuation is a metric space. Namely, the valuation induces a *norm* $|\cdot|: K \to \mathbb{R}$ by setting $|a| = \exp(-\mathrm{val}(a))$ for $a \in K^*$ and $|0| = 0$. The field K is a metric space with distance $|a - b|$ between two elements $a, b \in K$. In fact, this metric on K is an *ultrametric*, which means that

$$|a+b| \;\leq\; \max(|a|, |b|) \;\leq\; |a| + |b|.$$

This allows the use of analytical and topological methods to study K.

An important example is the field of *Puiseux series* in a variable t with complex coefficients. This field is denoted by $K = \mathbb{C}\{\{t\}\}$. It contains the field $\mathbb{C}(t)$ of rational functions and its algebraic closure $\overline{\mathbb{C}(t)}$. Indeed, every element in $\overline{\mathbb{C}(t)}$ can be expanded into a Puiseux series. The exponents in a Puiseux series are rational numbers that have a common denominator.

The *valuation* of a scalar c in K is the smallest exponent a of any term $c_a t^a$ with $c_a \neq 0$ that appears in the series expansion of c. We write $a = \mathrm{val}(c)$. This is an element in the value group $(\mathbb{Q}, +)$ of K. Here are two examples of scalars in the Puiseux series field K and their valuations:

$$c = \frac{1}{t^2 + 2t^3 + t^5} = t^{-2} - 2t^{-1} + 4 - 9t + 20t^2 - 44t^3 + 97t^4 - 214t^5 + 472t^6 - \cdots$$

has $\mathrm{val}(c) = -2$, while

$$c' = t^{2/7}\sqrt{1 - t^{2/3}} = t^{2/7} - \frac{1}{2}t^{20/21} - \frac{1}{8}t^{34/21} - \frac{1}{16}t^{16/7} - \frac{5}{128}t^{62/21} - \cdots$$

has $\mathrm{val}(c') = 2/7$.

It is known that the field K is algebraically closed [**38**, Theorem 2.1.5]. So every polynomial of degree d in $K[x]$ has d roots, counting multiplicities.

Example 7.3 (Puiseux series). Every cubic polynomial in $K[x]$ has three roots. For instance, the three roots of $f(x) = tx^3 - x^2 + 3tx - 2t^5$ are

$$\begin{gathered} t^{-1} - 3t - 9t^3 - 54t^5 + 2t^6 - 405t^7 + 18t^8 - 3402t^9 + 180t^{10} - 30618t^{11} + \cdots, \\ 3t + 9t^3 - \tfrac{2}{3}t^4 + 54t^5 - 2t^6 + \tfrac{10931}{27}t^7 - 18t^8 + 3402t^9 - \tfrac{43756}{243}t^{10} + 30618t^{11} + \cdots, \\ \tfrac{2}{3}t^4 + \tfrac{4}{27}t^7 + \tfrac{16}{243}t^{10} - \tfrac{8}{81}t^{12} + \tfrac{80}{2187}t^{13} - \tfrac{80}{729}t^{15} + \tfrac{448}{19683}t^{16} - \tfrac{224}{2187}t^{18} + \cdots. \end{gathered}$$

Such Puiseux series can be computed in a computer algebra system. The valuations of the three roots are -1, 1 and 4. These characterize the asymptotic behavior of the roots when t is a real number close to zero.

We now assume K is an algebraically closed field of characteristic zero with a valuation $K \to \mathbb{R} \cup \{\infty\}$. Let $f \in K[\mathbf{x}]$ be a polynomial in n variables. We define its *tropicalization*, denoted by $\mathrm{trop}(f)$, to be the polynomial over the tropical semiring obtained by replacing each coefficient in f by its valuation. In what follows we restrict to $n = 1$, but later we also take $n \geq 2$.

For instance, if f is the cubic in Example 7.3, then its tropicalization is

$$\mathrm{trop}(f) = 1 \odot x^{\odot 3} \oplus 0 \odot x^{\odot 2} \oplus 1 \odot x \oplus 5. \tag{7.1}$$

Definition 7.4. A *tropical polynomial* is a function that is the minimum of finitely many affine-linear functions. A real number u is said to be a *tropical root* of a given tropical polynomial if that minimum is attained at least twice when the affine-linear functions are evaluated at the argument u.

Example 7.5. If we evaluate (7.1) at the argument u, we obtain

$$\mathrm{trop}(f)(u) = \min\{1 + 3u,\ 0 + 2u,\ 1 + u,\ 5\}. \tag{7.2}$$

This is a tropical polynomial function. For instance, for the argument $u = 4$, the minimum above is attained twice, by $1+u$ and by 5. This means that 4 is a tropical root of $\mathrm{trop}(f)$. The other roots are $u = 1$ and $u = -1$. Note that the three tropical roots $-1, 1$ and 4 of $\mathrm{trop}(f)$ are precisely the valuations of the three classical roots of the cubic f in Example 7.3. In Theorem 7.7 below, we state the general result which explains this observation.

To evaluate a tropical polynomial, one takes the minimum of its tropical monomials. Tropical monomials are affine-linear functions. If $u = \mathrm{val}(c)$ for some $c \in K$ and u is not a tropical root of $\mathrm{trop}(f)$, then

$$\mathrm{val}(f(c)) \;=\; (\mathrm{trop}(f))(u). \tag{7.3}$$

In the following lemma we are claiming that two $\mathbb{R} \to \mathbb{R}$ functions agree.

Lemma 7.6. *Multiplication of polynomials is compatible with tropicalization. Namely, if $f, g \in K[x]$ then $\mathrm{trop}(fg) = \mathrm{trop}(f) \odot \mathrm{trop}(g)$.*

Proof. By [**38**, Lemma 2.1.12], we may assume that the value group of K is dense in $\mathbb{R}$. Consider a real number $u \in \mathbb{R}$ that is in the value group but is not a tropical root of $\mathrm{trop}(f)$, $\mathrm{trop}(g)$ or $\mathrm{trop}(fg)$. There exists a scalar $c \in K$ with $\mathrm{val}(c) = u$. A computation using (7.3) shows that

$$\mathrm{trop}(fg)(u) = \mathrm{val}((fg)(c)) = \mathrm{val}(f(c)) + \mathrm{val}(g(c)) = \mathrm{trop}(f)(u) \odot \mathrm{trop}(g)(u).$$

Hence the two functions agree on a dense set of arguments $u \in \mathbb{R}$. As the functions are continuous, we conclude that they are equal. $\square$

Lemma 7.6 holds also for polynomials in n variables, but in this section we stay with $n = 1$. The following facts relate classical root finding and tropical root finding. We continue to assume that K is algebraically closed.

Theorem 7.7. *Fix a univariate polynomial $f \in K[x]$ and let $g = \mathrm{trop}(f)$ be its tropicalization. If $c \in K \setminus \{0\}$ satisfies $f(c) = 0$, then $u = \mathrm{val}(c)$ is a tropical root of g. Conversely, every tropical root u of g arises from a zero c of f.*

Proof. Let $f(x) = \sum_{i=0}^{d} b_i x^i$ and suppose $f(c) = \sum_{i=0}^{d} b_i c^i$ is zero. Then $\infty = \mathrm{val}(f(c))$ but $\mathrm{val}(b_i c^i) < \infty$ for some i. Using [**38**, Lemma 2.1.1], this implies $\mathrm{val}(b_i c^i) = \mathrm{val}(b_j c^j) \leq \mathrm{val}(b_k c^k)$ for some $i \neq j$ and all $k \neq i, j$. This means that $u = \mathrm{val}(c)$ is a tropical root of $g = \mathrm{trop}(f)$.

Our proof of the second statement follows that of [**38**, Proposition 3.1.5]. We assume that u is a tropical root of $g = \mathrm{trop}(f)$. Since K is algebraically closed, we can factor $f(x) = \prod_{j=1}^{d} (a_j x - b_j)$. By Lemma 7.6, the tropicalization $g = \mathrm{trop}(f)$ is the classical sum of the d simple tropical functions

$$\mathrm{trop}(a_j x - b_j) \;=\; \mathrm{val}(a_j) \odot x \oplus \mathrm{val}(b_j) \;=\; \min\bigl\{\mathrm{val}(a_j) + x, \mathrm{val}(b_j)\bigr\}.$$

Each of these functions has precisely one tropical root, namely $\mathrm{val}(b_j) - \mathrm{val}(a_j)$, and each tropical root of their classical sum g must be one of these.

Let u be a tropical root of g. Then there is an index j such that $u = \mathrm{val}(b_j) - \mathrm{val}(a_j)$. If we set $c = b_j / a_j$ in K, then $f(c) = 0$ and $u = \mathrm{val}(c)$. $\square$

Fields with valuations provide a systematic way of speaking algebraically about logarithms. This explains the connection between classical arithmetic and tropical arithmetic. Here is a scenario that appears in number theory.

Example 7.8 (The p-adic valuation). For every prime number p, the field $K = \mathbb{Q}$ of rational numbers has a valuation val_p with value group $\mathbb{Z}$. Indeed, every rational number c can be written uniquely as $c = p^u \cdot \frac{q}{r}$ where $u \in \mathbb{Z}$ while q and r are relatively prime integers not divisible by p. We have $\mathrm{val}_p(c) = u$. The completion of $\mathbb{Q}$ with respect to the norm induced by val_p is the field of p-adic numbers. This field is important in number theory.

We shall return to fields and their valuations in our exploration of varieties in Section 7.3. First, however, let us develop some purely tropical machinery, in the more familiar setting of matrices and linear algebra.

7.2. Linear Algebra

Vectors and matrices make sense over the tropical semiring. For instance, the tropical scalar product in $\mathbb{R}^3$ of a row vector and a column vector is

$$\begin{aligned}(u_1, u_2, u_3) \odot (v_1, v_2, v_3)^T &= u_1 \odot v_1 \oplus u_2 \odot v_2 \oplus u_3 \odot v_3 \\ &= \min\{u_1 + v_1,\ u_2 + v_2,\ u_3 + v_3\}.\end{aligned}$$

Here is the product of a column vector and a row vector of length 3:

$$\begin{aligned}&(u_1, u_2, u_3)^T \odot (v_1, v_2, v_3) \\ &= \begin{pmatrix} u_1 \odot v_1 & u_1 \odot v_2 & u_1 \odot v_3 \\ u_2 \odot v_1 & u_2 \odot v_2 & u_2 \odot v_3 \\ u_3 \odot v_1 & u_3 \odot v_2 & u_3 \odot v_3 \end{pmatrix} = \begin{pmatrix} u_1 + v_1 & u_1 + v_2 & u_1 + v_3 \\ u_2 + v_1 & u_2 + v_2 & u_2 + v_3 \\ u_3 + v_1 & u_3 + v_2 & u_3 + v_3 \end{pmatrix}.\end{aligned} \tag{7.4}$$

Any matrix which can be expressed as such a product has *tropical rank* 1.

Fix a $d \times n$ matrix A. We may wish to find its image $\{A \odot \mathbf{x} : \mathbf{x} \in \mathbb{R}^n\}$ and solve linear systems $A \odot \mathbf{x} = \mathbf{b}$ for various right-hand sides $\mathbf{b}$. For an introduction to tropical linear systems see the books *Max-linear Systems* by Butkovič [7] and *Essentials of Tropical Combinatorics* by Joswig [**27**].

For a first application of tropical linear algebra, consider the problem of finding a shortest path in a weighted directed graph G with n nodes. Every directed edge (i, j) in G has an associated length d_{ij} which is a nonnegative real number. If (i, j) is not an edge of G then we set $d_{ij} = +\infty$. We represent G by its $n \times n$ *adjacency matrix* $D_G = (d_{ij})$, with zeros on the diagonal. The off-diagonal entries are the edge lengths d_{ij}. The matrix D_G need not be symmetric; we allow $d_{ij} \neq d_{ji}$. If G is an undirected graph, then we view G as a directed graph with two directed edges (i, j) and (j, i) for each undirected edge $\{i, j\}$. In that case, D_G is a symmetric matrix, where $d_{ii} = 0$ and $d_{ij} = d_{ji}$ is the distance between node i and node j.

Consider the $n \times n$ matrix with entries in $\mathbb{R}_{\geq 0} \cup \{\infty\}$ that results from tropically multiplying the given adjacency matrix D_G by itself $n - 1$ times:

$$D_G^{\odot(n-1)} = D_G \odot D_G \odot \cdots \odot D_G. \tag{7.5}$$

Proposition 7.9. *Let G be a weighted directed graph on n nodes with adjacency matrix D_G. The entry of the matrix $D_G^{\odot(n-1)}$ in row i and column j is the length of a shortest path from node i to node j in the graph G.*

Proof. Let $d_{ij}^{(r)}$ denote the minimum length of any path from node i to node j using at most r edges in G. We have $d_{ij}^{(1)} = d_{ij}$ for any two nodes i and j. Since the edge weights d_{ij} are assumed to be nonnegative, for each two nodes i and j, there exists a shortest path from i to j that visits each node of G at most once. Hence the length of a shortest path from i to j equals $d_{ij}^{(n-1)}$.

For $r \geq 2$ we have a recursive formula for the length of a shortest path:

$$d_{ij}^{(r)} \quad = \quad \min\{d_{ik}^{(r-1)} + d_{kj} \,:\, k = 1, 2, \ldots, n\}. \tag{7.6}$$

Using tropical arithmetic, this formula can be rewritten as

$$\begin{aligned} d_{ij}^{(r)} &= d_{i1}^{(r-1)} \odot d_{1j} \;\oplus\; d_{i2}^{(r-1)} \odot d_{2j} \;\oplus\; \cdots \;\oplus\; d_{in}^{(r-1)} \odot d_{nj}. \\ &= (d_{i1}^{(r-1)}, d_{i2}^{(r-1)}, \ldots, d_{in}^{(r-1)}) \odot (d_{1j}, d_{2j}, \ldots, d_{nj})^T. \end{aligned}$$

It follows by induction on r that $d_{ij}^{(r)}$ equals the entry in row i and column j of the $n \times n$ matrix $D_G^{\odot r}$. Indeed, the right-hand side of the recursive formula is the tropical product of row i in $D_G^{\odot(r-1)}$ and column j in D_G. That product is the (i,j) entry of $D_G^{\odot r}$. Applying this to $r = n-1$, we see that $d_{ij}^{(n-1)}$ is the entry in row i and column j of $D_G^{\odot(n-1)}$. This proves the claim. $\square$

The above algorithm belongs to the area known as *dynamic programming* in computer science. For us, it means performing the matrix multiplication

$$D_G^{\odot r} \quad = \quad D_G^{\odot(r-1)} \odot D_G \qquad \text{for } r = 2, \ldots, n-1.$$

We next consider the notion of the *tropical determinant.* This is the tropicalization of the classical determinant, here regarded as a polynomial of degree n in n^2 unknowns x_{ij}. Thus, we form the tropical sum over the tropical diagonal products obtained by taking all $n!$ permutations π of $\{1, 2, \ldots, n\}$:

$$\operatorname{tropdet}(X) \quad := \quad \bigoplus_{\pi \in S_n} x_{1\pi(1)} \odot x_{2\pi(2)} \odot \cdots \odot x_{n\pi(n)}. \tag{7.7}$$

Here S_n is the *symmetric group* of permutations of $\{1, 2, \ldots, n\}$. Evaluating the tropical determinant means solving the classical *assignment problem* from combinatorial optimization. Imagine a company that has n jobs and n workers, and each job needs to be assigned to exactly one of the workers. Let x_{ij} be the cost of assigning job i to worker j. The company wishes to find the cheapest assignment $\pi \in S_n$. The optimal total cost equals

$$\min\{x_{1\pi(1)} + x_{2\pi(2)} + \cdots + x_{n\pi(n)} \,:\, \pi \in S_n\}. \tag{7.8}$$

That minimum is the tropical determinant (7.7) of the matrix $X = (x_{ij})$:

Proposition 7.10. *The tropical determinant solves the assignment problem.*

In the assignment problem we seek the minimum over $n!$ quantities. This appears to require exponentially many operations. However, there is a polynomial-time method for this task, namely the *Hungarian algorithm.*

The square matrix X is called *tropically singular* if the minimum in (7.8) is attained at least twice. Following Definition 7.4, this is equivalent to saying that X is a tropical root of the tropical determinant (7.7). If Y is any rectangular matrix, we define the *tropical rank* of Y to be the size of the largest square submatrix that is not tropically singular. For rank 1, this is consistent with our earlier definition, given after (7.4). Namely, the matrix Y has tropical rank 1 if and only if it is the tropical product of a column vector and a row vector. For a detailed discussion of the tropical rank and other related notions of matrix rank, we refer to [**38**, §5.3].

Eigenvectors and eigenvalues of square matrices are central to linear algebra. The same is true in tropical linear algebra. We fix an $n \times n$ matrix $A = (a_{ij})$ over $\overline{\mathbb{R}} = \mathbb{R} \cup \{\infty\}$. An *eigenvalue* of A is a real number λ such that

$$A \odot \mathbf{v} = \lambda \odot \mathbf{v} \qquad \text{for some } \mathbf{v} \in \mathbb{R}^n. \tag{7.9}$$

We say that $\mathbf{v}$ is an *eigenvector* of the matrix A. The arithmetic operations in (7.9) are tropical. For instance, for $n = 2$, the left-hand side of (7.9) is

$$\begin{pmatrix} a_{11} & a_{12} \\ a_{21} & a_{22} \end{pmatrix} \odot \begin{pmatrix} v_1 \\ v_2 \end{pmatrix} = \begin{pmatrix} a_{11} \odot v_1 \oplus a_{12} \odot v_2 \\ a_{21} \odot v_1 \oplus a_{22} \odot v_2 \end{pmatrix} = \begin{pmatrix} \min\{a_{11} + v_1, a_{12} + v_2\} \\ \min\{a_{21} + v_1, a_{22} + v_2\} \end{pmatrix}.$$

The right-hand side of (7.9) equals

$$\lambda \odot \begin{pmatrix} v_1 \\ v_2 \end{pmatrix} = \begin{pmatrix} \lambda \odot v_1 \\ \lambda \odot v_2 \end{pmatrix} = \begin{pmatrix} \lambda + v_1 \\ \lambda + v_2 \end{pmatrix}.$$

Let $G(A)$ denote the directed graph with adjacency matrix A. Its nodes are labeled by $[n] = \{1, 2, \ldots, n\}$. There is an edge from node i to node j if and only if $a_{ij} < \infty$. The edge has length a_{ij}. In particular, $a_{ii} \neq \infty$ if and only if there is a loop at vertex i. The *normalized length* of a directed path $i_0, i_1, \ldots, i_k$ in $G(A)$ is $(a_{i_0 i_1} + a_{i_1 i_2} + \cdots + a_{i_{k-1} i_k})/k$, computed in classical arithmetic. If $i_k = i_0$ then the path is a *directed cycle*, and this quantity is the normalized length of the cycle. Recall that a directed graph is *strongly connected* if there is a directed path from any node to any other node.

Theorem 7.11. *Let A be an $n \times n$ matrix with entries in $\overline{\mathbb{R}}$ whose directed graph $G(A)$ is strongly connected. Then A has precisely one eigenvalue $\lambda(A)$. It equals the minimum normalized length of a directed cycle in $G(A)$.*

Proof. Let $\lambda = \lambda(A)$ be the minimum of the normalized lengths over all directed cycles in $G(A)$. We first prove that $\lambda(A)$ is the only possibility for an eigenvalue. Suppose that $\mathbf{z} \in \mathbb{R}^n$ is any eigenvector of A, and let γ be the corresponding eigenvalue. For any cycle $(i_1, i_2, \ldots, i_k, i_1)$ in $G(A)$ we have

$$a_{i_1 i_2} + z_{i_2} \geq \gamma + z_{i_1}, \; a_{i_2 i_3} + z_{i_3} \geq \gamma + z_{i_2},$$
$$a_{i_3 i_4} + z_{i_4} \geq \gamma + z_{i_3}, \; \ldots, \; a_{i_k i_1} + z_{i_1} \geq \gamma + z_{i_k}.$$

Adding the left-hand sides and the right-hand sides, we find that the normalized length of the cycle is greater than or equal to γ. In particular, we have $\lambda(A) \geq \gamma$. For the reverse inequality, start with any index i_1. Since $\mathbf{z}$ is an eigenvector with eigenvalue γ, there exists i_2 such that $a_{i_1 i_2} + z_{i_2} = \gamma + z_{i_1}$. Likewise, there exists i_3 such that $a_{i_2 i_3} + z_{i_3} = \gamma + z_{i_2}$. We continue in this manner until we reach an index i_l which was already in the sequence, say $i_k = i_l$ for $k < l$. By adding the equations along this cycle, we find that

$$\begin{aligned} &(a_{i_k i_{k+1}} + z_{i_{k+1}}) + (a_{i_{k+1} i_{k+2}} + z_{i_{k+2}}) + \cdots + (a_{i_{l-1} i_l} + z_{i_l}) \\ = \;\; &(\gamma + z_{i_k}) \; + \; (\gamma + z_{i_{k+1}}) \; + \; \cdots \; + \; (\gamma + z_{i_{l-1}}). \end{aligned}$$

We conclude that the normalized length of the cycle $(i_k, i_{k+1}, \ldots, i_l = i_k)$ in $G(A)$ is equal to γ. In particular, $\gamma \geq \lambda(A)$. This proves that $\gamma = \lambda(A)$.

It remains to prove the existence of an eigenvector. Let B be the matrix obtained from A by (classically) subtracting $\lambda(A)$ from every entry in A. All cycles in $G(B)$ have nonnegative length, and there exists a cycle of length zero. Using tropical matrix operations, we define

$$B^+ \; = \; B \oplus B^{\odot 2} \oplus B^{\odot 3} \oplus \cdots \oplus B^{\odot n}.$$

This matrix is known as the *Kleene plus* of the matrix B. The entry B^+_{ij} in row i and column j of B^+ is the length of a shortest path from node i to node j in the weighted directed graph $G(B)$. Here, we assume that a path contains some edges, so the shortest path from i to i may be strictly positive. Since $G(B)$ is strongly connected, we have $B^+_{ij} < \infty$ for all i and j.

Fix any node j that lies on a zero-length cycle of $G(B)$. Let $\mathbf{x} = B^+_{\cdot j}$ denote the jth column vector of the matrix B^+. We have $x_j = B^+_{jj} = 0$, as there is a path from j to itself of length zero and there are no negative-weight cycles. This implies $B^+ \odot \mathbf{x} \leq B^+_{\cdot j} = \mathbf{x}$. Since lengths of shortest paths obey the triangle inequality, we have $(B \odot \mathbf{x})_i = \min_l (B_{il} + x_l) = \min_l (B_{il} + B^+_{lj}) \geq B^+_{ij} = x_i$. In vector notation, we write this as $B \odot \mathbf{x} \geq \mathbf{x}$.

Since tropical linear maps preserve coordinatewise inequalities between vectors, we have $B^2 \odot \mathbf{x} \geq B \odot \mathbf{x}$, $B^3 \odot \mathbf{x} \geq B^2 \odot \mathbf{x}$, etc. Therefore, $B^+ \odot \mathbf{x} = B \odot \mathbf{x} \oplus B^2 \odot \mathbf{x} \oplus \cdots \oplus B^n \odot \mathbf{x} = B \odot \mathbf{x}$. This yields $\mathbf{x} \leq B \odot \mathbf{x} = B^+ \odot \mathbf{x} \leq \mathbf{x}$. Hence, $B \odot \mathbf{x} = \mathbf{x}$, so $\mathbf{x}$ is an eigenvector of B with eigenvalue 0. We conclude that $\mathbf{x}$ is an eigenvector with eigenvalue λ of our matrix A:

$$A \odot \mathbf{x} \;=\; (\lambda \odot B) \odot \mathbf{x} \;=\; \lambda \odot (B \odot \mathbf{x}) \;=\; \lambda \odot \mathbf{x}.$$

This completes the proof of Theorem 7.11. □

The eigenvalue of a tropical $n \times n$ matrix $A = (a_{ij})$ can be computed efficiently, using a *linear program* with $n+1$ decision variables $v_1, \ldots, v_n, \lambda$:

(7.10) Maximize λ subject to $a_{ij} + v_j \geq \lambda + v_i$ for all $1 \leq i, j \leq n$.

Proposition 7.12. *The unique eigenvalue $\lambda(A)$ of the given $n \times n$ matrix $A = (a_{ij})$ coincides with the optimal value λ^* of the linear program* (7.10).

Proof. See [**38**, Proposition 5.1.2]. □

We next determine the *eigenspace* of the matrix A, which is the set

$$\mathrm{Eig}(A) \;=\; \big\{ \mathbf{x} \in \mathbb{R}^n : A \odot \mathbf{x} = \lambda(A) \odot \mathbf{x} \big\}.$$

The set $\mathrm{Eig}(A)$ is closed under tropical scalar multiplication: If $\mathbf{x} \in \mathrm{Eig}(A)$ and $c \in \mathbb{R}$, then $c \odot \mathbf{x}$ is also in $\mathrm{Eig}(A)$. We can thus identify $\mathrm{Eig}(A)$ with its image in the quotient space $\mathbb{R}^n/\mathbb{R}\mathbf{1} \simeq \mathbb{R}^{n-1}$. Here $\mathbf{1} = (1, 1, \ldots, 1)$. This space is called the *tropical projective torus*; cf. [**27**, §1.4]. We saw that every eigenvector of the matrix A is also an eigenvector of the matrix $B = (-\lambda(A)) \odot A$ and vice versa. Hence the eigenspace $\mathrm{Eig}(A)$ is equal to

$$\mathrm{Eig}(B) \;=\; \big\{ \mathbf{x} \in \mathbb{R}^n : B \odot \mathbf{x} = \mathbf{x} \big\}.$$

Theorem 7.13. *Let B_0^+ be the submatrix of the Kleene plus B^+ given by the columns whose diagonal entry B_{jj}^+ is zero. The image of this matrix, in tropical arithmetic, equals the eigenspace:* $\mathrm{Eig}(A) = \mathrm{Eig}(B) = \mathrm{Image}(B_0^+)$.

Proof. See [**38**, Theorem 5.1.3]. □

Example 7.14. We demonstrate the computation of eigenvectors for $n = 3$. In our first example, the minimal cycle lengths are attained by the loops:

$$A = \begin{pmatrix} 3 & 4 & 4 \\ 4 & 3 & 4 \\ 4 & 4 & 3 \end{pmatrix} \quad \Rightarrow \quad \lambda(A) = 3 \quad \Rightarrow \quad B = B^+ = B_0^+ = \begin{pmatrix} 0 & 1 & 1 \\ 1 & 0 & 1 \\ 1 & 1 & 0 \end{pmatrix}.$$

The eigenspace is the tropical image of B. This image in $\mathbb{R}^3/\mathbb{R}\mathbf{1}$ is the hexagon with vertices $(0,1,1)$, $(0,0,1)$, $(1,0,1)$, $(1,0,0)$, $(1,1,0)$ and $(0,1,0)$. In our second example, the winner is the cycle $1 \to 2 \to 1$:

$$A = \begin{pmatrix} 3 & 1 & 4 \\ 1 & 3 & 2 \\ 4 & 4 & 3 \end{pmatrix} \Rightarrow \lambda(A) = 1 \Rightarrow B = \begin{pmatrix} 2 & 0 & 3 \\ 0 & 2 & 1 \\ 3 & 3 & 2 \end{pmatrix} \Rightarrow B^+ = \begin{pmatrix} 0 & 0 & 1 \\ 0 & 0 & 1 \\ 3 & 3 & 2 \end{pmatrix}.$$

The eigenspace of A is the tropical linear span of the first column of B^+:

$$\mathrm{Eig}(A) = \mathrm{Eig}(B) = \{c \odot (0,0,3)^T : c \in \mathbb{R}\} = \{(c,c,c+3)^T : c \in \mathbb{R}\}.$$

So here $\mathrm{Eig}(A)$ is just a single point in the tropical projective 2-torus $\mathbb{R}^3/\mathbb{R}\mathbf{1}$.

We computed the eigenspace of a square matrix as the image of another matrix. This motivates the study of images of tropical linear maps $\mathbb{R}^m \to \mathbb{R}^n$. Such images are *not* tropical linear spaces. They are known as *tropical polytopes.* Indeed, one defines *tropical convexity* in $\mathbb{R}^n/\mathbb{R}\mathbf{1}$ by taking tropical linear combinations. Tropical convexity is a rich and beautiful theory with many applications. For introductions see [**27**, Chapter 5] and [**38**, §5.2].

We give a brief illustration for $m = n = 3$. The image of a 3×3 matrix A is the set of all tropical linear combinations of three vectors in $\mathbb{R}^3$. We represent this set by its image in the plane $\mathbb{R}^3/\mathbb{R}\mathbf{1}$. That image is a *tropical triangle*, because it is the tropical convex hull of three points in the plane. This triangle is degenerate if the three points are tropically collinear in $\mathbb{R}^3/\mathbb{R}\mathbf{1}$. This happens when the minimum in the tropical determinant (7.7) is attained twice. In that case, the matrix A is called *tropically singular.*

Example 7.15. Consider the tropical triangle in $\mathbb{R}^3/\mathbb{R}\mathbf{1}$ given by the matrix

$$A = \begin{pmatrix} 0 & 0 & 2 \\ 0 & 3 & 1 \\ 1 & 0 & 0 \end{pmatrix} \qquad \text{or} \qquad A' = \begin{pmatrix} -1 & 0 & 2 \\ -1 & 3 & 1 \\ 0 & 0 & 0 \end{pmatrix}.$$

Each point in $\mathbb{R}^3/\mathbb{R}\mathbf{1}$ is represented uniquely by a column vector $(u, v, 0)$. This tropical triangle consists of the segment between $(-1,-1,0)$ and $(0,0,0)$, the segment between $(0,3,0)$ and $(0,1,0)$, the segment between $(2,1,0)$ and $(1,1,0)$, and the classical triangle with vertices $(0,0,0)$, $(0,1,0)$ and $(1,1,0)$.

There are five combinatorial types of tropical triangles. Similarly, there are 35 types of tropical quadrilaterals. These are shown in [**38**, Figure 5.2.4].

7.3. Tropical Varieties

The previous section explored tropical counterparts of concepts from linear algebra. In what follows we move on to nonlinear algebra. Our aim is to introduce the tropical counterparts of algebraic varieties. Our point of departure is the discussion of fields with valuation at the end of Section 7.1.

In what follows we assume that K is an algebraically closed field with a valuation whose value group is $\mathbb{Q}$. The Puiseux series field $K = \mathbb{C}\{\{t\}\}$ is our primary example. Another field, derived from Example 7.8 and of great interest in number theory, is the algebraic closure of the p-adic numbers.

Consider any polynomial in n variables $\mathbf{x} = (x_1, \ldots, x_n)$ over the field K:

$$(7.11) \qquad f \;=\; c_1 \mathbf{x}^{\mathbf{a}_1} \,+\, c_2 \mathbf{x}^{\mathbf{a}_2} \,+\, \cdots \,+\, c_s \mathbf{x}^{\mathbf{a}_s}.$$

The tropicalization of f is the following expression in tropical arithmetic:

$$\operatorname{trop}(f) \;=\; \operatorname{val}(c_1) \odot \mathbf{x}^{\odot \mathbf{a}_1} \,\oplus\, \operatorname{val}(c_2) \odot \mathbf{x}^{\odot \mathbf{a}_2} \,\oplus\, \cdots \oplus \operatorname{val}(c_s) \odot \mathbf{x}^{\odot \mathbf{a}_s}.$$

To evaluate the *tropical polynomial* $\operatorname{trop}(f)$ at a point $\mathbf{u} = (u_1, \ldots, u_n)$, we take the minimum of the s expressions in the tropical sum above:

$$\operatorname{val}(c_i) \odot \mathbf{u}^{\odot \mathbf{a}_i} \;=\; \operatorname{val}(c_i) \odot u_1^{\odot a_{i1}} \odot \cdots \odot u_n^{\odot a_{in}} \;=\; \operatorname{val}(c_i) + a_{i1} u_1 + \cdots + a_{in} u_n.$$

Here the index i runs over $\{1, \ldots, s\}$. If this minimum is attained at least twice, then $\mathbf{u}$ is a *tropical zero* of $\operatorname{trop}(f)$. The special case $n = 1$ appeared in Definition 7.4. The following result generalizes the first part of Theorem 7.7.

Proposition 7.16. *If* $\mathbf{z} = (z_1, \ldots, z_n) \in K^n$ *is a zero of a polynomial* f *in* $K[\mathbf{x}]$, *then its coordinatewise valuation* $\operatorname{val}(\mathbf{z}) = \big(\operatorname{val}(z_1), \ldots, \operatorname{val}(z_n)\big) \in \mathbb{Q}^n$ *is a tropical zero of the associated tropical polynomial* $\operatorname{trop}(f)$.

Proof. Let $u_i = \operatorname{val}(z_i)$ and $\mathbf{u} = (u_1, \ldots, u_n)$. The valuation of the term $c_i \mathbf{z}_i^{\mathbf{a}}$ in (7.11) equals $\operatorname{val}(c_i) \odot \mathbf{u}^{\odot \mathbf{a}_i}$. The sum of the s scalars $c_i \mathbf{z}_i^{\mathbf{a}}$ is zero in K, so the summands of lowest valuation must have a cancellation. In other words, the minimum valuation in the tropical sum is attained by two or more of the expressions $\operatorname{val}(c_i) \odot \mathbf{u}^{\odot \mathbf{a}_i}$. By definition, this means that the vector $\mathbf{u} \in \mathbb{Q}^n$ is a tropical zero of $\operatorname{trop}(f)$. $\square$

A celebrated result due to Kapranov states that the converse holds too. Namely, if $f \in K[\mathbf{x}]$ and $\mathbf{u} \in \mathbb{Q}^n$ is a tropical zero of $\operatorname{trop}(f)$, then there is a point $\mathbf{z} \in K^n$ such that $f(\mathbf{z}) = 0$ and $\operatorname{val}(\mathbf{z}) = \mathbf{u}$. We refer to [**38**, Theorem 3.1.3] for the proof and further details. For the $n = 1$ case see Theorem 7.7.

The element ∞ in the tropical semiring arises naturally from the arithmetic in the field K because $\operatorname{val}(0) = \infty$. Sometimes it is preferable to restrict tropical algebra to $\mathbb{R}$, or to $\mathbb{Q}$, thus excluding ∞. This is done by disallowing zero coordinates among the solutions of a polynomial system. For this, we set $K^* = K \backslash \{0\}$ and introduce the *algebraic torus* $(K^*)^n$. The ring of polynomial functions on $(K^*)^n$ is the *Laurent polynomial ring*

$$K[\mathbf{x}^{\pm}] \;:=\; K[x_1^{\pm 1}, x_2^{\pm 1}, \ldots, x_n^{\pm 1}].$$

The elements of $K[\mathbf{x}^{\pm}]$ are polynomials as in (7.11), but we now allow negative integers among the coordinates of the exponent vectors $\mathbf{a}_i$.

In what follows we fix the field of Puiseux series $K = \mathbb{C}\{\{t\}\}$. For the extension to other fields see [**38**, §2.4]. Given any vector $\mathbf{u} \in \mathbb{R}^n$, the *initial form* $\operatorname{in}_{\mathbf{u}}(f)$ is the subsum of terms $\overline{c_i} \mathbf{x}^{\mathbf{a}_i}$ in (7.11) for which $\operatorname{val}(c_i) \odot \mathbf{u}^{\odot \mathbf{a}_i}$ is minimal. Here $\overline{c_i}$ is the coefficient of the lowest-order term in the Puiseux series c_i. For instance, if $c = 3t + 9t^3 - \frac{2}{3}t^4 + \cdots \in K$ is the second scalar displayed in Example 7.3, then $\overline{c} = 3$. A *monomial* in the Laurent polynomial ring is a scalar times a product of variables raised to integer exponents.

Lemma 7.17. *For any Laurent polynomial $f \in K[\mathbf{x}^{\pm}]$ and any point $\mathbf{u} \in \mathbb{R}^n$, the following three conditions are equivalent:*

- *The initial form $\mathrm{in}_{\mathbf{u}}(f)$ is not a unit in $K[\mathbf{x}^{\pm}]$.*
- *The initial form $\mathrm{in}_{\mathbf{u}}(f)$ is not a monomial.*
- *The point $\mathbf{u}$ is a tropical zero of $\mathrm{trop}(f)$.*

Proof. Every monomial is invertible in the Laurent polynomial ring. To show the converse, we fix a lexicographic order on monomials. If $hg = 1$ then the product of the smallest monomials in the support of h and g must equal the product of the two largest. In particular, the smallest and the largest monomials appearing in h are the same, i.e. h is a monomial. We conclude that the first two items in Lemma 7.17 are equivalent. The equivalence of the last two items comes from the definition of tropical zeros. □

Fix any ideal I in $K[\mathbf{x}^{\pm}]$ and let $\mathcal{V}(I)$ be its variety in the algebraic torus $(K^*)^n$. We define the *tropical variety* of I to be the following subset of $\mathbb{R}^n$:

$$\mathrm{trop}(\mathcal{V}(I)) = \{ \mathbf{u} \in \mathbb{R}^n \ : \ \mathbf{u} \text{ is a tropical zero of } \mathrm{trop}(f) \text{ for all } f \in I \}.$$

We also refer to this set as the *tropicalization* of the variety $\mathcal{V}(I)$.

The study of tropical varieties is the subject of tropical algebraic geometry. Two important results in this area are the *Fundamental Theorem* [**38**, Theorem 3.2.3] and the *Structure Theorem* [**38**, Theorem 3.3.5]. The former extends Kapranov's Theorem. It states that the set of rational points in $\mathrm{trop}(\mathcal{V}(I))$ is the image of the classical variety $\mathcal{V}(I) \subset (K^*)^n$ under the coordinatewise valuation map. The latter states that the tropical variety $\mathrm{trop}(\mathcal{V}(I))$ is a balanced polyhedral complex. Furthermore, its dimension agrees with the dimension of $\mathcal{V}(I)$. Numerous concrete examples of such balanced polyhedral complexes are given in the textbooks [**27**] and [**38**].

Example 7.18. Fix $n = 9$ and let $\mathbf{x} = (x_{ij})$ be a 3×3 matrix whose entries are unknowns. Let I be the ideal in $K[\mathbf{x}^{\pm}]$ that is generated by the nine 2×2 minors of $\mathbf{x}$. Then $\mathcal{V}(I)$ is the 5-dimensional variety of 3×3 matrices of rank 1 in $(K^*)^{3\times 3}$. The tropical variety $\mathrm{trop}(\mathcal{V}(I))$ is the set of 3×3 matrices in (7.4), that is, matrices $\mathbf{u}$ of tropical rank 1. This is the linear subspace of dimension 5 in $\mathbb{R}^{3\times 3}$ defined by the tropical 2×2 determinants $u_{ij} \odot u_{kl} \oplus u_{il} \odot u_{kj}$. This minimum is attained twice if and only if $u_{ij} + u_{kl} = u_{il} + u_{kj}$. Every matrix $\mathbf{u} = (u_{ij})$ that satisfies these linear equations and has entries in $\mathbb{Q}$ arises as the valuation $\mathbf{u} = \mathrm{val}(\mathbf{z})$ of a rank-1 matrix $\mathbf{z} = (z_{ij})$ with entries in K^*. For instance, $\mathbf{z} = (t^{u_{ij}})$.

Consider the assignment problem in Proposition 7.10. The tropical variety $\mathrm{trop}(\mathcal{V}(I))$ represents scenarios where all six assignments for $n = 3$ have the same cost. The situation becomes more interesting when we pass from rank 1 to rank 2. Now only the two best assignments have the same cost.

To model this, let $J \subset K[\mathbf{x}^{\pm}]$ be the principal ideal generated by the determinant of $\mathbf{x}$. Then $\mathcal{V}(J)$ is a hypersurface of degree 3 in $(K^*)^{3\times 3}$. The tropical hypersurface $\mathrm{trop}(\mathcal{V}(J))$ is defined by the tropical determinant

$$\begin{aligned}\mathrm{tropdet}(\mathbf{x}) = \quad & x_{11} \odot x_{22} \odot x_{33} \oplus x_{11} \odot x_{23} \odot x_{32} \oplus x_{12} \odot x_{21} \odot x_{33} \\ & \oplus x_{12} \odot x_{23} \odot x_{31} \oplus x_{13} \odot x_{21} \odot x_{32} \oplus x_{13} \odot x_{22} \odot x_{31}.\end{aligned}$$

Thus $\mathrm{trop}(\mathcal{V}(J))$ is set of all 3×3 matrices $\mathbf{u} = (u_{ij})$ such that the minimum of the six terms is attained at least twice. For such a matrix, there is more than one optimal assignment of the three workers to the three jobs.

The set $\mathrm{trop}(\mathcal{V}(J))$ is a polyhedral fan of dimension 8 in the 9-dimensional ambient space $\mathbb{R}^{3\times 3}$. It is a cone with apex $\mathrm{trop}(\mathcal{V}(I)) \simeq \mathbb{R}^5$ over a 2-dimensional polyhedral complex. That complex consists of nine squares and six triangles, which are glued together as shown on the left in Figure 7.1.

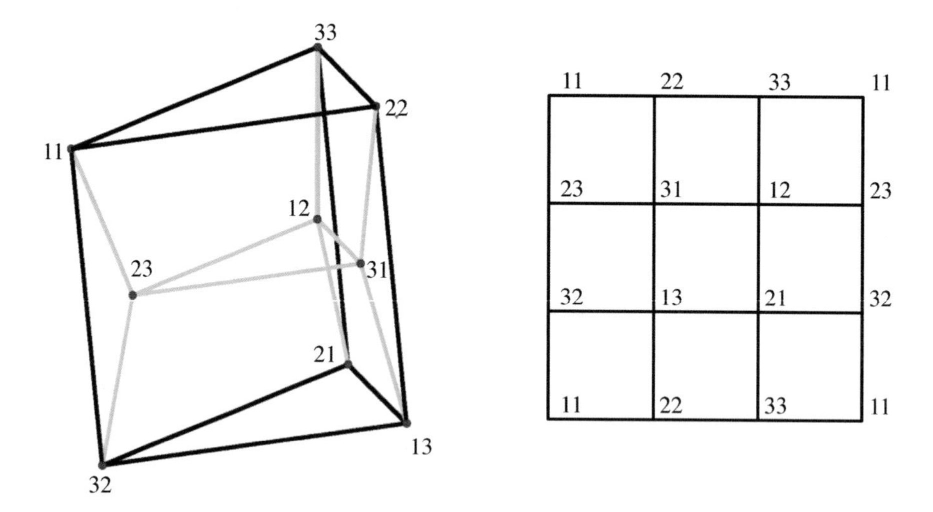

Figure 7.1. The combinatorial structure of the tropical hypersurface that is defined by the tropical 3×3 determinant.

The six triangles represent matrices $\mathbf{u}$ such that the minimum of the six terms in $\mathrm{tropdet}(\mathbf{u})$ is attained by two permutations in S_3 with the same sign. The nine squares on the right in Figure 7.1 are glued to form a torus. These represent matrices $\mathbf{u}'$ for which the minimum is attained by two permutations in S_3 with opposite signs. Concrete examples for the two cases are

$$\mathbf{u} = \begin{pmatrix} 0 & 0 & 1 \\ 1 & 0 & 0 \\ 0 & 1 & 0 \end{pmatrix} \quad \text{and} \quad \mathbf{u}' = \begin{pmatrix} 0 & 0 & 1 \\ 0 & 0 & 1 \\ 1 & 1 & 0 \end{pmatrix}.$$

Here are rank-2 matrices over K that map to $\mathbf{u}$ and $\mathbf{u}'$ under tropicalization:

$$\mathbf{z} = \begin{pmatrix} t+1 & t-1 & 2t \\ t & 1 & t+1 \\ 1 & t & t+1 \end{pmatrix} \quad \text{and} \quad \mathbf{z}' = \begin{pmatrix} 1 & 2 & t \\ 2 & 4 & 5t \\ 3t & 6t & 7 \end{pmatrix}.$$

The supports of the matrices $\mathbf{u} = \mathrm{trop}(\mathbf{z})$ and $\mathbf{u}' = \mathrm{trop}(\mathbf{z}')$ match the labels of the corresponding 2-cells in Figure 7.1. The matrix $\mathbf{u}$ has support $13, 21, 32$, which labels the bottom triangle on the left. The matrix $\mathbf{u}'$ has support $13, 23, 31, 32$, which labels the middle left square on the right.

We close with a remark on lifting Proposition 7.9 from tropical algebra to algebra over the field $K = \mathbb{C}\{\{t\}\}$. Given a directed graph G with rational edge weights d_{ij}, we now define a new adjacency matrix A_G. The entry of A_G in row i and column j equals $t^{d_{ij}}$ if (i,j) is an edge of G and equals 0 otherwise.

By construction, the valuation of the matrix A_G is the adjacency matrix D_G seen earlier in Section 7.1. Moreover, the tropical matrix power in (7.5) is the valuation of the corresponding power of the classical matrix A_G:

$$D_G^{\odot(n-1)} = (\mathrm{val}(A_G))^{\odot(n-1)} = \mathrm{val}\big(A_G^{n-1}\big). \tag{7.12}$$

Indeed, the (i,j) entry of A_G^{n-1} is the generating function for all paths. To be precise, this entry is the Puiseux polynomial $\sum_\ell c_\ell t^\ell$, where c_ℓ is the number of paths from node i to node j in the graph G that have length ℓ.

Exercises

(1) Let u, v and w be real numbers, and let x, y and z be variables. For $n \in \mathbb{N}$, what are the coefficients in the expansion of the expression

$$(u \odot x \oplus v \odot y \oplus w \odot z)^{\odot n}$$

in tropical arithmetic?

(2) Prove that tropical matrix multiplication is an associative operation.

(3) Draw the graph of the following function:

$$\mathbb{R} \to \mathbb{R}, \quad x \mapsto 1 \oplus 2 \odot x \oplus 3 \odot x^{\odot 2} \oplus 6 \odot x^{\odot 3} \oplus 10 \odot x^{\odot 4}.$$

What are the tropical zeros of this tropical polynomial?

(4) How would you define the tropical characteristic polynomial of a square matrix? Compute your polynomial for the 3×3 matrices in Example 7.14.

(5) Draw the graph of the $\mathbb{R}^2 \to \mathbb{R}$ function given by

$$(x, y) \mapsto 1 \oplus 2 \odot x \oplus 3 \odot y \oplus 6 \odot x \odot y \oplus 10 \odot x \odot y^{\odot 2}.$$

What are the tropical zeros of this tropical polynomial?

(6) Let G be the directed graph on n nodes with edge weights $d_{ij} = i \cdot j$ when $i \neq j$ and $d_{ii} = 0$, for $i, j \in \{1, \ldots, n\}$. Compute the tropical powers $D_G^{\odot i}$ of the matrix D_G for $i = 1, 2, \ldots, n-1$. What are their tropical ranks? Interpret the entries of these matrices in terms of paths.

(7) Take the graph G from above with $n = 5$, and fix the matrix $A_G = (t^{d_{ij}})$. Using classical arithmetic over the field $K = \mathbb{Q}(t)$, compute the powers A_G^i of the matrix A_G for $i < n$. What are the ranks of these matrices? Interpret the entries in terms of paths. Verify equation (7.12).

(8) Take the graph G from above with $n = 3$. Find the eigenvalues and eigenspaces of the classical matrix A_G. Find the tropical eigenvalue and tropical eigenspace of the matrix D_G. Do you see a relationship?

(9) Take the graph G from above with $n = 10$. Compute the determinant of A_G and the tropical determinant of D_G. Do you see a relationship?

(10) Take the graph G from above. The matrix D_G defines a tropical linear map from $\mathbb{R}^n$ to itself. Determine the image of this map for $n = 2, 3, 4$. Draw pictures in $\mathbb{R}^n/\mathbb{R}\mathbf{1} \simeq \mathbb{R}^{n-1}$. These are tropical polytopes.

(11) Consider the quartic polynomial $f(x) = t + t^2 x + t^3 x^2 + t^6 x^3 + t^{10} x^4$ in $K[x]$, where $K = \mathbb{C}\{\{t\}\}$. Identify its four roots. Write the first 10 terms of these Puiseux series. What are their valuations?

(12) Let J be the ideal generated by the determinant of a symmetric 3×3 matrix. It lives in a Laurent polynomial ring with six variables. Determine the tropical hypersurface $\text{trop}(\mathcal{V}(J))$. Write a discussion similar to Example 7.18. Draw the analogue of Figure 7.1 for symmetric matrices.

(13) Analyze the complexity of the algorithm described in Proposition 7.9. Can you improve the computation of $D_G^{\odot(n-1)}$? What happens if some edge weights of G are negative? What happens if G contains cycles of negative total weight? How can you detect if such a cycle exists?

(14) The Wikipedia page for *Tropical geometry* shows a tropical cubic curve. Find a tropical polynomial in two unknowns that defines this curve.

Chapter 8

Toric Varieties

"The art of doing mathematics consists in finding that special case which contains all the germs of generality", David Hilbert

Toric varieties are the simplest and most accessible varieties. They often arise in applications, both within mathematics and across the sciences. A toric variety is an irreducible variety that is parametrized by a vector of monomials. The relations between these monomials are binomials, i.e. polynomials with only two terms. Thus, an irreducible variety is toric if and only if its prime ideal is generated by binomials. Monomials and binomials correspond to points in an integer lattice, and we think of them as the lattice points in a polytope. Toric varieties appear prominently in optimization and statistics, thanks to the purely combinatorial description given above. This description also makes toric varieties perfect "model organisms" for algebraic geometers, who use toric varieties to test conjectures, teach geometric concepts, and compute invariants. For instance, the dimension and degree of a toric variety are the dimension and volume of the associated lattice polytope. This result will appear at the very end of this book, in Proposition 13.26.

8.1. The Affine Story

The adjective *toric* derives from the noun *torus*. We begin by introducing tori from an algebraic perspective. We fix an algebraically closed field K and the Laurent polynomial ring $K[\mathbf{x}^{\pm}] = K[x_1^{\pm 1}, \ldots, x_n^{\pm 1}]$. The corresponding variety $(K^*)^n$, having the ring of functions $K[\mathbf{x}^{\pm}]$, is the *algebraic torus* of dimension n over K. Here, $K^* = K\backslash\{0\}$ is the set of nonzero field elements.

The algebraic torus $(K^*)^n$ is a group under coordinatewise multiplication. The name torus comes from the special case where $n = 2$ and $K = \mathbb{C}$ is the field of complex numbers. Here, we have $(\mathbb{C}^*)^2 \simeq (\mathbb{R}_{\geq 0} \times \mathbb{S}^1)^2$, where $\mathbb{S}^1$ is the unit circle. Thus, the 2-dimensional algebraic torus $(\mathbb{C}^*)^2$ is the product of the topological torus $\mathbb{S}^1 \times \mathbb{S}^1$ and the contractible factor $\mathbb{R}_{\geq 0} \times \mathbb{R}_{\geq 0}$.

We recall from Section 7.3 that subvarieties of the algebraic torus $(K^*)^n$ are the objects one starts from when developing tropical algebraic geometry.

Definition 8.1 (Character of a torus). A *character* of the algebraic torus $T = (K^*)^n$ is a regular map $\chi : T \to K^*$ that is also a group morphism.

In Exercise 1 we shall see that characters are given by Laurent monomials

$$\mathbf{x}^{\mathbf{b}} = x_1^{b_1} x_2^{b_2} \cdots x_n^{b_n} \quad \text{where } \mathbf{b} \in \mathbb{Z}^n.$$

The characters of T are hence in bijection with $\mathbb{Z}^n$. Under this correspondence, multiplication of characters becomes addition in the group $(\mathbb{Z}^n, +)$:

$$\chi_1(\mathbf{x}) \cdot \chi_2(\mathbf{x}) = (\chi_1 + \chi_2)(\mathbf{x}).$$

A group isomorphic to $\mathbb{Z}^k$, for some k, is called a *lattice.* The lattice of characters of T is denoted by M_T or simply M. Since a subgroup of a free abelian group is free, any set of characters generates a sublattice $\tilde{M} \subset M$.

Let $\mathbf{a}_1, \ldots, \mathbf{a}_p$ be characters in $M_T \simeq \mathbb{Z}^n$. We write A for the $n \times p$ matrix whose columns are the vectors $\mathbf{a}_i$. The lattice $\tilde{M}$ generated by the characters $\mathbf{a}_i$ is the image of $\mathbb{Z}^p$ under left multiplication by the matrix A.

Proposition 8.2. *The image of T in $(K^*)^p$ under the map $f : \mathbf{x} \to \mathbf{x}^A = (\mathbf{x}^{\mathbf{a}_1}, \ldots, \mathbf{x}^{\mathbf{a}_p})$ is also a torus $\tilde{T}$. The character lattice of $\tilde{T}$ is equal to $\tilde{M}$.*

Proof. The monomial map $f : T \mapsto (K^*)^p$ induces the ring homomorphism

$$f^* : K[y_1^{\pm 1}, \ldots, y_p^{\pm 1}] \to K[x_1^{\pm 1}, \ldots, x_n^{\pm 1}], \quad y_i \mapsto \mathbf{x}^{\mathbf{a}_i}.$$

The image of f^* is the ring of functions on $\tilde{T} := \overline{f(T)}$. This image is the group algebra $K[\tilde{M}]$. By definition, this is the vector space over K with basis given by elements of $\tilde{M}$ and multiplication induced from addition in M_T. The lattice $\tilde{M}$ is isomorphic to the group $\mathbb{Z}^d$ for some integer $d \leq \min(n, p)$. Hence, the ring of functions on $\tilde{T}$ is isomorphic to $K[\mathbb{Z}^d]$, i.e. the closed image $\tilde{T}$ of the monomial map f is a torus with character lattice $\tilde{M} \simeq \mathbb{Z}^d$.

It remains to show that $f(T) = \tilde{T}$, i.e. the image is actually closed. For this we pick a point $p \in \tilde{T}$. It corresponds to a map of groups $e_p : \tilde{M} \to K^*$ given by evaluating characters at p. To prove that $p \in f(T)$, we need to extend this map to the whole lattice M_T. This may be done iteratively as follows. Pick $m \in M_T \backslash \tilde{M}$. If $km \notin \tilde{M}$ for all positive integers k, then we simply define $e_p(m) = 1$. Otherwise we pick the smallest

k such that $km \in \tilde{M}$. We define $e_p(m)$ as any kth root of $e_p(km)$ and extend $\tilde{M}$ by m. □

The d-dimensional torus $\tilde{T}$ lives in $(K^*)^p$. We are interested in its Zariski closure in the affine space K^p. Such an affine variety is a toric variety.

Definition 8.3. An *affine toric variety* is the closed image of a monomial map $(K^*)^n \to K^p$, $\mathbf{x} \mapsto (\mathbf{x}^{\mathbf{a}_1}, \ldots, \mathbf{x}^{\mathbf{a}_p})$, where $\mathbf{a}_i \in \mathbb{Z}^n$ and $K^* = K \backslash \{0\}$.

We specify a toric variety by an integer matrix $A \in \mathbb{Z}^{n \times p}$. The p columns of A represent characters of the torus $T = (K^*)^n$. Toric geometry relates the combinatorics of these lattice points to the geometry of the toric variety.

Example 8.4.

(1) Any affine space is a toric variety. The corresponding matrix A is the identity matrix.
(2) The cuspidal cubic curve $x^3 - y^2$ is a toric variety. It is the image of the map $z \mapsto (z^2, z^3)$ given by the matrix $A = \begin{pmatrix} 2 & 3 \end{pmatrix}$.

Proposition 8.5. *The dimension of the affine toric variety in Definition* 8.3 *is equal to the rank of the lattice* $\tilde{M}$ *that is spanned by* $\mathbf{a}_1, \ldots, \mathbf{a}_p$ *in* $\mathbb{Z}^n$.

Proof. We saw in the proof of Proposition 8.2 that the torus $\tilde{T}$ has dimension $d = \operatorname{rank}(\tilde{M})$. The toric variety is the Zariski closure of $\tilde{T}$. It has dimension d, since passing to the Zariski closure preserves dimension. □

We defined toric varieties as closures in K^p of subtori of the torus $(K^*)^p$. The coordinate ring of a toric variety is a monoid algebra $K[S]$. Here, S denotes the monoid generated by the p distinguished characters in M_T, in the notation of Proposition 8.2. Thus, S is the smallest set containing 0 and the chosen characters that is closed under addition in M_T.

Example 8.6.

(1) The cuspidal curve defined by the equation $x^3 - y^2$ equals $\operatorname{Spec} K[z^2, z^3]$. The underlying monoid is $S = \{0, 2, 3, 4, \ldots\}$.
(2) The affine line is the closure of the image of the map

$$K^* \ni x \mapsto x \in K.$$

Here the character lattice is $M = \mathbb{Z}$, the distinguished character corresponds to $1 \in M$, and the monoid is $S = \{0, 1, 2, 3, \ldots\}$.

There is a fundamental difference between the cuspidal curve and the affine line. The monoid for the cuspidal curve has a "hole", namely the character 1.

Figure 8.1. The cuspidal cubic curve.

Definition 8.7. A submonoid S in a lattice M is called *saturated* if and only if for any $x \in M$ and $k \in \mathbb{Z}_+$ the following implication holds:

$$kx \in S \ \Rightarrow\ x \in S.$$

An affine toric variety $X = \operatorname{Spec} K[S]$ for which S is saturated (in the lattice $\tilde{M}$ it generates) is called *normal.* For the algebraic definition of normality see [**3**, Chapter 5], where normal rings are called integrally closed. Nonnormal varieties are always singular. For curves, the two notions coincide. Example 8.6 shows one normal (i.e. smooth) curve and one nonnormal (i.e. singular) curve. The latter is seen in Figure 8.1.

We next discuss the prime ideal of the toric variety X. This is computed from the characters that define X. In general, given a variety defined as the Zariski closure of the image of a map, finding the defining equations is a hard problem, known as *implicitization.* We discussed this in Chapter 4. The implicitization problem greatly simplifies when the variety is toric. The prime ideal I_X of the toric variety X lives in the polynomial ring $K[y_1, \ldots, y_p]$. This *toric ideal* is the kernel of the restriction of f^* to this polynomial ring.

Lemma 8.8. *Let X be the toric variety defined by $A = (\mathbf{a}_1, \ldots, \mathbf{a}_p)$. Then*

(1) *any relation $\sum_i b_i\mathbf{a}_i = \sum_j c_j\mathbf{a}_j$ with nonnegative integer coefficients $b_i, c_j \in \mathbb{N}$ provides a binomial $\prod y_i^{b_i} - \prod y_j^{c_j}$ in the toric ideal I_X;*

(2) *every binomial in the ideal I_X is of the form described in* (1)*;*

(3) *the toric ideal I_X is generated by these binomials.*

Recall that a *binomial* is a polynomial with only two terms. Statement (2) is understood up to scaling—we can multiply the binomial by a constant.

Proof. Properties (1) and (2) follow from the fact that a polynomial in $y_1, \ldots, y_p$ vanishes on the toric variety X if and only if we obtain zero after substituting $y_i \mapsto \mathbf{x}^{\mathbf{a}_i}$ for all i. However, such a substitution turns monomials (in the variables y_j) into monomials (in the variables x_k). The fact that the monomials in $\mathbf{x}$ cancel is precisely encoded by the integral relations in (1). Property (3) follows similarly, by induction on number of terms of a polynomial in the ideal of X. In Exercise 2 the reader is asked to give a proof. For an argument using a monomial ordering see [**52**, Lemma 4.1]. $\square$

Example 8.9. Fix $n = 3$ and $p = 7$. To specify a toric variety, we choose characters $\mathbf{a}_1, \ldots, \mathbf{a}_7 \in \mathbb{Z}^3$. Let us take the column vectors of the matrix

$$A = \begin{pmatrix} 2 & 2 & 1 & 0 & 0 & 1 & 1 \\ 1 & 0 & 0 & 1 & 2 & 2 & 1 \\ 0 & 1 & 2 & 2 & 1 & 0 & 1 \end{pmatrix}.$$

The associated toric variety X is a threefold in K^7. The toric ideal I_X equals

$$\begin{aligned} \langle\, & y_1y_3 - y_2y_7\,,\; y_1y_4 - y_7^2\,,\; y_1y_5 - y_6y_7\,,\; y_2y_4 - y_3y_7\,,\; y_2y_5 - y_7^2\,, \\ & y_2y_6 - y_1y_7\,,\; y_3y_5 - y_4y_7\,,\; y_3y_6 - y_7^2\,,\; y_4y_6 - y_5y_7\,\rangle. \end{aligned} \tag{8.1}$$

Each of these nine binomials vanishes under the substitution $y_i \mapsto \mathbf{x}^{\mathbf{a}_i}$. Using the methods in Section 4.2, we can check that I_X is the desired prime ideal. The toric variety X has dimension 3 and lives in the affine space K^7.

The ideal I_X is homogeneous. Each of the nine binomials in (8.1) is homogeneous. This comes from the fact that the matrix A has column sums $(3,3,3,3,3,3,3)$. Geometrically speaking, the threefold X is a cone in K^7. We can therefore also regard X as a surface in the projective space $\mathbb{P}^6$. That surface is nonsingular and it has degree 6. This passage from appropriate matrices A to projective toric varieties will be our theme in Section 8.2.

Theorem 8.10. *The toric ideals I_X are precisely the prime ideals generated by binomials $\mathbf{y}^{\mathbf{b}} - \mathbf{y}^{\mathbf{c}}$. Every such ideal defines a toric variety X as above.*

Proof. Let I be a prime ideal generated by a set of binomials $\mathbf{y}^{\mathbf{b}_i} - \mathbf{y}^{\mathbf{c}_i}$ in p variables $y_1, \ldots, y_p$. By Hilbert's Basis Theorem, there is a finite subset of binomials that generates I. For each such generator, the nonnegative integer vectors $\mathbf{b}_i$ and $\mathbf{c}_i$ have disjoint support, since I is prime. We write the difference vectors $\mathbf{b}_i - \mathbf{c}_i$ as the columns of a matrix B that has p rows.

Let A be any integer matrix of format $n \times p$ whose rows span the kernel of B under left multiplication. Here, the kernel is understood as a $\mathbb{Z}$-module (a.k.a. an abelian group), so it can be computed using integer linear algebra (e.g. the Hermite normal form algorithm). We claim that the columns of B span the kernel A under right multiplication. This is clear over $\mathbb{Q}$, but it also holds over $\mathbb{Z}$ by our hypothesis that I is a prime ideal. Otherwise, there would exist a vector $\mathbf{b} - \mathbf{c}$ that is not in the column space of B but of which some integer multiple $k\mathbf{b} - k\mathbf{c}$ is in that column span. We pick the smallest possible k. The following binomial is in the binomial prime ideal I:

$$\mathbf{y}^{k\mathbf{b}} - \mathbf{y}^{k\mathbf{c}} = \left(\mathbf{y}^{\mathbf{b}} - \mathbf{y}^{\mathbf{c}}\right)\left(\mathbf{y}^{(k-1)\mathbf{b}} + \mathbf{y}^{(k-2)\mathbf{b}}\mathbf{y}^{\mathbf{c}} + \cdots + \mathbf{y}^{(k-1)\mathbf{c}}\right).$$

However, neither of the two factors is in I. In characteristic zero this is straightforward, as the second factor has coefficients not adding up to zero. In general, one can observe that if the second factor is in I, then we must have $\mathbf{y}^{(k-1)\mathbf{b}} - \mathbf{y}^{(k-1-i)\mathbf{b}}\mathbf{y}^{i\mathbf{c}} \in I$ for some positive i. This contradicts the choice of k. We obtain a contradiction to the hypothesis that I is prime.

We now take X to be the toric variety in K^p that is defined by the matrix A. The argument above shows that $I = I_X$, which gives the assertion in the theorem. For further details on this proof see [**12**, Proposition 1.1.11]. □

Definition 8.11. A *convex polyhedral cone* C in $\mathbb{R}^n$ is a subset of elements of the form $\lambda_1\mathbf{v}_1 + \cdots + \lambda_k\mathbf{v}_k$ where $\mathbf{v}_1, \ldots, \mathbf{v}_k \in \mathbb{R}^n$ are fixed and $\lambda_1, \ldots, \lambda_k$ range over $\mathbb{R}_{\geq 0}$. We call C *rational* if the vectors $\mathbf{v}_i$ can be chosen in $\mathbb{Q}^n$. In what follows we refer to rational convex polyhedral cones simply as *cones*.

Definition 8.12. A *face* F of a cone $C \subset \mathbb{R}^n$ is a subset of the form

$$F = \{\mathbf{c} \in C : \ell(\mathbf{c}) = 0\}, \tag{8.2}$$

where ℓ is a linear form that is nonnegative on C, i.e. $\ell(\mathbf{c}) \geq 0$ for all $\mathbf{c} \in C$. The *dimension* of a face of a cone is the dimension of the smallest linear space that contains it. If $\dim C = n = \dim F + 1$, then ℓ is unique up to a scalar. In this case, F is called a *facet* and the hyperplane it spans is the *supporting hyperplane* of C. We note that $\ell = 0$ gives $F = C$. Furthermore, any face of a cone is also a cone, and the relation "is a face of" is transitive.

Example 8.13. The orthant $C = \mathbb{R}^n_{\geq 0}$ is a cone. It has 2^n faces, ranging from the apex $\{0\}$ to the full cone C. There are $\binom{n}{i}$ faces of dimension i. Each of the n facets F arises by setting one coordinate to zero, so $F \simeq \mathbb{R}^{n-1}_{\geq 0}$.

If C is a convex polyhedral cone in $\mathbb{R}^n$, then its *f-vector* is the vector

$$f(C) = (f_1(C), f_2(C), \ldots, f_{n-1}(C)), \tag{8.3}$$

where $f_i(C)$ denotes the number of i-dimensional faces of C. For instance, the orthant in $\mathbb{R}^5$ has f-vector $(5, 10, 10, 5)$. The 3-dimensional cone spanned by the columns of the matrix A in Example 8.9 has f-vector $(6, 6)$.

By Proposition 8.2, the toric variety X is the closure in K^p of the torus $\tilde{T} \subset (K^*)^p$. The group $\tilde{T}$ acts both on itself and on K^p, and hence it also acts on its closure X. The *torus orbits* on X are the orbits of that action by $\tilde{T}$. We next provide a combinatorial and geometric description of the torus orbits. We assume that X is defined by an integer $n \times p$ matrix A as above. We write $C \subset \mathbb{R}^n$ for the cone that is generated by the p columns $\mathbf{a}_i$ of A.

Theorem 8.14. *The torus orbits in X are in bijection with the faces of the cone C. The orbit corresponding to a face F is* $\{\mathbf{y} \in X : y_i \neq 0 \Leftrightarrow \mathbf{a}_i \in F\}$.

The closure of this orbit is the toric variety Spec $K[F \cap A]$ *whose parametrization is* $(\mathbf{x}^{\mathbf{a}_i} : \mathbf{a}_i \in F)$. *The dimension of this orbit equals* $\dim(F)$. *Moreover, inclusion of orbit closures in* X *corresponds to inclusion of faces of* C.

Sketch of proof. For normal toric varieties, a proof is given in [**12**, §3.2]. However, normality is not needed. From the definition of F being a face, one can check directly that $\{\mathbf{y} \in X : y_i \neq 0 \iff \mathbf{a}_i \in F\}$ is a torus orbit. Indeed, using that K is algebraically closed, one constructs a point $\mathbf{t}$ in the torus such that $\mathbf{t} \cdot \mathbf{y} = \mathbf{p}_F$, where $(\mathbf{p}_F)_i = 1$ if $\mathbf{a}_i \in F$ and 0 otherwise. It remains to be shown that each point of X must lie in one of these orbits. This follows from the constraints on its support that are imposed by the fact that the binomials in I_X vanish at that point. □

Example 8.15. Let X be the toric threefold in K^7 given in Example 8.9. The cone C is spanned in $\mathbb{R}^3$ by the columns of the 3×7 matrix A. It is the cone over a hexagon, so it has $14 = 1+6+6+1$ faces. The variety X is the disjoint union of 14 torus orbits, as follows. The face $F = \{0\}$ corresponds to the origin in K^7. The 1-dimensional face $F = \mathbb{R}_{\geq 0}\{\mathbf{a}_1\}$ corresponds to the curve $\{(t,0,0,0,0,0,0) \in X : t \in K^*\}$. The 2-dimensional face $F = \mathbb{R}_{\geq 0}\{\mathbf{a}_1, \mathbf{a}_2\}$ corresponds to the surface $\{(t,u,0,0,0,0,0) \in X : t,u \in K^*\}$. Finally, $F = C$ corresponds to the 3-dimensional torus $\tilde{T} = X \cap (K^*)^7$.

In conclusion, the geometry of X is read off from the cone C representing it.

8.2. Varieties from Polytopes

Projective toric varieties are obtained from affine toric varieties that are cones. They can be defined as follows. Let $A = (\mathbf{a}_1, \mathbf{a}_2, \ldots, \mathbf{a}_p)$ be an integer $n \times p$ matrix of rank n that has the vector $(1,1,\ldots,1)$ in its row span. Let $I_A \subset K[y_1, y_2, \ldots, y_p]$ be the prime ideal of polynomial relations between the Laurent monomials $\mathbf{x}^{\mathbf{a}_1}, \mathbf{x}^{\mathbf{a}_2}, \ldots, \mathbf{x}^{\mathbf{a}_p}$. This is a toric ideal. According to Lemma 8.8, the ideal I_A is generated by the binomials $\mathbf{y}^{\mathbf{b}} - \mathbf{y}^{\mathbf{c}}$, where $\mathbf{b} - \mathbf{c}$ is in the kernel of A. Such binomials are homogeneous because $(1,1,\ldots,1)$ is in the row span of A. Hence, I_A defines a projective variety, denoted by X_A.

Definition 8.16. A *projective toric variety* is any projective variety in $\mathbb{P}^{p-1}$ of the form $X_A = \mathcal{V}(I_A)$ where I_A is a homogeneous toric ideal as above.

Definition 8.17. A *polytope* in $\mathbb{R}^n$ is the convex hull of a finite set of points. A polytope is a *lattice polytope* if it is the convex hull of points in $\mathbb{Z}^n$. Faces and facets of polytopes are defined as in (8.2), but now ℓ is affine-linear.

We write $P = \operatorname{conv}(A)$ for the convex hull of the column vectors $\mathbf{a}_i$ in $\mathbb{R}^n$. By construction, P is a polytope of dimension $n-1$ with $\leq p$ vertices.

For instance, in Example 8.9, the polytope P is a regular hexagon. Algebraic geometers know the surface X_A as the blow-up of $\mathbb{P}^2$ at three points.

Given a projective toric variety X_A, we associate to it the polytope $P = \mathrm{conv}(A)$. Conversely, any lattice polytope can be coordinatized so that it spans the affine hyperplane $\{y_1+y_2+\cdots+y_n = k\}$ for some $k, n \in \mathbb{Z}_+$. We then take $A = P \cap \mathbb{Z}^n$ and associate the projective toric variety $X_P := X_A$ with the polytope P. This variety lives in $\mathbb{P}^{p-1}$ where $p = |A|$.

The class of varieties X_A is strictly larger than the class of varieties X_P. The reason is that A can be a proper subset of $\mathrm{conv}(A) \cap \mathbb{Z}^n$. However, the projective toric varieties of most interest to us are the X_P for some polytope P.

Example 8.18. The *Veronese variety* and the *Segre variety* from classical algebraic geometry are two prominent examples of projective toric varieties.

We write Δ_{n-1} for the standard $(n-1)$-simplex, whose vertices are the unit vectors $\mathbf{e}_1, \ldots, \mathbf{e}_n$. Fix $k \in \mathbb{Z}_{>0}$ and $P = k\Delta_{n-1}$. Then $A = P \cap \mathbb{Z}^n$ consists of the nonnegative integer vectors with coordinate sum k. Hence $p = |A| = \binom{n+k-1}{k}$. The toric variety X_P is the *kth Veronese embedding* of $\mathbb{P}^{n-1}$. It has dimension $n-1$ and degree k^{n-1} in $\mathbb{P}^{p-1}$. Its toric ideal I_A consists of the polynomial relations between all monomials of degree k in n variables. For instance, if $n = k = 3$ then there are 10 such monomials:

$$A \quad = \quad \begin{pmatrix} 3 & 2 & 2 & 1 & 1 & 1 & 0 & 0 & 0 & 0 \\ 0 & 1 & 0 & 2 & 1 & 0 & 3 & 2 & 1 & 0 \\ 0 & 0 & 1 & 0 & 1 & 2 & 0 & 1 & 2 & 3 \end{pmatrix}.$$

The ideal defining this Veronese surface in $\mathbb{P}^9$ is generated by 27 quadrics:

$$I_A \;=\; \langle y_1y_4-y_2^2, y_1y_5-y_2y_3, y_1y_6-y_3^2, y_1y_7-y_2y_4, \ldots, y_7y_{10}-y_8y_9, y_8y_{10}-y_9^2\rangle.$$

The binomials encode convexity relations between the points in A.

Next, fix $n_1, n_2 \in \mathbb{Z}_{>0}$ and set $n = n_1 + n_2$ and $p = n_1n_2$. Let $P = \Delta_{n_1-1} \times \Delta_{n_2-1}$ and write $\mathbf{e}_i$ and $\mathbf{e}'_j$ for the unit vectors in $\mathbb{R}^{n_1}$ and $\mathbb{R}^{n_2}$ respectively. Then $A = P \cap \mathbb{Z}^n = \big\{\mathbf{e}_i + \mathbf{e}'_j : 1 \leq i \leq n_1 \text{ and } 1 \leq j \leq n_2\big\}$. We here deviate from our hypothesis, as the rank of A is $n-1$, not n. The variety X_P is the *Segre embedding* of the product $\mathbb{P}^{n_1-1} \times \mathbb{P}^{n_2-1}$ into $\mathbb{P}^{p-1}$. The points on X_P are $n_1 \times n_2$ matrices of rank 1, up to scaling. The toric ideal I_A is generated by the 2×2 minors of an $n_1 \times n_2$ matrix of unknowns. Each 2×2 minor corresponds to the square formed by four of the lattice points in P. Can you see this for $n_1 = n_2 = 3$, the case in Example 7.18?

In the algebraic geometry literature, it is often assumed that toric varieties are *normal*. This is motivated by the fact that normal toric varieties admit a nice intrinsic characterization, in terms of fans. In our setting, this hypothesis is generally not needed. Still, we present the relevant definition.

Definition 8.19. A lattice polytope P in $\mathbb{R}^n$ is *normal* if for any $k \in \mathbb{Z}$ and any $\mathbf{u} \in kP \cap \mathbb{Z}^n$, there exist $\mathbf{u}_1, \ldots, \mathbf{u}_k \in P \cap \mathbb{Z}^n$ such that $\mathbf{u} = \sum_{i=1}^k \mathbf{u}_i$. In this case, X_P is *projectively normal* in $\mathbb{P}^{p-1}$, i.e. its affine cone is normal.

Remark 8.20. In the current literature, the property in Definition 8.19 is called the *integer decomposition property* (IDP). It is invariant under lattice translation of P. If $0 \in P$ and $P \cap \mathbb{Z}^n$ generates $\mathbb{Z}^n$, then all reasonable definitions of normality and IDP coincide. One of these equivalent definitions says that the semigroup generated by $\{1\} \times (P \cap \mathbb{Z}^n) \subset \mathbb{Z}^{n+1}$ is saturated.

To appreciate the subtleties we are alluding to, consider the tetrahedron

$$P \;=\; \operatorname{conv}\{(0,0,0),(1,1,0),(1,0,1),(0,1,1)\}.$$

Then $(1,1,1) \in 2P$, but it is not a lattice sum of two points in P. However, note that the lattice points of P do not generate $\mathbb{Z}^3$, only the sublattice consisting of points with an even sum of coordinates. Inside that sublattice, P is just the standard simplex and satisfies the condition in Definition 8.19.

The simplices and products of simplices in Example 8.18 are normal. Hence the Veronese variety and the Segre variety are projectively normal. Exercise 6 gives an example of a 3-dimensional lattice polytope that is not normal. All lattice polytopes of dimensions 1 and 2 are normal. If P is normal, then the polyhedral fan that characterizes its toric variety X_P intrinsically is the *normal fan* of P. We refer to [**12**, §3.1] for the basic theory.

We now return to the setting where I_A is any homogeneous toric ideal, $X_A \subset \mathbb{P}^{p-1}$ is its toric variety, and $P = \operatorname{conv}(A)$ is not necessarily normal. Let T denote the subset of X_A consisting of all points with nonzero coordinates. This is a torus of dimension $n-1$. The torus acts on X_A with finitely many orbits. Theorem 8.14 extends essentially verbatim to the projective case.

Corollary 8.21. *The torus orbits in X_A are in bijection with the faces of the polytope P. The orbit corresponding to a face F is $\{\mathbf{y} \in X_A : y_i \neq 0 \iff \mathbf{a}_i \in F\}$. The closure of this orbit is the projective toric variety with parametrization $(\mathbf{x}^{\mathbf{a}_i} : \mathbf{a}_i \in F)$. The dimension of this orbit equals* $\dim(F)$. *Inclusion of orbit closures in X_A corresponds to inclusion of faces of P.*

Proof. We apply Theorem 8.14 to the affine toric variety defined by I_A in K^p. This is the affine cone over $X_A \subset \mathbb{P}^{p-1}$. Its orbits correspond to the faces of the cone C over the polytope P. Note that $\dim(C) = n = \dim(P) + 1$. Each i-dimensional face F of P corresponds to an $(i+1)$-dimensional face of C, namely the cone over F. Likewise, each i-dimensional orbit in X_A corresponds to an $(i+1)$-dimensional orbit of the affine cone over X_A. These bijections, for $i = 0, 1, \ldots, n-1$, establish the desired bijection for P and X_A. The only face of C that is missing in P is the origin $\{0\}$. Likewise, the cone point of the affine cone over X_A disappears in X_A. □

Example 8.22. Consider the Segre threefold $X_A = \mathbb{P}^1 \times \mathbb{P}^2$ in $\mathbb{P}^5$, given by $n_1 = 2$ and $n_2 = 3$ in Example 8.18. The toric ideal I_A is generated by the 2×2 minors of a 2×3 matrix of unknowns, and the polytope $P = \Delta_1 \times \Delta_2$ is a *triangular prism.* This 3-dimensional polytope has $21 = 6+9+5+1$ faces, one for each of the torus orbits on X_A. For instance, the five 2-dimensional orbits are given by setting one row or column of the 2×3 matrix to zero, and the 0-dimensional orbits in X_A are the matrices with one nonzero entry.

Corollary 8.21 establishes a combinatorial link between projective toric varieties and their lattice polytopes. In what follows we tighten this link to a geometric one. We now fix $K = \mathbb{C}$, the complex numbers. We argue that the geometry of the polytope coincides with the geometry of the toric variety. The key to this identification is the *moment map* from X_A onto P.

We work in the complex projective space $\mathbb{P}^{p-1}_{\mathbb{C}}$ with its homogeneous coordinates $\mathbf{y} = (y_1 : y_2 : \cdots : y_p)$. The following map onto $P = \mathrm{conv}(A)$ is defined via the usual Euclidean norm $|\cdot|$ on the complex plane $\mathbb{C} \simeq \mathbb{R}^2$:

$$(8.4) \qquad \mathbb{P}^{p-1}_{\mathbb{C}} \to \mathbb{R}^n, \quad \mathbf{y} \mapsto \frac{1}{\sum_{i=1}^p |y_i|} \sum_{i=1}^p |y_i| \cdot \mathbf{a}_i.$$

This map is well-defined because the image is invariant under scaling of the vector $\mathbf{y}$, and $\sum_{i=1}^p |y_i|$ is always positive. Its image lies in the polytope P, since we are taking convex combinations of the points $\mathbf{a}_i$ in $\mathbb{R}^n$.

Definition 8.23. The *algebraic moment map* $\mu_A : X_A \to \mathbb{R}^n$ is defined as the restriction of (8.4) from the ambient space $\mathbb{P}^{p-1}_{\mathbb{C}}$ to the toric variety X_A. Let $X_{A,\mathbb{R}}$ be the subset of real points in X_A. Its subset of nonnegative (resp. positive) points is denoted by $X_{A,\geq 0}$ (resp. $X_{A,>0}$). These are semialgebraic sets in the real projective space $\mathbb{P}^{p-1}_{\mathbb{R}}$. To be precise, the *positive toric variety* $X_{A,>0}$ consists of all positive solutions, up to scaling, of the binomial equations in I_A, and similarly for the *nonnegative toric variety* $X_{A,\geq 0}$.

The complex projective toric variety X_A maps naturally onto its nonnegative part $X_{A,\geq 0}$ under the coordinatewise absolute value map

$$(8.5) \qquad (y_1 : y_2 : \cdots : y_p) \;\mapsto\; (|y_1| : |y_2| : \cdots : |y_p|).$$

The fibers of this map are real tori. The fiber over each point in $X_{A,>0}$ is homeomorphic to the topological torus $(\mathbb{S}^1)^{n-1}$. This torus is a subgroup of the complex torus $T \simeq (\mathbb{C}^*)^{n-1}$. We may think of (8.5) as the quotient map

$$X_A \longrightarrow X_A/(\mathbb{S}^1)^{n-1} \simeq X_{A,\geq 0}.$$

See [**12**, Proposition 12.2.3] for a formal statement for normal toric varieties. The algebraic moment map μ_A factors through the quotient map (8.5).

Theorem 8.24. *The restriction of the algebraic moment map μ_A to the nonnegative toric variety $X_{A,\geq 0}$ is a homeomorphism onto the polytope P.*

Proof. This can be found in many sources. A good place to start is Sottile's article [**50**, Theorem 8.4] on toric methods in geometric modeling. □

Corollary 8.25. *If the linear system of equations $A\mathbf{y} = \mathbf{b}$ has a nonnegative solution $\mathbf{y} \in \mathbb{R}^p_{\geq 0}$, then it has a unique such solution $\hat{\mathbf{y}}$ in the toric variety X_A.*

Proof. We identify X_A with the affine cone over the projective toric variety defined by the $n \times p$ matrix A. The algebraic moment map μ_A lifts uniquely, by scaling, to a map from this affine cone to the cone C over the polytope P. The linear system $A\mathbf{y} = \mathbf{b}$ has a nonnegative solution if and only if $\mathbf{b}$ lies in C. Theorem 8.24 implies that the point $\mathbf{b}$ has a unique preimage $\hat{\mathbf{y}} = \mu_A^{-1}(\mathbf{b}) \in X_{A,\geq 0}$ under the moment map, which is the desired solution. □

Example 8.26. Let A be the $(n_1{+}n_2) \times (n_1 n_2)$ matrix for the polytope $P = \Delta_{n_1-1} \times \Delta_{n_2-1}$ as in Examples 8.18 and 8.22. This matrix A represents the linear map that takes an $n_1 \times n_2$ matrix $\mathbf{y}$ to the vector $\mathbf{b}$ of its row and column sums. The polytopes $\{\mathbf{y} \in \mathbb{R}^{n_1 \times n_2}_{\geq 0} : A\mathbf{y} = \mathbf{b}\}$ are known as *transportation polytopes.* The points in the Segre variety X_A are the $n_1 \times n_2$ matrices $\mathbf{y}$ of rank 1. In this case, Corollary 8.25 has the following interpretation: *Every transportation polytope contains a unique rank-*1 *matrix* $\hat{\mathbf{y}}$.

Example 8.26 has important consequences in statistics. We saw this already for $n_1 = n_2 = m$ in Example 2.5. Suppose the nonnegative matrix $\mathbf{y}$ has entries that sum to 1. Then $\mathbf{y}$ is a joint distribution of two random variables that have n_1 and n_2 states. The nonnegative variety $X_{A,\geq 0}$ is the independence model for these two random variables. The map A computes the sufficient statistics $\mathbf{b} = A\mathbf{y}$, i.e. the column vector of row sums and the row vector of column sums. The product of these vectors is the rank-1 matrix $\hat{\mathbf{y}} = \mu_A^{-1}(\mathbf{b})$. This is the *maximum likelihood estimate* for the empirical distribution $\mathbf{y}$ with respect to the independence model.

Example 8.27 ($n_1 = n_2 = 2$)**.** The independence model for two binary random variables is a quadratic surface in $\mathbb{P}^3_{\geq 0} = \Delta_3$. This is the nonnegative part $X_{A,\geq 0}$ of the Segre quadric $X_A = \mathbb{P}^1 \times \mathbb{P}^1 \subset \mathbb{P}^3$. That surface meets the boundary of the tetrahedron Δ_3 in four edges that form a 4-cycle.

The moment map μ_A projects the tetrahedron onto a square. The 4-cycle is mapped onto the boundary of the square. The surface $X_{A,>0}$ is mapped bijectively onto the interior of the square. Figure 8.2 illustrates this scenario.

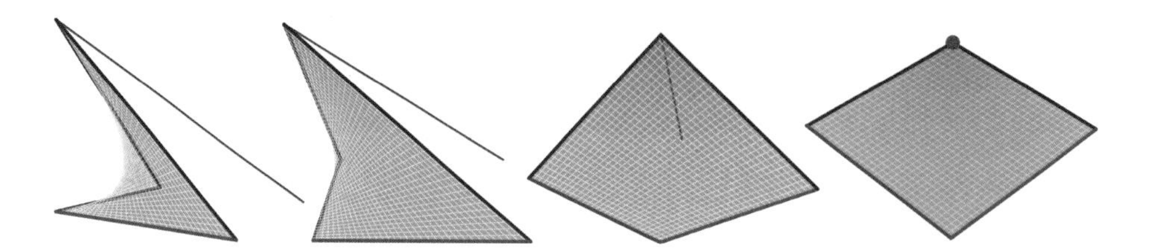

Figure 8.2. Nonnegative part of the Segre quadric from different points of view. The red line represents the direction of the projection given by the moment map that identifies the quadric with a square.

Example 8.27 is an instance of a general construction in algebraic statistics. Projective toric varieties X_A correspond to a class of statistical models, referred to as *toric models* in [**43**] and as *log-linear models* in [**57**]. The inverse moment map μ_A^{-1} is the maximum likelihood estimator for the model $X_{A,\geq 0}$. Given a point $\mathbf{b}$ in the model polytope $P = \mathrm{conv}(A)$, the estimate $\mu_A^{-1}(\mathbf{b})$ is the *Birch point* in $X_{A,\geq 0}$. This distribution best explains the data with sufficient statistic $\mathbf{b}$. See [**43**, Proposition 1.9] and [**57**, Corollary 7.3.9].

Toric varieties are ubiquitous in applications. One explanation for this is the following observation, which connects this chapter to the previous one.

Proposition 8.28. *Let X be an irreducible variety over a field K as in Section* 7.3. *If X is toric then its tropicalization* $\mathrm{Trop}(X)$ *is a linear space.*

Proof. If $X = \mathcal{V}(I_A)$ then every point in X has the form $(\mathbf{x}^{\mathbf{a}_1}, \ldots, \mathbf{x}^{\mathbf{a}_p})$. The images of these points under coordinatewise valuation are $\mathbf{u}A$ where $\mathbf{u} = \mathrm{val}(\mathbf{x})$ runs over $\mathbb{Q}^n$. This implies that $\mathrm{Trop}(X)$ is the row space of A. □

In fact, the converse to this proposition also holds, with a slightly more inclusive definition of toric variety. Informally speaking, toric varieties are precisely those varieties that become linear spaces under taking logarithms.

8.3. The World Is Toric

The occurrence of toric structures in an application can be either obvious or hidden. A typical example of the former is log-linear models in statistics. These are obviously toric, as seen around Example 8.27. In this section we discuss some scenarios where the toric structure is hidden and needs to be unearthed, often by a nontrivial choice of coordinates. Our style in this section is extremely informal. We briefly visit four areas in which toric varieties arise. Under each header we focus on one concrete instance of a toric variety $X_A \subset \mathbb{P}^{p-1}$. The broader context is discussed alongside that example.

Chemical Reactions. Three chemical species $\sigma_1, \sigma_2, \sigma_3$ can form four chemical complexes $3\sigma_1, 3\sigma_2, 3\sigma_3, \sigma_1+\sigma_2+\sigma_3$. Each complex can react so as to transform into any other complex. We introduce unknowns c_1, c_2, c_3 for the species concentrations and K_1, K_2, K_3, K_4 to encode rate constants.

This chemical reaction system is modeled by the *toric balancing ideal*

$$I_A \;=\; \left(\left\langle 2 \times 2 \text{ minors of } \begin{pmatrix} K_1 & K_2 & K_3 & K_4 \\ c_1^3 & c_2^3 & c_3^3 & c_1c_2c_3 \end{pmatrix} \right\rangle \,:\, (c_1c_2c_3)^\infty \right).$$

This toric ideal has 10 minimal generators, six from the 2×2 minors, plus

$$c_1^2K_2K_3 - c_2c_3K_4^2, \quad c_2^2K_1K_3 - c_1c_3K_4^2, \quad c_3^2K_1K_2 - c_1c_2K_4^2, \quad \underline{K_1K_2K_3 - K_4^3}.$$

The variety $X_A = \mathcal{V}(I_A)$ is a threefold of degree 13 in $\mathbb{P}^6$. The underlying 4×7 matrix A is found with the integer linear algebra method in the proof of Theorem 8.10. The polytope $P = \operatorname{conv}(A)$ is a triangular prism. One triangular face is Δ_2 with vertices labeled by c_1, c_2, c_3. The other triangular face is $3\Delta_2$ with vertices K_1, K_2, K_3 and centroid K_4. The underlined cubic generates the *moduli ideal*, which identifies the *toric dynamical systems.*

The mathematical theory of chemical reaction network systems with mass action kinetics is an important domain of application of nonlinear algebra. For an introduction we refer to the textbook by Dickenstein and Feliu [**17**]. The term "toric dynamical systems" was coined in the article [**13**]. In the chemical literature, these are known as complex balancing mass action systems. The toric ideals above were introduced in [**13**, §2].

Gaussian Maximum Likelihood Estimation. Let $n = 5$ and $p = 10$, and consider the integer matrix

$$A \;=\; \begin{pmatrix} 1 & 1 & 1 & 1 & 0 & 0 & 0 & 0 & 0 & 0 \\ 1 & 0 & 0 & 0 & 1 & 1 & 1 & 0 & 0 & 0 \\ 0 & 1 & 0 & 0 & 1 & 0 & 0 & 1 & 1 & 0 \\ 0 & 0 & 1 & 0 & 0 & 1 & 0 & 1 & 0 & 1 \\ 0 & 0 & 0 & 1 & 0 & 0 & 1 & 0 & 1 & 1 \end{pmatrix}. \tag{8.6}$$

The columns of A are labeled $y_{01}, y_{02}, \ldots, y_{34}$. These are our coordinates for $\mathbb{P}^9$. The polytope $P = \operatorname{conv}(A)$ is the *second hypersimplex* of dimension 4. It has f-vector $(10, 30, 30, 10)$ but is not self-dual. Its toric ideal I_A has 10 quadratic generators $y_{ij}y_{kl} - y_{ik}y_{jl}$, and X_A is a fourfold of degree 11 in $\mathbb{P}^9$. The f-vector of a polytope records the number of its faces, as in (8.3).

This toric variety of the second hypersimplex of dimension m arises when studying Gaussian distributions on $\mathbb{R}^m$ with structured covariance matrix Σ.

Consider the model given by prescribing all off-diagonal entries to be equal. Thus, for $m = 4$, we are interested in the linear space of symmetric matrices

$$\Sigma = \begin{bmatrix} \sigma_1 & \sigma_0 & \sigma_0 & \sigma_0 \\ \sigma_0 & \sigma_2 & \sigma_0 & \sigma_0 \\ \sigma_0 & \sigma_0 & \sigma_3 & \sigma_0 \\ \sigma_0 & \sigma_0 & \sigma_0 & \sigma_4 \end{bmatrix}. \tag{8.7}$$

Given a sample covariance matrix S, one seeks to maximize the log-likelihood

$$\ell(\Sigma) \;=\; -\log\,\det(\Sigma) - \mathrm{trace}(S\Sigma^{-1}) \;=\; \log\,\det(K) - \mathrm{trace}(SK).$$

This function is convex in the *concentration matrix* $K = \Sigma^{-1}$. For that reason, we study the set of matrices K whose inverses have the structure (8.7). This is a nonlinear projective variety of dimension m in $\mathbb{P}^{\binom{m}{2}}$. Defining polynomials are obtained by equating off-diagonal entries in the adjoint of K. These are complicated expressions with many terms of degree $m-1$.

We find that this is a toric variety, after the linear change of coordinates

$$K = \begin{bmatrix} y_{01}+y_{12}+y_{13}+y_{14} & -y_{12} & -y_{13} & -y_{14} \\ -y_{12} & y_{02}+y_{12}+y_{23}+y_{24} & -y_{23} & -y_{24} \\ -y_{13} & -y_{23} & y_{03}+y_{13}+y_{23}+y_{34} & -y_{34} \\ -y_{14} & -y_{24} & -y_{34} & y_{04}+y_{14}+y_{24}+y_{34} \end{bmatrix}.$$

For any m, this is the reduced Laplacian matrix of the complete graph on $m+1$ nodes. It follows from [**56**, Theorem 1.2] that the variety of matrices K whose inverses are constant away from the diagonal equals the toric variety of the second hypersimplex. Using the coordinates y_{ij}, the inverse of the matrix Σ in (8.7) satisfies the 10 quadratic binomials in I_A. The article [**56**] establishes this toric structure for a larger class of Gaussian models, one for each rooted tree, thus contributing to likelihood inference for such models.

Phylogenetics. Group-based models in phylogenetics are varieties that become toric after a linear change of coordinates. The nonlinear algebra of this transformation was pioneered in [**54**]. For the relevant background from molecular biology we refer to [**43**, Chapter 4]. The following case study is taken from [**54**, Example 3]. The *Cavender-Farris-Neyman model*, also known as the *binary Jukes-Cantor model*, for the claw tree $K_{1,3}$ is a group-based model for three binary random variables. Its eight joint probabilities are

$$\begin{aligned} p_{000} &= \pi_0\alpha_0\beta_0\gamma_0 + \pi_1\alpha_1\beta_1\gamma_1, & p_{001} &= \pi_0\alpha_0\beta_0\gamma_1 + \pi_1\alpha_1\beta_1\gamma_0\,, \\ p_{010} &= \pi_0\alpha_0\beta_1\gamma_0 + \pi_1\alpha_1\beta_0\gamma_1, & p_{011} &= \pi_0\alpha_0\beta_1\gamma_1 + \pi_1\alpha_1\beta_0\gamma_0\,, \\ p_{100} &= \pi_0\alpha_1\beta_0\gamma_0 + \pi_1\alpha_0\beta_1\gamma_1, & p_{101} &= \pi_0\alpha_1\beta_0\gamma_1 + \pi_1\alpha_0\beta_1\gamma_0\,, \\ p_{110} &= \pi_0\alpha_1\beta_1\gamma_0 + \pi_1\alpha_0\beta_0\gamma_1, & p_{111} &= \pi_0\alpha_1\beta_1\gamma_1 + \pi_1\alpha_0\beta_0\gamma_0. \end{aligned}$$

Here π_0 and $\pi_1 = 1-\pi_0$ give the root distribution. The other model parameters $\alpha_0 = 1-\alpha_1$, $\beta_0 = 1-\beta_1$ and $\gamma_0 = 1-\gamma_1$ are the transition probabilities from the root to the three leaves. When all parameters are nonnegative, the model is a 4-dimensional semialgebraic subset of the probability simplex Δ_7. The *Fourier transform* gives a change of coordinates in the parameter space:

$$\begin{array}{llll} \pi_0 = \frac{1}{2}(r_0 + r_1), & \pi_1 = \frac{1}{2}(r_0 - r_1), & \alpha_0 = \frac{1}{2}(a_0 + a_1), & \alpha_1 = \frac{1}{2}(a_0 - a_1), \\ \beta_0 = \frac{1}{2}(b_0 + b_1), & \beta_1 = \frac{1}{2}(b_0 - b_1), & \gamma_0 = \frac{1}{2}(c_0 + c_1), & \gamma_1 = \frac{1}{2}(c_0 - c_1). \end{array}$$

It also gives a linear change of coordinates in the probability space:

$$p_{ijk} \quad = \quad \frac{1}{8}\sum_{r=0}^{1}\sum_{s=0}^{1}\sum_{t=0}^{1}(-1)^{ir+js+kt}\cdot y_{rst}.$$

After these coordinate changes, the parametrization is now toric:

$$\begin{array}{llll} y_{000} = r_0a_0b_0c_0, & y_{001} = r_1a_0b_0c_1, & y_{010} = r_1a_0b_1c_0, & y_{011} = r_0a_0b_1c_1, \\ y_{100} = r_1a_1b_0c_0, & y_{101} = r_0a_1b_0c_1, & y_{110} = r_0a_1b_1c_0, & y_{111} = r_1a_1b_1c_1. \end{array}$$

This corresponds to a matrix $A \in \{0,1\}^{8\times 8}$ of rank 5. The toric ideal equals

$$I_A \quad = \quad \langle y_{001}y_{110} - y_{000}y_{111},\, y_{010}y_{101} - y_{000}y_{111},\, y_{100}y_{011} - y_{000}y_{111}\rangle.$$

Hence X_A is a complete intersection of codimension 3 and degree 8 in $\mathbb{P}^7$. The study of such phylogenetic models is an active area of research.

Paths and Signatures. Let $n = 6$ and $p = 19$, and consider the monomial map

$$\begin{array}{ll} y_{ijk} = a_ia_ja_k & \text{for } 1 \leq i \leq j \leq k \leq 3, \\ z_{k;ij} = a_kb_{ij} & \text{for } k = 1,2,3 \text{ and } 1 \leq i < j \leq 3. \end{array}$$

This defines a toric variety X_A of dimension 5 and degree 24 in $\mathbb{P}^{18}$. The matrix $A \in \{0,1\}^{6\times 19}$ has rows indexed by $a_1, a_2, a_3, b_{12}, b_{13}, b_{23}$ and $19 = 10 + 9$ columns indexed by $y_{111}, y_{112}, \ldots, y_{333}$ and $z_{1;12}, z_{1;13}, \ldots, z_{3;23}$. The toric ideal I_A is generated by 81 binomial quadrics, namely the 2×2 minors of

$$(8.8) \qquad \begin{pmatrix} y_{111} & y_{112} & y_{113} & y_{122} & y_{123} & y_{133} & z_{1;12} & z_{1;13} & z_{1;23} \\ y_{112} & y_{122} & y_{123} & y_{222} & y_{223} & y_{233} & z_{2;12} & z_{2;13} & z_{2;23} \\ y_{113} & y_{123} & y_{133} & y_{223} & y_{233} & y_{333} & z_{3;12} & z_{3;13} & z_{3;23} \end{pmatrix}.$$

Let $\tilde{X}_A$ be the join of X_A with $\mathbb{P}^7$. This is a 13-dimensional toric variety of degree 24 in $\mathbb{P}^{26}$. It is defined by the same ideal I_A but now in 27 variables. We replace these by the entries of a $3\times 3\times 3$ tensor $\sigma = (\sigma_{ijk})$ as follows:

$$(8.9) \qquad \begin{array}{lllll} y_{ijk} & = & \sigma_{kij} + \sigma_{ikj} + \sigma_{ijk} & + & \sigma_{kji} + \sigma_{jki} + \sigma_{jik}, \\ z_{k;ij} & = & \frac{1}{2}(\sigma_{kij} + \sigma_{ikj} + \sigma_{ijk}) & - & \frac{1}{2}(\sigma_{kji} + \sigma_{jki} + \sigma_{jik}). \end{array}$$

The resulting variety $\mathcal{U}_{3,3}$ is the universal variety of third-order signature tensors of arbitrary paths in $\mathbb{R}^3$. Such tensors play an important role in *stochastic analysis*, especially in the Hairer-Lyons theory of rough paths.

A natural generalization of the universal variety is the *rough Veronese variety*, which was shown to be toric by Colmenarejo et al. in [**9**]. This variety is a variant of the classical Veronese variety, but adapted to the study of rough paths. For an introduction to this theory see [**9**, §1] and the references therein. Returning to the example above, in the notation of [**9**, §2], we have $\mathcal{U}_{3,3} = \mathcal{R}_{3,3,3} = \tilde{X}_A \subset \mathbb{P}^{26}$ and $\mathcal{R}_{3,3,2} = X_A \subset \mathbb{P}^{18}$. The equations defining these varieties are obtained by substituting (8.9) into (8.8). This 3×10 matrix has rank ≤ 1 for the signatures of all paths in $\mathbb{R}^3$.

Exercises

(1) Prove that every character χ of the torus $T = (K^*)^n$ is given by a Laurent monomial $\mathbf{x}^{\mathbf{b}} = x_1^{b_1} x_2^{b_2} \cdots x_n^{b_n}$ for some integer vector $\mathbf{b}$ in $\mathbb{Z}^n$.

(2) Show that every polynomial in the ideal I_A of an affine toric variety is a K-linear combination of binomials. This gives statement (3) in Lemma 8.8.

(3) Describe the ideal of the Segre variety $\mathbb{P}^{a_1-1} \times \cdots \times \mathbb{P}^{a_s-1}$ inside $\mathbb{P}^{a_1 \cdots a_s - 1}$. What is the degree of this toric variety? Describe its lattice polytope P.

(4) There is a natural bijection between (convex, rational, polyhedral) cones in $\mathbb{R}^d$ and finitely generated saturated monoids in $\mathbb{Z}^d$. Prove this fact.

(5) Determine the toric ideal I_A and the toric variety X_A for the matrix

$$A = \begin{pmatrix} 0 & 0 & 1 & 1 & 1 \\ 0 & 1 & 0 & 1 & 1 \\ 0 & 0 & 0 & 2 & 3 \\ 1 & 1 & 1 & 1 & 1 \end{pmatrix}.$$

(6) Let A be as in Exercise 5. Show that the 3-dimensional lattice polytope $P = \mathrm{conv}(A)$ is not normal. Draw this polytope and find its f-vector.

(7) Prove that every 2-dimensional lattice polytope is normal.

(8) Prove the following result: For any k-dimensional lattice polytope P, the scaled polytope $(k-1)P$ is normal. Hint: For $k = 2$, this is the previous exercise. Start with the case $k = 3$. Consult [**12**, §2.2] if needed.

(9) Determine the number of lattice points in kP where P is the polytope in Exercises 5 and 6. Show that this number is a cubic polynomial in k. This is known as the *Ehrhart polynomial* of the polytope P.

(10) Prove the following theorem due to Mumford in the case of toric varieties. Let X be a projective toric variety. For r large enough, the rth Veronese reembedding $v_r(X)$ of X is defined by quadratic equations.

(11) Compute an explicit Gröbner basis for the toric ideal I_A, where A is the matrix in Example 8.9. Is your initial monomial ideal $\mathrm{in}(I_A)$ radical?

(12) Compute the inverse of the matrix Σ in (8.7), and verify that its entries satisfy the 10 quadratic binomials given by the second hypersimplex.

(13) Let A be the 3×7 matrix in Example 8.9. Can you give a formula for the inverse moment map μ_A^{-1}? Is there an expression in radicals?

(14) (a) Compute the number of points of a projective toric variety X_P over a finite field in terms of the f-vector of the associated polytope P.

(b) Assuming that X_P is smooth and $K = \mathbb{C}$, use the Weil conjectures (which are theorems, thanks to Grothendieck and Deligne) to give a formula for Betti numbers of X_P, again in terms of the f-vector.

(15) Give an example of a toric threefold X_A in $\mathbb{P}^6$ that has degree 11. Draw the polytope $P = \mathrm{conv}(A)$. Can you arrange for X_A to be smooth?

(16) Verify Theorem 8.24 for the case where P is the triangular prism. This corresponds to the Segre variety with $n_1 = 2$ and $n_2 = 3$ in Example 8.18.

(17) Section 8.3 discusses four toric varieties that arise in applications. Pick two of them and study the corresponding polytopes. Find their f-vectors.

Chapter 9

Tensors

"The name TensorFlow derives from the operations that neural networks perform on multidimensional data arrays, which are referred to as tensors",
Wikipedia

Tensors are ubiquitous in many branches of modern mathematics. They are higher-dimensional analogues of matrices. Just as matrices are basic objects in *linear algebra*, tensors are fundamental to *nonlinear algebra*. One reason why they appear so late in this book is that we already saw them in disguise: Homogeneous polynomials are symmetric tensors. In this chapter we show that basic attributes of matrices, such as eigenvectors and rank, can also be defined for tensors. However, their behavior is far more interesting in the tensor context. We also discuss applications of tensors, focusing on a central open algorithmic problem: How fast can one multiply two matrices? As always, linear algebra is our door to nonlinear algebra. Further, the new nonlinear tools will be applied to revisit fundamental questions in linear algebra.

9.1. Eigenvectors

In this section we extend the familiar concepts of eigenvectors, rank and singular values from matrices to the setting of tensors. We start by reviewing some basics of linear algebra, beginning with the study of symmetric matrices. Recall that symmetric matrices uniquely represent quadratic forms.

For instance, consider the following quadratic form in three variables:

$$Q = 2x^2 + 7y^2 + 23z^2 + 6xy + 10xz + 22yz. \tag{9.1}$$

This quadratic form is represented by a symmetric 3×3 matrix as follows:

$$(9.2) \qquad Q \;=\; \begin{pmatrix} x & y & z \end{pmatrix} \begin{pmatrix} 2 & 3 & 5 \\ 3 & 7 & 11 \\ 5 & 11 & 23 \end{pmatrix} \begin{pmatrix} x \\ y \\ z \end{pmatrix}.$$

The gradient of the quadratic form Q is the vector of its partial derivatives. Thus, the gradient is a vector of linear forms. It defines a linear map from K^3 to itself. Up to multiplication by 2, this is the linear map one usually associates with a square matrix. For the quadratic form in (9.1) we have

$$\nabla Q \;=\; \begin{pmatrix} \partial Q/\partial x \\ \partial Q/\partial y \\ \partial Q/\partial z \end{pmatrix} \;=\; 2 \cdot \begin{pmatrix} 2 & 3 & 5 \\ 3 & 7 & 11 \\ 5 & 11 & 23 \end{pmatrix} \begin{pmatrix} x \\ y \\ z \end{pmatrix}.$$

In this section, the field K is usually $\mathbb{R}$ or $\mathbb{C}$. We call $\mathbf{v} \in K^n \backslash \{0\}$ an *eigenvector* of Q if $\mathbf{v}$ is mapped to a scalar multiple of $\mathbf{v}$ by the gradient map:

$$(\nabla Q)(\mathbf{v}) \;=\; \lambda \cdot \mathbf{v} \quad \text{for some } \lambda \in K.$$

Just like in the earlier chapters, it is convenient to replace the n-dimensional affine space K^n with the $(n-1)$-dimensional projective space $\mathbb{P}^{n-1}$. Thus, two nonzero vectors are identified if they are parallel. From a nonzero quadratic form Q we obtain a rational self-map of the projective space:

$$(9.3) \qquad \nabla Q \,:\, \mathbb{P}^{n-1} \dashrightarrow \mathbb{P}^{n-1}.$$

The dashed arrow means that this map may not be defined everywhere. Algebraic geometers call this a *rational map*. If Q is rank-deficient then the linear map has a kernel. It consists of all points at which the gradient ∇Q vanishes. These are the *base points* of the map (9.3). If Q has full rank, then ∇Q is a regular map $\mathbb{P}^{n-1} \to \mathbb{P}^{n-1}$, i.e. it is defined on all of $\mathbb{P}^{n-1}$. We conclude our discussion with the following remark on the gradient map:

Remark 9.1. The eigenvectors of Q are the fixed points ($\lambda \neq 0$) and base points ($\lambda = 0$) of the gradient map ∇Q in (9.3). These points $\mathbf{v}$ live in $\mathbb{P}^{n-1}$.

Symmetric $n \times n$ matrices often appear in statistics. Consider n real-valued random variables $X_1, \ldots, X_n$. Their *covariance matrix* is the matrix Σ whose (i,j) entry is $\text{cov}[X_i, X_j] = E[(X_i - E[X_i])(X_j - E[X_j])]$. We note that Σ is positive semidefinite, i.e. all its eigenvalues are nonnegative.

An $n \times n$ matrix usually has n linearly independent *eigenvectors*, provided the underlying field K is algebraically closed. If the matrix is real and symmetric, then its eigenvectors have real coordinates and are *orthogonal*. For a rectangular matrix, one considers pairs of *singular vectors*, defined below, one on the left and one on the right. The number of these singular vector pairs is equal to the smaller of the two matrix dimensions.

Eigenvectors and singular vectors are familiar from linear algebra, where they are introduced in concert with *eigenvalues* and *singular values.* Numerical linear algebra is the foundation of applied mathematics and scientific computing. Specifically, the concept of eigenvectors, and numerical algorithms for computing them, became a key technology during the 20th century.

Singular vectors are defined for rectangular matrices. We review their definition through the lens of Remark 9.1. We begin with the observation that each rectangular matrix uniquely represents a bilinear form, e.g.

$$B = 2ux+3uy+5uz+3vx+7vy+11vz = \begin{pmatrix} u & v \end{pmatrix} \begin{pmatrix} 2 & 3 & 5 \\ 3 & 7 & 11 \end{pmatrix} \begin{pmatrix} x \\ y \\ z \end{pmatrix}. \tag{9.4}$$

The gradient of the bilinear form defines an endomorphism on the direct sum of an m-dimensional space and an n-dimensional space. This fuses left multiplication and right multiplication by our matrix into a single linear map. For the bilinear form in (9.4), the gradient is the vector of linear forms

$$\nabla B = \Big(\big(\frac{\partial B}{\partial x}, \frac{\partial B}{\partial y}, \frac{\partial B}{\partial z}\big), \big(\frac{\partial B}{\partial u}, \frac{\partial B}{\partial v}\big) \Big). \tag{9.5}$$

The associated endomorphism has the form $\nabla B : K^3 \oplus K^2 \to K^3 \oplus K^2$. This gradient map takes the pair $\big((x,y,z),(u,v)\big)$ to the pair in (9.5), i.e. to

$$(\,(2u+3v, 3u+7v, 5u+11v)\,, \,(2x+3y+5z, 3x+7y+11z)\,)\,.$$

More generally, let B be an $m \times n$ matrix over K. Consider the equations

$$B\mathbf{x} = \lambda\mathbf{y} \quad \text{and} \quad B^T\mathbf{y} = \lambda\mathbf{x}, \tag{9.6}$$

where λ is a scalar, $\mathbf{x}$ is a nonzero vector in K^n, and $\mathbf{y}$ is a nonzero vector in K^m. These are our unknowns. Given a solution to (9.6), we see that $\mathbf{x}$ is an eigenvector of B^TB, $\mathbf{y}$ is an eigenvector of BB^T, and λ^2 is a common eigenvalue of these two symmetric matrices. Assuming $K = \mathbb{R}$, this eigenvalue's nonnegative square root $\lambda \geq 0$ is a *singular value* of B. Associated to λ are the *right singular vector* $\mathbf{x}$ and the *left singular vector* $\mathbf{y}^T$. In analogy to Remark 9.1, the process of solving (9.6) has the following dynamical interpretation.

Remark 9.2. Each singular vector pair $(\mathbf{x}, \mathbf{y})$ of a rectangular matrix B is mapped to a scalar multiple $(\lambda\mathbf{x}, \lambda\mathbf{y})$ under the gradient map ∇B of the associated bilinear form. Here λ is the singular value, which can be zero. Up to scaling, the gradient map is a self-map on a product of projective spaces:

$$\nabla B : \mathbb{P}^{n-1} \times \mathbb{P}^{m-1} \dashrightarrow \mathbb{P}^{n-1} \times \mathbb{P}^{m-1},$$

$$(\mathbf{x}, \mathbf{y}) \mapsto \Big(\big(\frac{\partial B}{\partial x_1}, \ldots, \frac{\partial B}{\partial x_n}\big), \big(\frac{\partial B}{\partial y_1}, \ldots, \frac{\partial B}{\partial y_m}\big) \Big).$$

To be precise, each pair $(\mathbf{x}, \mathbf{y})$ of singular vectors gives rise to either a base point or a fixed point of the rational map ∇B between the product of projective spaces. Conversely, suppose that $(\mathbf{x}, \mathbf{y}) \in \mathbb{P}^{n-1} \times \mathbb{P}^{m-1}$ is a fixed point or a base point. In order to lift $(\mathbf{x}, \mathbf{y})$ to a pair of singular vectors, some compatibility is required. For fixed points, we can take an arbitrary vector representative of $\mathbf{x}$ and the corresponding representative of $\mathbf{y}$. But for a base point $(\mathbf{x}, \mathbf{y})$, this lifting is possible if and only if the vector is actually in the kernel of the linear map $K^{m+n} \to K^{m+n}$ defined by ∇B.

Working over $\mathbb{R}$, we have the following orthogonal decompositions:

$$Q = O \cdot \mathrm{diag} \cdot O^T \quad \text{and} \quad B = O_1 \cdot \mathrm{diag} \cdot O_2.$$

Here diag represents diagonal matrices, the latter one being rectangular. The matrices O, O_1 and O_2 are orthogonal, i.e. $OO^T = \mathrm{Id}$, $O_1O_1^T = \mathrm{Id}$ and $O_2O_2^T = \mathrm{Id}$. The entries of diag are respectively the eigenvalues of Q and the singular values of B. The columns of O are the eigenvectors of Q. The columns of O_1 and the rows of O_2 give all pairs of singular vectors of B. Note that O_1 and O_2 are both square matrices, but of different sizes. The possibly unmatched rows of O_2 are in the kernel of B. The formulas above are known as the *spectral decomposition* and the *singular value decomposition*.

For $K = \mathbb{C}$ there is a version of the spectral decomposition for *Hermitian matrices* Q. A Hermitian matrix Q is equal to its conjugate transpose: $Q_{ij} = \overline{Q_{ji}}$. The matrix O is now *unitary*, i.e. its inverse is equal to its conjugate transpose. The eigenvalues of Q remain real. Analogously, for the singular value decomposition one needs to replace the orthogonal matrices by unitary matrices. The singular values are nonnegative also in this case.

We summarize our brief review of linear algebra in the following points:

- Symmetric matrices Q represent quadratic forms.
- Rectangular matrices B represent bilinear forms.
- Their gradients ∇Q and ∇B specify the linear maps one usually identifies with the matrices Q and $B \oplus B^T$.
- The *eigenvectors* and *singular vectors* are fixed points of these maps.
- These fixed points are computed via *spectral* and *singular value decompositions*:

$$Q = O \cdot \mathrm{diag} \cdot O^T \quad \text{and} \quad B = O_1 \cdot \mathrm{diag} \cdot O_2.$$

In the age of big data, the role of matrices is increasingly played by *tensors*, that is, multidimensional arrays of numbers. Principal component analysis tells us that the eigenvectors of a covariance matrix $Q = BB^T$ give directions in which the data B is most spread. One hopes to identify similar features for tensor data. This has encouraged engineers and scientists to spice up their linear algebra tool box with a pinch of algebraic geometry.

The spectral theory of tensors is the theme of the following discussion. This theory was pioneered around 2005 by Lek-Heng Lim and Liqun Qi. We refer to the textbook [**44**] for background and context. Our aim is to generalize familiar notions, such as rank, eigenvectors and singular vectors, from matrices to tensors. Specifically, we address the following two questions. The answers to these two questions are provided in Examples 9.7 and 9.13.

Question 9.3. How many eigenvectors does a $3 \times 3 \times 3$ tensor have?

Question 9.4. How many singular vector triples does a $3{\times}3{\times}3$ tensor have?

A *tensor* is a d-dimensional array $T = (t_{i_1 i_2 \cdots i_d})$. Here the entries $t_{i_1 i_2 \cdots i_d}$ are elements in the ground field K. The set of all tensors of format $n_1{\times}n_2{\times}\cdots{\times}n_d$ forms a vector space of dimension $n_1 n_2 \cdots n_d$ over K. For $d = 1, 2$ we get vectors and matrices. A tensor has *rank* 1 if it is the outer product of d vectors, written $T = \mathbf{u} \otimes \mathbf{v} \otimes \cdots \otimes \mathbf{w}$ or, in coordinates,

$$t_{i_1 i_2 \cdots i_d} \;=\; u_{i_1} v_{i_2} \cdots w_{i_d}.$$

The problem of *tensor decomposition* is the following. We wish to express a given tensor T as a sum of rank-1 tensors, using as few summands as possible. The minimal number of rank-1 summands needed to represent T is the *rank* of T. We note that the rank of any tensor is always finite. We will discuss this topic in detail in the next section.

An $n{\times}n{\times}\cdots{\times}n$ tensor $T = (t_{i_1 i_2 \cdots i_d})$ is called *symmetric* if it is unchanged upon permuting the indices. The space $\mathrm{Sym}_d(\mathbb{R}^n)$ of such symmetric tensors has dimension $\binom{n+d-1}{d}$. It is identified with the space of homogeneous polynomials of degree d in n variables, written as

$$T \;=\; \sum_{i_1,\ldots,i_d=1}^{n} t_{i_1 i_2 \cdots i_d} \cdot x_{i_1} x_{i_2} \cdots x_{i_d}. \tag{9.7}$$

Example 9.5. A tensor T of format $3{\times}3{\times}3$ has 27 entries. If T is a symmetric tensor, then it has at most 10 distinct entries, one for each coefficient of the associated cubic polynomial in three variables. This polynomial defines a cubic curve in the projective plane $\mathbb{P}^2$, as indicated in Figure 9.1.

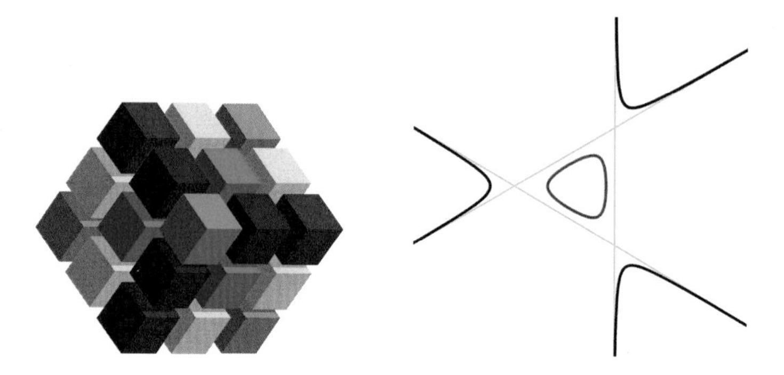

Figure 9.1. A symmetric 3×3×3 tensor represents a cubic curve in $\mathbb{P}^2$.

Symmetric tensor decomposition writes T as a sum of powers of linear forms:

$$(9.8) \qquad T \;=\; \sum_{j=1}^{r} \lambda_j \mathbf{v}_j^{\otimes d} \;=\; \sum_{j=1}^{r} \lambda_j (v_{1j}x_1 + v_{2j}x_2 + \cdots + v_{nj}x_n)^d,$$

where $\mathbf{v}_j = (v_{1j}, \ldots, v_{nj})$ and the λ_j are scalars. As before, the gradient of T defines a map $\nabla T : K^n \to K^n$, but this map is now nonlinear. A vector $\mathbf{v} \in K^n$ is called an *eigenvector* of T if $(\nabla T)(\mathbf{v}) = \lambda \cdot \mathbf{v}$ for some $\lambda \in K$.

Eigenvectors of tensors arise naturally in optimization. Consider the problem of maximizing a real homogeneous polynomial T over the unit sphere in $\mathbb{R}^n$. If λ denotes a Lagrange multiplier, one sees that the eigenvectors of T are the critical points of this optimization problem. One can check the values of T at these points to find global maxima and minima.

We find it convenient to replace K^n by the projective space $\mathbb{P}^{n-1}$. The gradient map is then a rational map from this projective space to itself:

$$\nabla T \,:\, \mathbb{P}^{n-1} \dashrightarrow \mathbb{P}^{n-1}.$$

The eigenvectors of T are *fixed points* ($\lambda \neq 0$) and *base points* ($\lambda = 0$) of ∇T. Thus the spectral theory of tensors is closely related to the study of dynamical systems on $\mathbb{P}^{n-1}$. The matrix case ($d = 2$) appeared in (9.3). By the Spectral Theorem of linear algebra, a real quadratic form T has a real decomposition (9.8) with $d = 2$. Here r is the rank, the λ_j are the eigenvalues of T, and the eigenvectors $\mathbf{v}_j = (v_{1j}, v_{2j}, \ldots, v_{nj})$ are orthonormal. We can compute the eigenvectors numerically by *power iteration*, namely by applying ∇T until an approximate fixed point is reached. This will be a dominant eigenvector, and the other eigenvectors are then found by an appropriate inductive scheme.

For $d \geq 3$, one can still use power iteration to compute eigenvectors of T. However, the eigenvectors are usually not the vectors $\mathbf{v}_i$ in the low-rank decomposition (9.8). One exception arises when the symmetric tensor is *odeco*, or orthogonally decomposable. This means that T has the form (9.8), where $r = n$ and $\{\mathbf{v}_1, \mathbf{v}_2, \ldots, \mathbf{v}_r\}$ is an orthogonal basis of $\mathbb{R}^n$. These basis vectors are the attractors of the dynamical system ∇T, provided $\lambda_j > 0$.

The following gives a count of the eigenvectors of a symmetric tensor.

Theorem 9.6 (Cartwright-Sturmfels [**8**]). *If K is algebraically closed, then the number of eigenvectors of a general tensor $T \in \mathrm{Sym}_d(K^n)$ equals*

$$\frac{(d-1)^n - 1}{d-2} \quad = \quad \sum_{i=0}^{n-1} (d-1)^i.$$

Example 9.7 ($n = d = 3$). The Fermat cubic $T = x^3 + y^3 + z^3$ is an odeco tensor. Its gradient map is the regular map that squares each coordinate: $\nabla T : \mathbb{P}^2 \to \mathbb{P}^2$, $(x : y : z) \mapsto (x^2 : y^2 : z^2)$. This dynamical system has $7 = 1 + 2 + 2^2$ fixed points, of which only the first three are attractors:

$$(1:0:0), (0:1:0), (0:0:1),\ (1:1:0), (1:0:1), (0:1:1),\ (1:1:1).$$

We conclude that T has seven eigenvectors, as predicted by Theorem 9.6.

It is known that all eigenvectors can be real for suitable tensors. This was proved by Khozhasov [**29**] using the theory of *harmonic polynomials*. For $n = 3$, we can construct tensors with all eigenvectors real by the following geometric method. Let T be a product of d linear forms in x, y, z, defining d lines in $\mathbb{P}^2_{\mathbb{R}}$. The $\binom{d}{2}$ vertices of the line arrangement are base points of ∇T. One can show that each of the $\binom{d}{2} + 1$ regions contains one fixed point. This accounts for all $1 + (d-1) + (d-1)^2$ eigenvectors, which are hence real.

Example 9.8. Let $d = 4$ and fix the quartic $T = xyz(x+y+z)$, which is a symmetric $3 \times 3 \times 3 \times 3$ tensor. Its curve in $\mathbb{P}^2$ is an arrangement of four lines. All $13 = 6 + 7$ eigenvectors of T are real. The 6 vertices of the arrangement are the base points of ∇T. Each of the 7 regions contains one fixed point.

For special tensors T, two of the eigenvectors in Theorem 9.6 may coincide. This corresponds to vanishing of the *eigendiscriminant*, a big polynomial in the $t_{i_1 i_2 \cdots i_d}$. In the matrix case ($d = 2$), this is the discriminant of the characteristic polynomial of an $n \times n$ matrix [**53**, §7.5]. For $3 \times 3 \times 3$ tensors, the eigendiscriminant is a polynomial of degree 24 in 27 unknowns.

Theorem 9.9 (Abo-Seigal-Sturmfels [**1**]). *The eigendiscriminant is an irreducible homogeneous polynomial of degree $n(n-1)(d-1)^{n-1}$ in the $t_{i_1 i_2 \cdots i_d}$.*

Example 9.10 ($n = 2$). The eigendiscriminant of a binary form $T(x, y)$ of degree d is the discriminant of $x\frac{\partial T}{\partial y} - y\frac{\partial T}{\partial x}$, so it has degree $2d - 2$ in T.

Singular value decomposition is a central notion in linear algebra and its applications. Remark 9.2 casts the singular vector pairs of a matrix as fixed points of a self-map of a product of two projective spaces. Consider now a d-dimensional tensor T in $K^{n_1 \times \cdots \times n_d}$. It corresponds to a multilinear form. The *singular vector tuples* of T are the fixed points of the gradient map

$$\nabla T \,:\, \mathbb{P}^{n_1-1} \times \cdots \times \mathbb{P}^{n_d-1} \dashrightarrow \mathbb{P}^{n_1-1} \times \cdots \times \mathbb{P}^{n_d-1}.$$

Example 9.11. The trilinear form $T = x_1y_1z_1 + x_2y_2z_2$ is interpreted as a $2{\times}2{\times}2$ tensor. The gradient ∇T of this trilinear form is the rational map

$$\begin{array}{ccc} \mathbb{P}^1 \times \mathbb{P}^1 \times \mathbb{P}^1 & \dashrightarrow & \mathbb{P}^1 \times \mathbb{P}^1 \times \mathbb{P}^1, \\ \big((x_1 : x_2), (y_1 : y_2), (z_1 : z_2)\big) & \mapsto & \big((y_1z_1 : y_2z_2), (x_1z_1 : x_2z_2), (x_1y_1 : x_2y_2)\big). \end{array}$$

This map has six fixed points, namely $\big((1{:}0),(1{:}0),(1{:}0)\big)$, $\big((0{:}1),(0{:}1),(0{:}1)\big)$, $\big((1{:}1),(1{:}1),(1{:}1)\big)$, $\big((1{:}1),(1{:}{-1}),(1{:}{-1})\big)$, $\big((1{:}{-1}),(1{:}1),(1{:}{-1})\big)$, and $\big((1{:}{-1}),(1{:}{-1}),(1{:}1)\big)$. These are the singular vector triples of the tensor T.

Here is a formula for the expected number of singular vector tuples.

Theorem 9.12 (Friedland and Ottaviani [**20**])**.** *For a general $n_1 \times \cdots \times n_d$ tensor T over an algebraically closed field K, the number of singular vector tuples is the coefficient of the monomial $z_1^{n_1-1} \cdots z_d^{n_d-1}$ in the polynomial*

$$\prod_{i=1}^{d} \frac{(\widehat{z_i})^{n_i} - z_i^{n_i}}{\widehat{z_i} - z_i} \quad \textit{where} \quad \widehat{z_i} = z_1 + \cdots + z_{i-1} + z_{i+1} + \cdots + z_d.$$

We end our study of the spectral theory of tensors by answering Question 9.4.

Example 9.13. Let $d = n_1 = n_2 = n_3 = 3$. The polynomial in Theorem 9.12 is

$$(\widehat{z_1}^2 + \widehat{z_1}z_1 + z_1^2)(\widehat{z_2}^2 + \widehat{z_2}z_2 + z_2^2)(\widehat{z_3}^2 + \widehat{z_3}z_3 + z_3^2) \;=\; \cdots + \mathbf{37} z_1^2 z_2^2 z_3^2 + \cdots.$$

Therefore, a general $3{\times}3{\times}3$ tensor has exactly 37 triples of singular vectors. Likewise, a general $3{\times}3{\times}3{\times}3$ tensor has 997 quadruples of singular vectors.

9.2. Tensor Rank

There are many ways to define the rank of an $a \times b$ matrix M over a field K:

(1) the smallest integer r such that all $(r+1) \times (r+1)$ minors vanish;

(2) the dimension of the image of the induced linear map $K^a \to K^b$;

(3) the dimension of the image of the induced linear map $K^b \to K^a$;

(4) the smallest r such that $M = UW$ for $U \in K^{a \times r}$ and $W \in K^{r \times b}$.

The first point implies that matrices of rank at most r form a variety. The last point implies that a matrix of rank r is a sum of r matrices of rank 1. This is also true for symmetric matrices: A symmetric matrix of rank r is a sum of r symmetric rank-1 matrices. Another fact is that a real matrix of rank r also has rank r when viewed over $\mathbb{C}$. This seems obvious, but a priori it is not clear why there is no shorter decomposition into rank-1 matrices with entries in $\mathbb{C}$. Our aim is to study these issues for arbitrary tensors.

In this section we work in a tensor product $V_1 \otimes V_2 \otimes \cdots \otimes V_d$ of finite-dimensional vector spaces V_i over a field K. This is the space of all tensors of format $(\dim V_1) \times \cdots \times (\dim V_d)$. By fixing a basis, we may identify $V_i \simeq K^{n_i}$ and thus $V_1 \otimes V_2 \otimes \cdots \otimes V_d \simeq K^{n_1 \times n_2 \times \cdots \times n_d}$. A nonzero tensor T in this space has rank 1 if it is the outer product of d vectors, i.e. $T = \mathbf{u} \otimes \mathbf{v} \otimes \cdots \otimes \mathbf{w}$. In coordinates, this means that the entries of T factor as

$$t_{i_1 i_2 \cdots i_d} \;=\; u_{i_1} v_{i_2} \cdots w_{i_d}.$$

Tensors of rank at most 1 form an affine variety. It is the affine cone over the Segre variety $\mathbb{P}^{n_1-1} \times \cdots \times \mathbb{P}^{n_d-1}$ in $\mathbb{P}^{n_1 \cdots n_d - 1}$. In fact, from Chapters 2 and 8 we know the equations of this variety. They are binomial quadrics, namely the 2×2 minors of all *flattenings* of T. By flattenings we mean representations of the tensor T as a matrix. We shall now explain this.

To begin with, the flattenings of the tensor T have the following invariant description. Let I be any nonempty, proper subset of $[d] = \{1, 2, \ldots, d\}$. The corresponding flattening is the linear map defined by T as follows:

$$K^{\prod_{i\in I} n_i} = \bigotimes_{i \in I} V_i^* \longrightarrow \bigotimes_{j \in [d]\setminus I} V_j = K^{\prod_{j \in [d]\setminus I} n_j}, \qquad (9.9)$$

$$e_{\mathbf{i}}^* \mapsto \sum_{\mathbf{j}} t_{\mathbf{ij}} e_{\mathbf{j}}.$$

Here $e_{\mathbf{i}}^*$ is a fixed basis vector of $K^{\prod_{i\in I} n_i} = \bigotimes_{i\in I} V_i^*$, $e_{\mathbf{j}}$ is any basis vector of $\bigotimes_{j\in[d]\setminus I} V_j = K^{\prod_{j\in[d]\setminus I} n_j}$, and $t_{\mathbf{ij}}$ is the entry of T corresponding to given $\mathbf{i}$ and $\mathbf{j}$. Thus a tensor T has rank 1 if and only if all $2^d - 2$ flattenings of T are matrices of rank 1. In fact, it is enough to check d flattenings corresponding to $|I| = 1$. There is a similar result for tensors of rank 2, due to Landsberg, Manivel and Raicu, but for all higher ranks only one direction is true: The rank of T is bounded below by that of any flattening.

Example 9.14. A tensor $T = (t_{ijk}) \in V_1 \otimes V_2 \otimes V_3$ induces the linear map

$$V_1^* \to V_2 \otimes V_3\,, \quad \mathbf{e}_i^* \mapsto (t_{ijk})_{j,k} = \sum_{j,k} t_{ijk} \cdot \mathbf{f}_j \otimes \mathbf{g}_k, \qquad (9.10)$$

where $(\mathbf{e}_i)$, $(\mathbf{f}_j)$ and $(\mathbf{g}_k)$ are respectively bases of V_1, V_2 and V_3. This is the case where $I = \{1\}$ in (9.9). We think of (9.10) as an $n_1 \times n_2 n_3$ matrix.

The transpose of this matrix is the $n_2 n_3 \times n_1$ matrix that corresponds to $I = \{2,3\}$ in (9.9). Thus, a three-way tensor has three distinct flattenings, up to transposition.

We conclude that rank-1 tensors behave in a very nice way. However, arbitrary tensors exhibit rather strange properties. Recall that the rank of a tensor T is the minimal r such that T is the sum of r rank-1 tensors. For instance, the three-way tensors of rank ≤ 2 are the tensors of the form

$$T \quad = \quad \mathbf{a} \otimes \mathbf{b} \otimes \mathbf{c} + \mathbf{d} \otimes \mathbf{e} \otimes \mathbf{f}. \tag{9.11}$$

We shall see that the set of these tensors is not Zariski closed in $K^{n_1 \times n_2 \times n_3}$.

Example 9.15. Let $d = 3$ and $V_1 = V_2 = V_3 = \mathbb{C}^2$ with basis $\{\mathbf{e}_0, \mathbf{e}_1\}$. The following $2 \times 2 \times 2$ tensor is known in quantum physics as the *W-state*:

$$W \quad = \quad \mathbf{e}_0 \otimes \mathbf{e}_0 \otimes \mathbf{e}_1 \; + \; \mathbf{e}_0 \otimes \mathbf{e}_1 \otimes \mathbf{e}_0 \; + \; \mathbf{e}_1 \otimes \mathbf{e}_0 \otimes \mathbf{e}_0. \tag{9.12}$$

This representation shows that W has rank at most 3. In fact, $\operatorname{rk} W = 3$, as the reader is asked to prove in Exercise 9. To do so, equate W with T in (9.11). This gives an inconsistent system of eight cubic equations in the 12 unknown coordinates $a_0, a_1, b_0, \ldots, f_1$ of the vectors $\mathbf{a}, \mathbf{b}, \ldots, \mathbf{f}$ in (9.11).

However, there exist rank-2 tensors arbitrarily close to W. We have

$$\begin{aligned}
&\frac{1}{\epsilon}\big((\mathbf{e}_0 + \epsilon\mathbf{e}_1) \otimes (\mathbf{e}_0 + \epsilon\mathbf{e}_1) \otimes (\mathbf{e}_0 + \epsilon\mathbf{e}_1) - \mathbf{e}_0 \otimes \mathbf{e}_0 \otimes \mathbf{e}_0\big) \\
&\quad = W \, + \, \epsilon(\mathbf{e}_1 \otimes \mathbf{e}_1 \otimes \mathbf{e}_0 + \mathbf{e}_1 \otimes \mathbf{e}_0 \otimes \mathbf{e}_1 + \mathbf{e}_0 \otimes \mathbf{e}_1 \otimes \mathbf{e}_1) \, + \, \epsilon^2 \mathbf{e}_1 \otimes \mathbf{e}_1 \otimes \mathbf{e}_1.
\end{aligned}$$

This is an identity for all $\epsilon \neq 0$. In particular, we have

$$\lim_{\epsilon \to 0} \frac{1}{\epsilon}\big((\mathbf{e}_0 + \epsilon\mathbf{e}_1) \otimes (\mathbf{e}_0 + \epsilon\mathbf{e}_1) \otimes (\mathbf{e}_0 + \epsilon\mathbf{e}_1) - \mathbf{e}_0 \otimes \mathbf{e}_0 \otimes \mathbf{e}_0\big) \; = \; W.$$

We conclude that the W-state is a tensor of rank 3, but it can be approximated with arbitrary precision by a sequence of tensors of rank 2.

Definition 9.16. The *border rank* $\operatorname{brk}(T)$ of a complex tensor T is the smallest r such that T lies in the closure of the set of tensors of rank r.

The notion of border rank requires a topology on the space of tensors. The geometric locus of tensors of border rank $\leq r$ is the closure of the locus of tensors of rank $\leq r$. Note that we may realize tensors of rank $\leq r$ as the image of a polynomial map. Hence, over the complex numbers, by Corollary 4.20 of Chevalley's Theorem, it does not matter whether we take the Euclidean topology or the Zariski topology—the closures coincide. However, the situation is different over the real numbers. To prove this, we shall use the hyperdeterminant, denoted by Det, from Example 4.10.

Lemma 9.17. *The hyperdeterminant of the rank-2 tensor in* (9.11) *is*

$$\operatorname{Det}(T) \quad = \quad (a_0 d_1 - a_1 d_0)^2 (b_0 e_1 - b_1 e_0)^2 (c_0 f_1 - c_1 f_0)^2. \tag{9.13}$$

Proof. We have an explicit formula for $\mathrm{Det}(T)$ as a homogeneous polynomial of degree 4 in the eight tensor entries t_{ijk}. If we substitute $t_{ijk} = a_ib_jc_k+d_ie_jf_k$ and factor, then we obtain the above product of degree 12. $\square$

Corollary 9.18. *Let* $T \in \mathbb{R}^{2\times 2\times 2}$. *If* T *has real rank* ≤ 2, *then* $\mathrm{Det}(T) \geq 0$.

Proof. Our hypothesis says that T has a representation (9.11) over $\mathbb{R}$. The expression (9.13) for $\mathrm{Det}(T)$ is a square in $\mathbb{R}$. It is hence nonnegative. $\square$

Example 9.19. Let $i = \sqrt{-1}$. The following tensor has rank 2 in $\mathbb{C}^{2\times 2\times 2}$:

$$T \;=\; \frac{1}{2}\big((\mathbf{e}_1 + i\mathbf{e}_0)^{\otimes 3} \;+\; (\mathbf{e}_1 - i\mathbf{e}_0)^{\otimes 3}\big) \;=\; \mathbf{e}_1 \otimes \mathbf{e}_1 \otimes \mathbf{e}_1 \;-\; W.$$

Note that this tensor is real. By substituting its coefficients into the formula for the hyperdeterminant in Example 4.10, we find that $\mathrm{Det}(T) = -4 < 0$. Corollary 9.18 implies that the real rank of T is ≥ 3. Thus the tensor T has the property that its complex rank is strictly smaller than its real rank.

Exercise 10 states that the set of rank-2 tensors is Zariski dense in the space of $2 \times 2 \times 2$ tensors. This holds for any infinite field K. If $K = \mathbb{C}$ then they are also dense in the Euclidean topology. In fact, a tensor T has complex rank ≤ 2 if $\mathrm{Det}(T) \neq 0$. This property is specific to tensors of format $2\times 2\times 2$. For larger formats, the situation is much more complicated.

Note that the W-state has rank 3 and satisfies $\mathrm{Det}(W) = 0$. If T is a $2 \times 2 \times 2$ tensor that satisfies $\mathrm{Det}(T) = 0$ and all three flattenings of T have rank 2, then T always has complex rank 3. If exactly one flattening has rank 1, then T has complex rank 2. If two flattenings have rank 1, then all three have rank 1 and T has complex rank 1. If $K = \mathbb{R}$ then we must distinguish the two cases $\mathrm{Det}(T) > 0$ and $\mathrm{Det}(T) < 0$. In the former case, T has real rank 2. In the latter case, T has real rank 3.

Our discussion has the following interpretation in projective geometry. Tensors of rank 1 form the Segre threefold $X = \mathbb{P}^1 \times \mathbb{P}^1 \times \mathbb{P}^1$ in $\mathbb{P}^7$. By Exercise 10, the secant variety of X, described in Definition 9.20 below, fills $\mathbb{P}^7$. However, the tangential variety of X, which is the union of all tangent spaces to X, has dimension 6. It is the hypersurface $\{\mathrm{Det}(T) = 0\}$ in $\mathbb{P}^7$. The W-state is a point on this hyperdeterminant hypersurface. The line $\{\lambda \mathbf{e}_0 \otimes \mathbf{e}_0 \otimes \mathbf{e}_0 + W \,:\, \lambda \in K\}$ crosses that hypersurface transversally. If $K = \mathbb{R}$ then the real rank on that line depends only on the sign of λ.

To conclude, unlike in the case of matrices or rank-1 tensors:

- Tensors of rank at most r may form a nonclosed set.
- A real tensor can have smaller rank when viewed as a complex tensor.
- Real tensors of bounded real border rank form semialgebraic sets.

We have described rank-1 tensors as the Segre product of projective spaces. It is natural to ask for a geometric description of tensors of rank at most r.

Definition 9.20 (Secant variety)**.** Let X be any projective variety in $\mathbb{P}^n$. The *kth secant variety* of X is the closure of the set of k-secant planes to X:

$$\sigma_k(X) \quad := \quad \overline{\bigcup_{p_1,\ldots,p_k \in X} \langle p_1, \ldots, p_k \rangle}. \tag{9.14}$$

Note that $X = \sigma_1(X) \subset \sigma_2(X) \subset \cdots \subset \sigma_{\dim\langle X\rangle}(X) = \langle X\rangle$. These containments are strict until $\sigma_r(X)$ equals the linear span $\langle X\rangle$ of the variety X.

If X is the Segre variety, then the union in (9.14) is the set of tensors of rank $\leq r$. Its closure $\sigma_r(X)$ is the set of tensors of border rank $\leq r$. It is a major open problem to determine the prime ideal of $\sigma_r(X)$. This would provide a test for a tensor to have border rank r. The simplest equations for $\sigma_r(X)$ are the $(r+1)\times(r+1)$ minors of the various flattenings, as in (9.9). These have degree $r+1$. No polynomials of degree $\leq r$ vanish on $\sigma_r(X)$.

If X is the Veronese variety, then we obtain the notion of *symmetric rank*. A symmetric tensor T is in X if the following equivalent conditions hold:

(1) The rank of T as a tensor is 1.

(2) $T = \lambda \mathbf{v} \otimes \mathbf{v} \otimes \cdots \otimes \mathbf{v}$ for some vector $\mathbf{v}$ and scalar λ.

(3) T, as a polynomial, is a power of a linear form times a constant.

Given a symmetric tensor T, the *symmetric border rank* of T is the smallest positive integer r such that $T \in \sigma_r(X)$. The rank of T is a lower bound for the symmetric rank of T, and ditto for the border rank.

It was a longstanding question, known as Comon's conjecture, whether the rank of a symmetric tensor is always equal to its symmetric rank. It turns out that the answer is no. A counterexample was constructed by Shitov in [**48**]. The border rank analogue of Comon's conjecture remains open.

It is easy to prove that general tensors have high rank and high border rank, but it is extremely hard to find explicit examples. In particular, it is not known how to provide examples of complex $n \times n \times n$ tensors T of rank greater than $3n$. Explicit constructions of tensors of rank close to $3n$ can be found in [**2**] and of border rank above $2n$ in [**33**]. By Exercise 12, a general tensor in $K^{n\times n\times n}$ has border rank quadratic in n.

We next offer a case study on tensor ranks for the case in Figure 9.1.

Example 9.21 ($3\times 3\times 3$ tensors)**.** Fix an algebraically closed field K. Tensors of format $3\times 3\times 3$ are points in the projective space $\mathbb{P}^{26}$. The 6-dimensional Segre variety $X = \mathbb{P}^2 \times \mathbb{P}^2 \times \mathbb{P}^2$ consists of all tensors of rank 1. Tensors of border rank at most 2 form the secant variety $\sigma_2(X)$, which has dimension 13. Its ideal is generated by the 3×3 minors of the three flattenings. These flattenings are 3×9 matrices, like $[A \,|\, B \,|\, C]$ where $A = (t_{ij1})$, $B = (t_{ij2})$ and $C = (t_{ij3})$ are matrices obtained as the slices in our tensor. These 3×3 minors span a space of cubics that has dimension 222. With another computation one also verifies that $\sigma_2(X)$ has degree 783.

The variety $\sigma_3(X)$ of tensors of border rank at most 3 has dimension 20. Its ideal is generated by a collection of quartic polynomials, namely the entries of the 3×3 matrices $A \cdot \mathrm{adj}(B) \cdot C - C \cdot \mathrm{adj}(B) \cdot A$, where we allow all possible ways of slicing the tensor. What is the degree of this variety?

Finally, there is the variety $\sigma_4(X)$ of $3 \times 3 \times 3$ tensors of border rank ≤ 4. This is a hypersurface of degree 9 in $\mathbb{P}^{26}$. Its defining polynomial is known as the *Strassen invariant*. The Strassen invariant can be computed as

$$\det(B)^2 \cdot \det\bigl(A \cdot B^{-1}C \;-\; C \cdot B^{-1}A\bigr).$$

The expression has 9216 terms and is independent of the choice of slicing. The fifth secant variety $\sigma_5(X)$ is equal to $\mathbb{P}^{26}$. In other words, the set of tensors of rank ≤ 5 is dense in the space of all $3 \times 3 \times 3$ tensors.

We refer to the book by Landsberg [**32**] for further details and much more information on ideals defining the varieties of tensors of bounded rank.

We now restrict the rank stratification to the space of symmetric tensors.

Example 9.22 (Ternary cubics)**.** Symmetric $3 \times 3 \times 3$ tensors T are ternary cubics, that is, homogeneous polynomials of degree 3 in three variables. We regard them as points in $\mathbb{P}^9 = \mathbb{P}(\mathrm{Sym}_3(K^3))$. Their ranks coincide with their symmetric ranks, i.e. Comon's conjecture is true in this tiny case [**45**].

The three flattenings $[A \,|\, B \,|\, C]$ in Example 9.21 are now all equal. After removing redundant columns, this becomes a 3×6 matrix, known as the *Hankel matrix* or *catalecticant*. The ideal of 2×2 minors of the Hankel matrix is generated by the 27 binomial quadrics seen for $A = 3\Delta_2$ in Example 8.18. Its variety is the Veronese surface $X \simeq \mathbb{P}^2$ whose points in $\mathbb{P}^9$ are the cubics of rank 1. The secant variety $\sigma_2(X)$ has dimension 5, its points are cubics of border rank ≤ 2, and it is defined by the 3×3 minors of the Hankel matrix.

Finally, the variety $\sigma_3(X)$ of cubics of border rank ≤ 3 is a quartic hypersurface in $\mathbb{P}^9$. Its defining polynomial is the classical *Aronhold invariant*. This has 25 terms and can be obtained by specializing any of the entries of

$$A \cdot \mathrm{adj}(B) \cdot C \;-\; C \cdot \mathrm{adj}(B) \cdot A. \tag{9.15}$$

We have already discussed the distinction between complex rank and real rank. A further refinement of the latter is the notion of *nonnegative rank*. It is very important in applications, e.g. in statistics, where one deals with probabilities. A tensor $T = (t_{i_1 i_2 \cdots i_d})$ is called *nonnegative* if its entries

$t_{i_1 i_2 \cdots i_d}$ are all nonnegative. The *nonnegative rank* of a nonnegative tensor T is the minimal number r of nonnegative rank-1 tensors that sum to T. In general, the nonnegative rank is larger than the real rank, even for matrices.

9.3. Matrix Multiplication

The multiplication of two matrices is a bilinear operation. In this section we identify this operation with a very special tensor. We will use this connection to explain how tensors may be regarded as computational problems, tensor decompositions as algorithms, and tensor rank as a complexity measure.

Determining the rank of a tensor is an important computational problem in nonlinear algebra. In general one cannot hope for an efficient solution, as the problem is NP-hard [**26**]. However, special cases are of particular interest. The most well-known and important instance is the matrix multiplication tensor.

Let $\mathrm{Mat}_{a,b} \simeq K^{a\times b}$ be the space of $a \times b$ matrices over a field K. The operation of matrix multiplication is a bilinear map $\mathrm{Mat}_{a,b} \times \mathrm{Mat}_{b,c} \to \mathrm{Mat}_{a,c}$. Bilinear maps from a Cartesian product of vector spaces are in bijection with linear maps from the tensor product of those spaces; see Exercise 14. Hence, matrix multiplication is an element of the vector space

$$\mathrm{Hom}\big(\mathrm{Mat}_{a,b} \otimes \mathrm{Mat}_{b,c}\,,\, \mathrm{Mat}_{a,c}\big) \;\simeq\; \mathrm{Mat}^*_{a,b} \otimes \mathrm{Mat}^*_{b,c} \otimes \mathrm{Mat}_{a,c}.$$

This is a canonical isomorphism. We write $M_{a,b,c}$ for the *matrix multiplication tensor*. This third-order tensor is a special element of the tensor space on the right-hand side. To simplify notation we write $M_n := M_{n,n,n}$ for the tensor that represents the multiplication of two square matrices.

Let $\{\mathbf{e}_{ij}\}$, $\{\mathbf{f}_{jk}\}$ and $\{\mathbf{g}_{ik}\}$ be the standard bases of the spaces $\mathrm{Mat}^*_{a,b}$, $\mathrm{Mat}^*_{b,c}$ and $\mathrm{Mat}_{a,c}$. Thus $\mathbf{g}_{ik}$ is the $a \times c$ matrix whose entries are zero except for a 1 in row i and column k. The other two bases are dual to such matrix units. The matrix multiplication tensor has the following representation:

$$M_{a,b,c} \;=\; \sum_{i=1}^{a}\sum_{j=1}^{b}\sum_{k=1}^{c} \mathbf{e}_{ij} \otimes \mathbf{f}_{jk} \otimes \mathbf{g}_{ik}. \tag{9.16}$$

Another representation is suggested in Exercise 13.

Example 9.23. Consider the tensor M_2 that represents multiplication of 2×2 matrices. Fixing the ordered basis $(\mathbf{e}_{00}, \mathbf{e}_{01}, \mathbf{e}_{10}, \mathbf{e}_{11})$ for $\mathrm{Mat}_{2,2} \simeq K^4$, we can write M_2 explicitly as a $4\times 4\times 4$ tensor with entries in $\{0,1\}$. Among the 64 entries in this tensor, there are precisely 8 ones and 56 zeros.

The rank-1 decomposition of the tensor $M_{a,b,c}$ given in (9.16) can be interpreted as an *algorithm* for computing the product of the two matrices:

- To carry out matrix multiplication, one needs to add abc partial results labelled by (i,j,k) in $\{1,\ldots,a\} \times \{1,\ldots,b\} \times \{1,\ldots,c\}$.
- In step (i,j,k) one multiplication is performed. Namely, one multiplies the (i,j) entry of the first matrix by the (j,k) entry of the second matrix. The product is added to the (i,k) entry of the output matrix.

This is the familiar classical algorithm for multiplying two matrices. It performs $abc-1$ additions and abc multiplications. For $a=b=c=n$, its running time is $O(n^3)$. We note that the number of multiplications is *exactly* equal to the number of rank-1 tensors appearing in the decomposition.

What if we represent $M_{a,b,c}$ in a different way? Could it be that the number of multiplications we need is smaller than abc? Equivalently, is the rank of $M_{a,b,c}$ smaller than abc? Half a century ago, Volker Strassen set out on a quest to prove that this is not possible. He quickly realized that the case of arbitrary a, b and c is extremely hard and focused on the first nontrivial case of $a=b=c=2$. For that tensor, he discovered a most surprising formula:

$$\begin{aligned} M_2 = {} & (\mathbf{e}_{11}+\mathbf{e}_{22}) \otimes (\mathbf{f}_{11}+\mathbf{f}_{22}) \otimes (\mathbf{g}_{11}+\mathbf{g}_{22}) \\ & + (\mathbf{e}_{21}+\mathbf{e}_{22}) \otimes \mathbf{f}_{11} \otimes (\mathbf{g}_{21}-\mathbf{g}_{22}) \\ & + \mathbf{e}_{11} \otimes (\mathbf{f}_{12}-\mathbf{f}_{22}) \otimes (\mathbf{g}_{12}+\mathbf{g}_{22}) \\ & + \mathbf{e}_{22} \otimes (\mathbf{f}_{21}-\mathbf{f}_{11}) \otimes (\mathbf{g}_{11}+\mathbf{g}_{21}) \\ & + (\mathbf{e}_{11}+\mathbf{e}_{12}) \otimes \mathbf{f}_{22} \otimes (\mathbf{g}_{12}-\mathbf{g}_{11}) \\ & + (\mathbf{e}_{21}-\mathbf{e}_{11}) \otimes (\mathbf{f}_{11}+\mathbf{f}_{12}) \otimes \mathbf{g}_{22} \\ & + (\mathbf{e}_{12}-\mathbf{e}_{22}) \otimes (\mathbf{f}_{21}+\mathbf{f}_{22}) \otimes \mathbf{g}_{11}. \end{aligned} \tag{9.17}$$

Thus the rank of the matrix multiplication tensor M_2 is strictly less than $2^3=8$. In fact, the rank and border rank of M_2 are both exactly 7. The latter is a highly nontrivial statement. We are not aware of any easy proof. Even showing that the rank of M_2 is not equal to 6 is a challenging exercise.

Why would such a decomposition be interesting? It furnishes an algorithm for multiplying 2×2 matrices that adds *seven* partial results. We describe only the first two, as the reader can reconstruct the other five:

(1) Add the $(1,1)$ entry to the $(2,2)$ entry of the first matrix and multiply by the sum of the $(1,1)$ and $(2,2)$ entries of the second matrix. Retain this result in the $(1,1)$ and $(2,2)$ entries of the first partial result.

(2) Add the $(2,1)$ entry of the first matrix to the $(2,2)$ entry and multiply by the $(1,1)$ entry of the second matrix. Put the result in the $(2,1)$ entry and negated $(2,2)$ entry of the second partial result.

Computing each partial result requires only one multiplication. Although we have decreased the number of multiplications, we increased the number of additions (and subtractions) to 21. Why should this be exciting? The reason is that multiplication of 2×2 matrices is not our final aim.

We would like to multiply very large matrices. Consider two 512×512 matrices. How do we multiply them? We may regard our matrices as 2×2 matrices with entries that are 256×256 matrices and apply Strassen's algorithm! We would have to add a lot of 256×256 matrices, but we only need to perform *seven* multiplications of such matrices. Further, these multiplications may be done *recursively* by applying the same algorithm, reducing to multiplication of 128×128 matrices, etc. Anyone who has tried multiplying or adding very large matrices knows that it is beneficial to trade multiplication even for many additions. This is in fact a theorem: The complexity of the (optimal) algorithm for multiplying matrices is governed by the rank of M_n.

The asymptotics of these quantities is measured by the constant

$$\begin{aligned} \omega &= \inf\{\tau : \text{the complexity of multiplying two } n \times n \text{ matrices is } O(n^\tau)\} \\ &= \inf\{\tau : \text{rank of } M_n = O(n^\tau)\}. \end{aligned}$$

This quantity is known as the *exponent of matrix multiplication*. The naive algorithm shows that $\omega \leq 3$. However Strassen's algorithm, as described above, gives $\omega \leq \log_2 7$. As matrices are of size n^2, we also know that $\omega \geq 2$.

The central conjecture in this field says that the lower bound is attained:

Conjecture 9.24. *The constant ω is equal to* 2.

The conjecture would imply that it is not much harder to multiply very large matrices than to add them (or even output the result)! At this point we note that our story is really relevant to scientific computing. Strassen's algorithm is implemented and used in practice to multiply large matrices.

A careful reader might now have an idea of how to proceed with a proof of Conjecture 9.24. As Strassen looked at 2×2 matrices, we should focus on larger matrices, say 3×3. The disappointing fact is that despite many attempts, no one knows either the rank or the border rank of the $9 \times 9 \times 9$ tensor M_3. For the current best estimates we refer to [**34, 35, 49**].

For each fixed n, deciding whether the rank (resp. border rank) of M_n is $\leq r$ means deciding whether M_n belongs to the image (resp. closed image) of a particular polynomial map. Thus, the methods of Chapter 4 apply. However, as tensor spaces are high-dimensional, this process is impossible to carry out on a computer, even for $n = 3$. What one can use instead is representation theory, as described in Chapter 10. The optimal estimates for ω are beyond the scope of this book. Currently we know that $2 \leq \omega < 2.38$. It is fascinating that the upper bounds are based on the border rank and nonconstructive methods—one proves the existence of an algorithm without explicitly providing it.

In general, we lack methods to show that a tensor has high rank or high border rank. To prove that $\omega > 2$ we would need to show that the rank of the tensor $M_n \in \mathbb{C}^{n^2} \otimes \mathbb{C}^{n^2} \otimes \mathbb{C}^{n^2}$ grows superlinearly with the dimension n^2 of the space of matrices. However, we currently cannot even prove that any (explicit) given tensor has rank greater than $3n^2$. Some methods of obtaining bounds for the rank of the tensor will be discussed in Section 10.3.

Exercises

(1) Fix the quadratic form Q in (9.1). Compute all the maxima and minima of Q on the unit 2-sphere. Find all fixed points of the gradient map $\nabla Q : \mathbb{P}^2 \to \mathbb{P}^2$. How are these two questions related?

(2) Compute all fixed points of the map $\nabla B : \mathbb{P}^2 \times \mathbb{P}^1 \to \mathbb{P}^2 \times \mathbb{P}^1$ given by the bilinear form B in (9.4). What are the singular vectors?

(3) Consider the $2 \times 2 \times 2 \times 2$ tensor defined by the multilinear form $T = x_1y_1z_1w_1 + x_2y_2z_2w_2$. Determine all quadruples of singular vectors of T.

(4) For $n = 2, 3, 4$, pick random symmetric tensors of formats $n \times n \times n$ and $n \times n \times n \times n$ with entries in $\mathbb{R}$. Compute all eigenvectors of your tensors.

(5) Prove Theorem 9.6.

(6) Find an explicit real $3 \times 3 \times 3 \times 3$ tensor with precisely 13 *real* eigenvectors.

(7) Find the number of singular vector tuples for your tensors in Exercise 4.

(8) Compute the eigendiscriminants for symmetric tensors of formats 2×2, $2 \times 2 \times 2$ and $2 \times 2 \times 2 \times 2$. Write them explicitly as homogeneous polynomials in the entries of an unknown tensor of each format.

(9) Prove that the rank of the W-state equals 3. Hint: Show that the polynomial system $W = T$ described in Example 9.15 has no solution.

(10) Show that the Zariski closure of the set of tensors of rank 2 in $\mathbb{R}^2 \otimes \mathbb{R}^2 \otimes \mathbb{R}^2$ is the whole space. Hint: Use the Jacobian of the parametrization.

(11) Find the equations of the tangential variety to $\mathbb{P}^1 \times \mathbb{P}^1 \times \mathbb{P}^2$ in $\mathbb{P}^{11}$.

(12) Prove that in $\mathbb{C}^n \otimes \mathbb{C}^n \otimes \mathbb{C}^n$:
   (a) there exists a tensor of border rank at least $\frac{1}{3}n^2$;
   (b) every tensor has rank at most n^2.

(13) Linear maps from V_1 to V_2 are identified with tensors in $V_1^* \otimes V_2$. The composition of linear maps in $V_1 \to V_2 \to V_3$ may be regarded as a map

$$(V_1^* \otimes V_2) \times (V_2^* \otimes V_3) \to (V_1^* \otimes V_3).$$

Hence, the matrix multiplication tensor $M_{\dim V_1, \dim V_2, \dim V_3}$ belongs to

$$(V_1^* \otimes V_2)^* \otimes (V_2^* \otimes V_3)^* \otimes (V_1^* \otimes V_3) = (V_1 \otimes V_1^*) \otimes (V_2 \otimes V_2^*) \otimes (V_3 \otimes V_3^*).$$

   (a) Explain why $M_{\dim V_1, \dim V_2, \dim V_3}$ is an element of the space on the right. Do not refer to the basis, only to the linear maps $V_i \to V_i$. Hint: The identity map is a distinguished element in $V_i^* \otimes V_i$.
   (b) Provide a natural isomorphism $\mathrm{Mat}_{a,b}^* \simeq \mathrm{Mat}_{b,a}$.
   (c) The tensor $M_{a,b,c}$ can also be identified with a trilinear map

$$\mathrm{Mat}_{a,b} \times \mathrm{Mat}_{b,c} \times \mathrm{Mat}_{c,a} \to K.$$

   Describe this trilinear map without referring to coordinates.

(14) Show that the following four vector spaces are naturally isomorphic:
   - $V_1 \otimes \cdots \otimes V_k$;
   - k-linear maps $V_1^* \times \cdots \times V_k^* \to \mathbb{C}$;
   - $(k-1)$-linear maps $V_1^* \times \cdots \times V_{k-1}^* \to V_k$;
   - linear maps $V_1^* \otimes \cdots \otimes V_{k-1}^* \to V_k$.

(15) The matrix multiplication tensor $M_{2,2,3}$ has format $4 \times 6 \times 6$. Write this tensor explicitly in coordinates. What do you know about its rank?

(16) Expand the Aronhold invariant and the Strassen invariant in monomials.

(17) Compute the ideal of the secant variety $\sigma_2(X)$ where $X = \mathbb{P}^1 \times \mathbb{P}^2 \times \mathbb{P}^2$ is the Segre variety in $\mathbb{P}^{17}$. Can you answer the same question for $\sigma_3(X)$?

(18) How can you test whether a complex $4 \times 4 \times 4$ tensor has rank ≤ 4?

(19) How many singular vector triples does a general $3 \times 4 \times 5$ tensor have?

Chapter 10

Representation Theory

"*Reality favors symmetry*", Jorge Luis Borges

Symmetry is the key to many applications and computations. While this is true across the mathematical sciences, it is especially pertinent in nonlinear algebra. In its most basic form, symmetry is expressed via the action of a group acting linearly on a vector space. The study of such actions is the subject of *representation theory*. For instance, the symmetric group on three letters acts on the plane by the rotations and reflections that fix a regular triangle with centroid at the origin. The map that takes each group element to its associated matrix is the representation of the group. The matrix representations of the groups we study here can be simultaneously block-diagonalized. The blocks are irreducible representations. Identifying these blocks is tantamount to exploiting symmetry in explicit computations. Our objective in this chapter is to offer a first glimpse of representation theory.

10.1. Groups, Representations and Characters

The most important groups we study in this chapter are the following:

- $\mathrm{GL}(V) = \mathrm{GL}(\dim V)$, the group of linear automorphisms of a finite-dimensional vector space V. This group has the structure of an algebraic variety, given by Exercise 8 in Chapter 2.
- $\mathrm{SL}(V) = \mathrm{SL}(\dim V)$, the group of linear automorphisms of V that preserve volume and orientation. This is the algebraic variety defined by the polynomial equation $\det A = 1$.

- S_n, the group of permutations of a set with n elements. This is an algebraic variety consisting of $n!$ distinct points in $\mathrm{GL}(n)$, namely the $n \times n$ permutation matrices.

The groups that we consider have two structures: of an abstract group and of an algebraic variety. We note that basic group operations, such as taking the inverse or acting by group elements, are in fact morphisms of algebraic varieties. We call such groups *algebraic*. Thus, we restrict our attention to algebraic groups and morphisms between them. These are maps that are both group morphisms and morphisms of algebraic varieties. In what follows, we work over an algebraically closed field K of characteristic zero.

In general, the following strategy for studying a mathematical object can be very powerful. *We examine all maps from (resp. to) this object to (resp. from) another basic object that we know well.* This general approach can be seen as the motivation for studying homotopy, homology or the theory of embeddings. For groups, we obtain the following central definition.

Definition 10.1. *A representation* of a group G is a morphism $G \to \mathrm{GL}(V)$. We will always assume that V is finite-dimensional.

Given a representation $\rho : G \to \mathrm{GL}(V)$, every group element $g \in G$ induces a linear map $\rho(g) : V \to V$. It is useful to think of a representation as a map $G \times V \to V$ with the notation

$$gv \;:=\; \rho(g)(v) \;\in\; V.$$

The following compatibilities hold for all $\lambda \in K$, $v, v_1, v_2 \in V$ and $g, g_1, g_2 \in G$:

$$(g_1 g_2)v \;=\; g_1(g_2 v) \quad \text{and} \quad g(\lambda v_1 + v_2) \;=\; \lambda g v_1 + g v_2. \tag{10.1}$$

If these hold, then we say that the group G *acts on* the vector space V. Often, the action is clear from the context, and we just call V a *representation* of G.

Example 10.2. Consider the regular triangle in $\mathbb{R}^2$ with vertices $(0,1)$, $(\frac{\sqrt{3}}{2}, -\frac{1}{2})$ and $(-\frac{\sqrt{3}}{2}, -\frac{1}{2})$. The group S_3 permutes the vertices of this triangle. This induces a representation $S_3 \to \mathrm{GL}(2)$. Explicitly, we have

$$(12) \mapsto \begin{pmatrix} \frac{1}{2} & \frac{\sqrt{3}}{2} \\ \frac{\sqrt{3}}{2} & -\frac{1}{2} \end{pmatrix}, \qquad (23) \mapsto \begin{pmatrix} -1 & 0 \\ 0 & 1 \end{pmatrix}.$$

These two transpositions generate S_3. This example generalizes to higher dimensions by the action of S_n on a regular $(n-1)$-dimensional simplex.

Example 10.3. The groups $\mathrm{GL}(n)$ and $\mathrm{SL}(n)$ act (by linear changes of coordinates) on the space $V = K[x_1, \ldots, x_n]_k \simeq K^{\binom{n+k-1}{k}}$ of homogeneous polynomials of degree k in n variables. Using the monomial basis on V, the representation ρ maps a small matrix of size $n \times n$ to a large matrix with rows and columns indexed by monomials of degree k. The entries of the large matrix are homogeneous polynomials of degree k in the entries of the small matrix. We recommend working this out for $n = k = 2$. This representation ρ of $\mathrm{GL}(n)$ plays an important role in *invariant theory*, the topic of Chapter 11.

The polynomial representations in Example 10.3 can be restricted to any subgroup G of $\mathrm{GL}(n)$, and this will turn V into a representation of G.

Example 10.4. Let $n = 2$ and $k = 3$, and consider the group of rotation matrices

$$G = \left\{ \begin{pmatrix} \sin(\theta) & \cos(\theta) \\ -\cos(\theta) & \sin(\theta) \end{pmatrix} : \theta \in \mathbb{R} \right\}.$$

This is an algebraic group, isomorphic to the circle $\{s^2 + c^2 = 1\}$ in $\mathbb{R}^2$. For some computations, it is advantageous to use the rational parametrization

$$s = \sin(\theta) = \frac{1-t^2}{1+t^2}, \quad c = \cos(\theta) = \frac{2t}{1+t^2}.$$

The representation of G on the space of binary cubics $V = \mathbb{R}[x_1, x_2]_3$ is

$$G \to \mathrm{GL}(V), \begin{pmatrix} s & c \\ -c & s \end{pmatrix} \mapsto \begin{pmatrix} s^3 & -cs^2 & c^2s & -c^3 \\ 3cs^2 & s^3-2c^2s & c^3-2cs^2 & 3c^2s \\ 3c^2s & 2cs^2-c^3 & s^3-2c^2s & -3cs^2 \\ c^3 & c^2s & cs^2 & s^3 \end{pmatrix}. \tag{10.2}$$

This is the formula for the usual monomial basis of V. Note that the determinant of the 4×4 matrix on the right is equal to $(s^2 + c^2)^6 = 1$. Since we want our field K to be algebraically closed, from now on we replace $\mathbb{R}$ by $\mathbb{C}$.

A *morphism* f between representations $\rho_1 : G \to \mathrm{GL}(V_1)$ and $\rho_2 : G \to \mathrm{GL}(V_2)$ is a linear map $f : V_1 \to V_2$ that is compatible with the group action:

$$f(\rho_1(g)(v)) = \rho_2(g)(f(v)) \quad \text{for all } g \in G \text{ and } v \in V_1.$$

This can also be written as $f(gv) = gf(v)$. The kernel and cokernel of f are also representations of G. This is the topic of Exercise 3.

Our first aim is to describe the basic building blocks of representations. A *subrepresentation* of a representation V of a group G is a linear subspace $W \subset V$ such that the action of G restricts to W, i.e.

$$gw \in W \text{ for all } w \in W \text{ and } g \in G.$$

For any representation V, the subspaces $\{0\}$ and V are subrepresentations.

Definition 10.5. A representation V is called *irreducible* if and only if 0 and V are its only subrepresentations.

We next show that the only morphism between two nonisomorphic irreducible representations is trivial.

Lemma 10.6 (Schur's Lemma). *Let V_1 and V_2 be irreducible representations of a group G. If $f : V_1 \to V_2$ is a morphism of representations, then either f is an isomorphism or $f = 0$. Further, any two such isomorphisms between V_1 and V_2 differ by a scalar multiple.*

Proof. Both $\ker f$ and $\operatorname{im} f$ are representations. As V_1 is irreducible, either $\ker f = V_1$ or f is injective. In the latter case, $\operatorname{im} f \simeq V_1$ is a subrepresentation of V_2, and hence f is also surjective, i.e. it is an isomorphism. For the last part, consider two isomorphisms f_1 and f_2. We may assume that f_1 is the identity on V_1. If v is the eigenvector of f_2 with eigenvalue $\lambda \in K$, then

$$f_2(v) = \lambda v = \lambda f_1(v).$$

Consider the morphism of representations $f := f_2 - \lambda f_1$. Clearly, $v \in \ker f$. Hence, by the first part, $f_2 - \lambda f_1$ is the zero map, and so $f_2 = \lambda f_1$. □

In the following theorem we assume that the group G is finite.

Theorem 10.7 (Maschke's Theorem). *Let V be a representation of a finite group G. There exists a direct sum decomposition*

$$V = \bigoplus V_i$$

where each V_i is an irreducible representation of G.

Remark 10.8. We recall our assumption that the field has characteristic zero and is algebraically closed. This makes representation theory well behaved. If the characteristic is finite, the situation is much more complicated.

Proof of Theorem 10.7. By induction on the dimension, it is enough to prove the following statement: If W is a subrepresentation of V, then there exists a subrepresentation W' such that $V = W \oplus W'$.

We fix a representation $\rho : G \to \mathrm{GL}(V)$ where G is finite. Let $\pi : V \to W$ be any linear (surjective) projection. We define a linear map $\tilde{\pi} : V \to W$ by

$$\tilde{\pi} = \frac{1}{|G|} \sum_{g \in G} \rho(g)_{|W} \circ \pi \circ \rho(g)^{-1}.$$

We note that $\tilde{\pi}$ is a morphism of representations, and it is also a surjective projection from V to W. Hence, $V = W \oplus \ker \tilde{\pi}$. □

Remark 10.9. The existence of a decomposition into irreducible components holds not only for finite groups. It also holds for $\mathrm{GL}(n)$, $\mathrm{SL}(n)$ and $\mathrm{SO}(n)$. A proof similar to the one above is known as the *unitarian trick*. It was introduced by Hurwitz and generalized by Weyl.

A representation of an arbitrary group that allows such a decomposition is called *semi-simple* or *completely reducible*. If all representations of G are completely reducible, then the group G is called *reductive*. This property will be essential in our discussion of invariant theory in Chapter 11. The group G being reductive is equivalent to the existence of a Reynolds operator as in Lemma 11.3. Remark 10.9 says that $\mathrm{GL}(n)$, $\mathrm{SL}(n)$ and $\mathrm{SO}(n)$ are reductive.

Example 10.10. Let $K = \mathbb{C}$ and consider the representation $V = \mathbb{C}^4$ of the circle group $G = \mathrm{SO}(2)$ in Example 10.4. The 4×4 matrix in (10.2) equals

$$M \cdot D(\theta) \cdot M^{-1},$$

where M is the matrix whose columns are the eigenvectors, namely

$$M \quad = \quad \begin{pmatrix} 1 & 1 & 1 & 1 \\ 3i & -3i & -i & i \\ 3 & -3 & 1 & 1 \\ -i & i & -i & i \end{pmatrix} \qquad \text{with } i = \sqrt{-1},$$

and $D(\theta)$ is the diagonal matrix whose entries are the eigenvalues. We have

$$D(\theta) \;=\; \mathrm{diag}\big(\, i \cdot \exp(3i\theta),\, (-i) \cdot \exp(-3i\theta),\, i \cdot \exp(-i\theta),\, (-i) \cdot \exp(i\theta) \,\big).$$

This shows that V equals $\mathbb{C}^1 \oplus \mathbb{C}^1 \oplus \mathbb{C}^1 \oplus \mathbb{C}^1$ as a representation of G.

Example 10.11. The additive group $G = (K, +)$ is not reductive. Indeed, let us consider the following representation:

$$G \ni a \mapsto \begin{pmatrix} 1 & a \\ 0 & 1 \end{pmatrix} \in \mathrm{GL}(2).$$

The subspace of K^2 spanned by the first basis vector is invariant under the action. However, it does not allow an invariant complement (see Exercise 13).

The decomposition into irreducible representations in Maschke's Theorem is not unique. The following example makes this clear.

Example 10.12. Any group G acts on any vector space V trivially by $gv = v$. Every subspace of V is a subrepresentation. The irreducible subrepresentations are the 1-dimensional subspaces of V. Hence, any decomposition into 1-dimensional subspaces $V = K^1 \oplus K^1 \oplus \cdots \oplus K^1$ is a decomposition into irreducible representations. There is no distinguished one.

As we will see, the reason for nonuniqueness is the fact that distinct V_i's appearing in the decomposition may be isomorphic. Let us therefore group the isomorphic V_i's together. Then we obtain the direct sum

$$V = \bigoplus_j V_j^{\oplus a_j}, \tag{10.3}$$

where $V_{j_0} \simeq V_{j_1}$ if and only if $j_0 = j_1$. The subrepresentations $V_j^{\oplus a_j}$ are called the *isotypic components* of V. The number a_j is the *multiplicity* of the irreducible representation V_j in V.

Corollary 10.13 (of Schur's Lemma). *The isotypic components and multiplicities of a semi-simple representation V are well-defined, i.e. they do not depend on the choice of the decomposition into irreducible representations.*

Proof. Consider two decompositions of a semi-simple representation:

$$V = \bigoplus_j V_j^{\oplus a_j} = \bigoplus_k V_k^{\oplus b_k}.$$

Allowing a_j and b_k to be equal to zero, we may assume that all irreducible representations occur and that the indexing in both sums $\bigoplus$ is the same.

First we prove that for a given irreducible representation V_i we have $a_i = b_i$. The restriction of the identity gives us an injective map

$$m : V_i^{\oplus a_i} \to \bigoplus_k V_k^{\oplus b_k}.$$

We claim that the composition of m with the projection

$$\pi_s : \bigoplus_k V_k^{\oplus b_k} \to V_s^{\oplus b_s}$$

equals zero, unless $s = i$. Indeed, if the map $V_i^{\oplus a_i} \to V_s^{\oplus b_s}$ is nonzero, it induces a nonzero map between some V_i and V_s. By Schur's Lemma such a map may exist only if $i = s$. Hence, $\operatorname{im}(m) \subset V_i^{\oplus b_i}$. In particular, by a dimension count, $a_i \leq b_i$. Analogously, $b_i \leq a_i$, i.e. the multiplicities do not depend on the decomposition. Further, the composition $\pi_s \circ m$ is the identity if $s = i$ and is zero if $s \neq i$. It follows that $\operatorname{im}(m) = V_i^{\oplus b_i}$. Thus, the isotypic components are mapped to (the same) isotypic components. □

Our next aim is to understand the irreducible representations of a given group G. The following definition provides us with the most important tool.

Definition 10.14 (Character). Let $\rho : G \to \mathrm{GL}(V)$ be a representation. The *character* $\chi_\rho = \chi_V$ of ρ is the function $G \to K$ that is obtained by composing the representation ρ with the trace function Tr on square matrices:

$$\chi_\rho(g) = \mathrm{Tr}(\rho(g)).$$

Recall that the *trace* of a matrix is the sum of its diagonal elements, or the sum of its eigenvalues. The character χ_ρ is an invariant of a representation ρ.

See Remark 10.16 for the relationship to the characters in Chapter 8. Properties of the trace function imply the following facts about characters:

- If $V = \bigoplus V_i$ then $\chi_V = \sum \chi_{V_i}$.
- If $g_1, g_2 \in G$ are conjugate, then $\chi(g_1) = \chi(g_2)$ for any character χ.
- If V_1 and V_2 are representations with characters χ_1 and χ_2, then their tensor product $V_1 \otimes V_2$ is a representation with character $\chi_1\chi_2$.
- We have $\chi_V(e) = \dim V$, where $e \in G$ is the identity element.

We now set $K = \mathbb{C}$. For a finite group G, we define the following scalar product on the finite-dimensional vector space $\mathbb{C}^G$ of functions from G to $\mathbb{C}$:

$$\langle \chi_1, \chi_2 \rangle \; := \; \frac{1}{|G|} \sum_{g \in G} \chi_1(g)\overline{\chi_2(g)}. \tag{10.4}$$

We learn from Serre's book [**46**, Chapter 2] that the characters of all irreducible representations of G are orthonormal with respect to this scalar product. In particular, the characters of irreducible representations are linearly independent elements in $\mathbb{C}^G$. Hence, we can find the multiplicities a_j in the isotypic decomposition $V = \bigoplus_j V_j^{a_j}$ by decomposing the character:

$$\chi_V \; = \; \textstyle\sum_j a_j \chi_j.$$

For any finite group G there are finitely many irreducible representations. The sum of the squares of their dimensions equals the order of the group [**46**, §2.5, Corollary 2]. A *class function* is a function $G \to \mathbb{C}$ that is constant on conjugacy classes. The characters of the irreducible representations form a basis for the space of class functions on G. One represents these characters in a table, called the *character table.* This makes the decomposition of an arbitrary representation easy, if we know its character. The character table is a square matrix, since the number of conjugacy classes (row labels) equals the number of irreducible representations (column labels).

Example 10.15. Consider the group S_3 of permutations of three elements. There are three conjugacy classes: the class of the identity (with one element), the class of a 3-cycle (with two elements), and the class of any transposition (with three elements). Hence, there are three irreducible representations. The first is the trivial representation $gv = v$, the second is the sign representation $gv = (\operatorname{sgn} g)v$, and the third is the 2-dimensional representation, given by the symmetries of a regular triangle. Each column in the table below represents a function $S_3 \to \mathbb{C}$ that is constant on conjugacy classes. We present the character table for the symmetric group S_3:

	Trivial representation	Sign repr.	2-dimensional repr.
Identity	1	1	2
Cycles (ijk)	1	1	-1
Transpositions (ij)	1	-1	0

The reader should check that these functions are orthonormal with respect to the inner product (10.4). In fact, one builds the character table of a finite group by exploiting the orthonormality of the columns. In this manner, one obtains the 5×5 character table for S_4 and the 7×7 character table for S_5.

10.2. Invertible Matrices and Permutations

In this section we study the representations of the groups $\mathrm{GL}(n)$ and $\mathrm{SL}(n)$. We cannot represent their characters by tables as there are infinitely many conjugacy classes. However, we can represent each character χ by its values on the Zariski dense subset of diagonalizable matrices. Hence, we fix the algebraic torus $T = (K^*)^n \subset \mathrm{GL}(n)$ of diagonal matrices, and we restrict the character χ to T. As χ is constant on any conjugacy class and any diagonalizable matrix is conjugate to an element of T, the function $\chi|_T$ characterizes χ. Therefore, given any representation W of $\mathrm{GL}(n)$, we restrict the group and regard W as a representation of T. By Exercise 1 and Corollary 10.13 we know that, as a representation of T, the space W decomposes into irreducible 1-dimensional representations with multiplicities. By Exercise 1 from Chapter 8, the character of each such 1-dimensional representation is a morphism $\mathbf{t} \mapsto \mathbf{t}^{\mathbf{b}}$ from T to K^*. Hence, it may be identified with $\mathbf{b} \in \mathbb{Z}^n$.

In conclusion, we obtain the decomposition into 1-dimensional spaces

$$W = \bigoplus_{\mathbf{b} \in \mathbb{Z}^n} W_{\mathbf{b}}^{a_{\mathbf{b}}}, \tag{10.5}$$

where $W_{\mathbf{b}}$ is the span of a nonzero vector $w \in W$ that is mapped to $\mathbf{t}^{\mathbf{b}} w$ by the torus element $\mathbf{t} = (t_1, \ldots, t_n)$. Thus w is a common eigenvector of all matrices in the representation of T. The isotypic component $W_{\mathbf{b}}^{a_{\mathbf{b}}}$ for the T-action is the span of all such vectors w for fixed $\mathbf{b}$. These $W_{\mathbf{b}}^{a_{\mathbf{b}}}$ are called *weight spaces*. The characters $\mathbf{b}$ of T for which $a_{\mathbf{b}} \neq 0$ are called *weights*.

Remark 10.16. Let T be the torus of diagonal matrices $\mathbf{t} = \mathrm{diag}(t_1, \ldots, t_n)$ in $\mathrm{GL}(n)$. If χ is a character of $\mathrm{GL}(n)$, then its restriction to T is the function $\chi|_T : T \to K$, $\mathbf{t} \mapsto \mathrm{Tr}(\rho(\mathbf{t}))$. Here Tr denotes the trace of a (large) square matrix. The restricted character $\chi|_T$ equals

$$\chi|_T(\mathbf{t}) = \sum_{\mathbf{b} \in \mathbb{Z}^n} a_{\mathbf{b}} \mathbf{t}^{\mathbf{b}}.$$

This Laurent polynomial in $t_1, \ldots, t_n$ is invariant under permutation of its n unknowns. This means that it is a symmetric function.

Example 10.17. Following Example 10.3, we consider the action of $\mathrm{GL}(n)$ on the space of homogeneous polynomials in n variables of degree k. Let χ be its character. Then $\chi|_T$ is the *complete symmetric polynomial* of degree k, i.e. $\chi|_T(\mathbf{t})$ is the sum of all monomials $\mathbf{t}^{\mathbf{a}}$ where $\mathbf{a} \in \mathbb{N}^n$ and $|\mathbf{a}| = k$.

Example 10.18. The group $\mathrm{GL}(n)$ acts naturally on the kth exterior power $V = \bigwedge^k K^n$. Write ρ for this representation and χ for its character. We fix the standard basis for V given by $\{e_{i_1} \wedge \cdots \wedge e_{i_k} : 1 \leq i_1 < \cdots < i_k \leq n\}$. The image $\rho(g)$ of an $n \times n$ matrix $g = (g_{ij})$ is the *kth compound matrix*, or *kth exterior power*, whose entries are the $k \times k$ minors of g. We note that the determinant of $\rho(g)$ equals $\det(g)^{\binom{n-1}{k-1}}$. The restricted character $\chi|_T(\mathbf{t})$ is the kth elementary symmetric polynomial in $t_1, \ldots, t_n$.

For a concrete example, let $k = 2$. Then $\rho(g)$ is an $\binom{n}{2} \times \binom{n}{2}$ matrix. Its rows and columns are labeled by ordered pairs from $\{1, 2, \ldots, n\}$. The entry in row $(i < j)$ and column $(k < l)$ equals $g_{ik}g_{jl} - g_{il}g_{jk}$. We have $\det(\rho(g)) = \det(g)^{n-1}$ and $\chi|_T(\mathbf{t}) = \sum_{i<j} t_i t_j$. For $k = 1$ we have $\rho(g) = g$, so $\chi|_T(\mathbf{t}) = t_1 + t_2 + \cdots + t_n$. For $k = n$, we get the 1-dimensional representation where $\rho(g)$ is the 1×1 matrix with entry $\det(g)$, so we have $\chi|_T(\mathbf{t}) = t_1 t_2 \cdots t_n$. This representation is trivial when restricted to $\mathrm{SL}(n)$.

Let ρ be any representation of $\mathrm{GL}(n)$. We fix the lexicographic order on the set of weights $\mathbf{b}$ that occur in ρ. Of particular importance is the *highest weight*. The corresponding eigenvectors $w \in W_{\mathbf{b}}$ in (10.5) are called *highest weight vectors*. They span the *highest weight space*. In Example 10.17, the highest weight is $(k, 0, \ldots, 0) \in \mathbb{Z}^n$, and a highest weight vector is the monomial x_1^k. In Example 10.18, the highest weight is $(1, \ldots, 1, 0, \ldots, 0)$, and a highest weight vector is $e_1 \wedge \cdots \wedge e_k$. In both cases, the highest weight space is 1-dimensional. We note that the highest weight vector does not depend on n, provided it exists (e.g. $n \geq k$ in the exterior power case).

Example 10.19 (Adjoint representation)**.** The space of $n \times n$ matrices M is a representation of $\mathrm{GL}(n)$ under the action by conjugation: $\rho(g)(M) := gMg^{-1}$. This is called the *adjoint representation* of $\mathrm{GL}(n)$. For $\mathrm{SL}(n)$ one considers the same action but on the space of traceless matrices. This is the *adjoint representation* of $\mathrm{SL}(n)$. Intersecting the torus of diagonal matrices in $\mathrm{GL}(n)$ with $\mathrm{SL}(n)$, we obtain an $(n-1)$-dimensional torus, with $t_1 \cdots t_n = 1$. We use the convention $t_n = t_1^{-1} \cdots t_{n-1}^{-1}$. The weights of the adjoint representation are t_i/t_j. These are known as *roots*. The highest weight t_1/t_n is represented by $(2, 1, \ldots, 1) \in \mathbb{Z}^{n-1}$. For $i \neq j$ the weight space corresponding to t_i/t_j is 1-dimensional and spanned by a matrix with one nonzero (i, j) entry. In particular, the highest weight space is 1-dimensional. Further, all diagonal matrices are invariant with respect to the torus action contributing to an n-dimensional space of weight $0 \in \mathbb{Z}^{n-1}$.

Theorem 10.20. *Every irreducible representation of* $\mathrm{SL}(V)$ *is determined (up to isomorphism) by its highest weight, and the highest weight space is* 1*-dimensional. A weight* $(a_1, \ldots, a_{n-1}) \in \mathbb{Z}^{n-1}$ *is the highest weight for some irreducible representation if and only if* $a_1 \geq a_2 \geq \cdots \geq a_{n-1} \geq 0$.

Proof. For the proof we refer to [**21**, Chapter 15]. □

Here is a tool for building irreducible representations from highest weights:

Definition 10.21. A *Young diagram* with k rows is a nonincreasing sequence of k positive integers. This is also known as a *partition*, and it is usually presented in the following graphical form, e.g. for a sequence $(2, 1, 1)$:

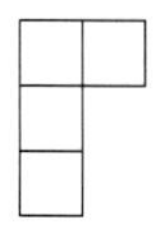

This Young diagram encodes the weight $(2, 1, 1)$. For $\mathrm{SL}(4)$ it represents the adjoint representation. For $\mathrm{SL}(5)$, this Young diagram does *not* represent the adjoint representation, but the highest weight in both cases is the same.

Theorem 10.20 says that the irreducible representations of $\mathrm{SL}(n)$ are in bijection with Young diagrams having at most $n - 1$ rows. Representations of $\mathrm{GL}(n)$ are similar; first, every irreducible representation V of $\mathrm{GL}(n)$ is also an irreducible representation of $\mathrm{SL}(n)$, so it has a Young diagram λ.

However, different representations of $\mathrm{GL}(n)$ give the same representation of $\mathrm{SL}(n)$ if they differ by a power of the determinant. Precisely, consider a representation $\rho : \mathrm{SL}(n) \to \mathrm{GL}(V)$ with associated Young diagram λ. Then we have the following representations of $\mathrm{GL}(n)$ for any $a \in \mathbb{Z}$:

$$\rho_a(g) \ := \ (\det g)^a \cdot (\sqrt[n]{\det g})^{|\lambda|} \cdot \rho\Big(\frac{1}{\sqrt[n]{\det g}} \cdot g\Big).$$

Here, the argument of ρ is in $\mathrm{SL}(n)$ and $|\lambda|$ is the number of boxes in λ. The irreducible representations of $\mathrm{GL}(n)$ are in bijection with pairs of a Young diagram with at most $n - 1$ rows and an integer $a \in \mathbb{Z}$. The 1-dimensional representation $g \mapsto \det(g)$ of $\mathrm{GL}(n)$ corresponds to $a = 1$ and the empty Young diagram. Equivalently, it may be represented by a Young diagram with one column and n rows. Thus, for $a \geq 0$, the representation ρ_a is often associated to a Young diagram λ' obtained by extending λ with a columns of height n. For a vector space U, the representation of $\mathrm{GL}(U)$ corresponding to a Young diagram λ with at most n rows is denoted by $S^\lambda(U)$.

Given a Young diagram λ, we write χ_λ for the character of the irreducible representation $S^\lambda(U)$. This is a symmetric polynomial in $\mathbf{t} = (t_1, \ldots, t_n)$, known as the *Schur polynomial* of λ. Schur polynomials include the complete symmetric polynomials in Example 10.17, for $\lambda = (n)$, and the elementary symmetric polynomials in Example 10.18, for $\lambda = (1, 1, \ldots, 1)$.

The next result gives an explicit formula for the Schur polynomials.

Proposition 10.22. *The Schur polynomial for λ is the following ratio of $n \times n$ determinants. If λ has less than n rows, then we extend it by zeros:*

$$\chi_\lambda(\mathbf{t}) \quad = \quad \frac{\det\big(t_i^{\lambda_j + n - j} \big)_{1 \le i,j \le n}}{\det\big(t_i^{n-j} \big)_{1 \le i,j \le n}}.$$

We can find the decomposition (10.3) of a representation V into irreducibles by writing the character χ_V as a linear combination of Schur polynomials χ_λ with nonnegative integer coefficients a_j. These coefficients are the multiplicities. This expression is unique because the Schur polynomials form a $\mathbb{Z}$-linear basis for the ring of symmetric polynomials in n variables.

Example 10.23. Let $n = 3$. The Schur polynomial for $\lambda = (\lambda_1, \lambda_2, \lambda_3)$ is

$$\chi_\lambda(\mathbf{t}) \quad = \quad \frac{1}{(t_1 - t_2)(t_1 - t_3)(t_2 - t_3)} \cdot \det \begin{pmatrix} t_1^{\lambda_1+2} & t_1^{\lambda_2+1} & t_1^{\lambda_3} \\ t_2^{\lambda_1+2} & t_2^{\lambda_2+1} & t_2^{\lambda_3} \\ t_3^{\lambda_1+2} & t_3^{\lambda_2+1} & t_3^{\lambda_3} \end{pmatrix}.$$

From this, we compute the three Schur polynomials of degree $|\lambda| = 3$:

$$\begin{aligned} \chi_{(3,0,0)} &= t_1^3 + t_1^2 t_2 + t_1 t_2^2 + t_2^3 + t_1^2 t_3 + t_1 t_2 t_3 + t_2^2 t_3 + t_1 t_3^2 + t_2 t_3^2 + t_3^3, \\ \chi_{(2,1,0)} &= (t_1 + t_2)(t_1 + t_3)(t_2 + t_3), \\ \chi_{(1,1,1)} &= t_1 t_2 t_3. \end{aligned}$$

The action of $\mathrm{GL}(3)$ on $U = K^3$ induces an action on the space $U^{\otimes 3} \simeq K^{27}$ of $3 \times 3 \times 3$ tensors. As characters are multiplicative under tensor product,

$$\chi_{U^{\otimes 3}} \quad = \quad (t_1 + t_2 + t_3)^3 \quad = \quad \chi_{(3,0,0)} \; + \; 2 \cdot \chi_{(2,1,0)} \; + \; \chi_{(1,1,1)}. \tag{10.6}$$

From this decomposition into Schur polynomials, we obtain the following decomposition of the triple tensor product into irreducible representations:

$$U^{\otimes 3} \quad = \quad S^{(3)}(U) \; \oplus \; \big(S^{(2,1)}(U) \oplus S^{(2,1)}(U) \big) \; \oplus \; S^{(1,1,1)}(U). \tag{10.7}$$

The first summand is the space of symmetric tensors, the last summand is the space of antisymmetric tensors, and the middle summand consists of two copies of the adjoint representation of $\mathrm{SL}(U)$, seen in Example 10.19.

The irreducible representations $S^\lambda(U)$ of $\mathrm{SL}(U)$ give rise to nice projective varieties. The group $\mathrm{SL}(U)$ acts also on the projective space $\mathbb{P}(S^\lambda(U))$. The latter action has a unique closed orbit, namely the orbit of the highest weight vector. Here are two examples of such *highest weight orbits*:

(1) Consider the highest weight vector $x_1^k = [e_1 \cdots e_1]$ in $S^k(U)$. Its orbit in $\mathbb{P}(S^k(U))$ is the kth Veronese embedding of $\mathbb{P}(U)$.

(2) The orbit of $[e_1 \wedge \cdots \wedge e_k] \in \mathbb{P}(\bigwedge^k(U))$ is the Grassmannian $G(k, U)$ in its Plücker embedding. Here $\lambda = (1, \ldots, 1)$ as in Example 10.18.

The study of highest weight orbits provides us with a unified approach to homogeneous varieties. It can also be used to build nice representations.

Example 10.24. Fix a Young diagram λ and let $k\lambda$ be obtained by scaling each row by k. Given a homogeneous variety X in $\mathbb{P}(S^\lambda(U))$, we can take the kth Veronese map v_k of this projective space. The linear span of $v_k(X)$ is $S^{k\lambda}(U)$. A special case of this construction is point (1) where $X = \mathbb{P}(U)$.

Next, we present a beautiful connection between the finite groups S_n and the Lie groups $\mathrm{SL}(n)$ or $\mathrm{GL}(n)$. This is the *Schur-Weyl duality*. Our description follows [**21**, Chapter 4]. Let us begin by going back to irreducible representations of S_n. Their characters form a basis of class functions. Hence the number of irreducible representations equals the number of conjugacy classes. Each conjugacy class can be encoded by the lengths of cycles in a decomposition of a permutation into cycles. These cycle lengths can be further represented by a Young diagram with n boxes, where the first row represents the length of the longest cycle and the last row the length of the shortest cycle. Thus, the number of irreducible representations of S_n equals the number of Young diagrams with n boxes.

Example 10.25. The symmetric group S_3 has three conjugacy classes:

- the identity $(1)(2)(3)$ with the Young diagram $(1,1,1)$;
- transpositions, e.g. $(12)(3)$, with the Young diagram $(2,1)$;
- 3-cycles, e.g. (123), with the Young diagram (3).

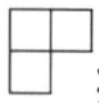

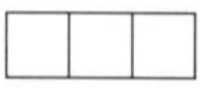

We shall exhibit a natural bijection between Young diagrams λ with n boxes and irreducible representations S_λ of S_n. Fix a vector space U and consider the n-fold tensor power $U^{\otimes n}$. There are two groups acting on it: $\mathrm{GL}(U)$, on each factor; and S_n, by permuting factors. Schur-Weyl duality provides a simultaneous decomposition of $U^{\otimes n}$ with respect to both groups.

Theorem 10.26 (Schur-Weyl duality). *Suppose* $\dim(U) \geq n$. *Then*

$$U^{\otimes n} = \bigoplus_{|\lambda|=n} S_\lambda \otimes S^\lambda(U), \tag{10.8}$$

where the sum is over all Young diagrams λ *with precisely* n *boxes and the* S_λ *are irreducible representations of the permutation group* S_n.

When $n = 2$ and $\dim U \geq 2$, we obtain $U^{\otimes 2} = S^2(U) \oplus \bigwedge^2 U$, as there are only two irreducible representations of S_2, both 1-dimensional. This recovers the fact that every square matrix is uniquely the sum of a symmetric matrix and a skew-symmetric matrix. The S_2 action on the matrix space $U^{\otimes 2}$ is transposition, which acts trivially on $S^2(U)$ and changes the sign on $\bigwedge^2 U$.

The $n = 3$ case is the first interesting one. The three irreducible representations S_λ of S_3 in Example 10.15 correspond to the three outer summands in (10.7). Note that $\dim(S_\lambda) = 2$ for $\lambda = (2,1)$. The middle summand in (10.7) is the 16-dimensional space $S_{(2,1)} \otimes S^{(2,1)}(U)$.

One can use (10.8) to define and construct all irreducible representations S_λ of the symmetric group S_n. These are known as *Specht modules*. By Schur-Weyl duality, the dimension of the Specht module S_λ equals the multiplicity a_λ of $S^\lambda(U)$ in $U^{\otimes n}$. This is seen from the decomposition of $U^{\otimes n}$ as a $\mathrm{GL}(U)$-representation into isotypic components:

$$U^{\otimes n} = \bigoplus_{\lambda} (S^\lambda(U))^{a_\lambda}.$$

The multiplicities a_λ can be found using Schur polynomials as in (10.6). For each isotypic component $(S^\lambda(U))^{a_\lambda}$ consider the highest weight space. This is the space of eigenvectors of the torus action with weight λ. The permutation group S_n acts on the highest weight space. This representation of S_n is irreducible. The Specht module S_λ is what we need for (10.8).

Let us return to the familiar $n = 2$ example of decomposing matrices into symmetric and skew-symmetric ones. The highest weight vector $e_1 e_1 = e_1 \otimes e_1$ of $S^2(U)$ is invariant with respect to transposition, i.e. it provides the trivial representation of the two-element group S_2. The highest weight vector $e_1 \wedge e_2 = \frac{1}{2}(e_1 \otimes e_2 - e_2 \otimes e_1)$ of $\bigwedge^2(U)$ changes sign when transposed, i.e. it provides the sign representation of the two-element group S_2.

Example 10.27 ($n = 3$). We consider $\lambda = (2,1)$ for $U^{\otimes 3}$. The isotypic component $(S^{(2,1)}(U))^2$ in the middle of (10.7) has a 2-dimensional subspace S_λ of highest weight vectors. One possible basis of this space consists of the tensors $e_{112} + e_{211} - 2e_{121}$ and $e_{121} + e_{211} - 2e_{112}$, where $e_{ijk} := e_i \otimes e_j \otimes e_k$.

10.3. Exploiting Symmetry

Representation theory is useful for many questions concerning tensors. Properties and varieties that occur naturally tend to be invariant under group actions on tensors, and this can be exploited for algorithms and applications.

We begin by showing how to obtain *lower bounds* for the border rank of a tensor. This has many applications, notably in complexity theory. One well-known instance is the matrix multiplication tensor M_n, which we saw in Section 9.3. Bounds on the rank and border rank of M_n translate into complexity bounds for optimal algorithms for multiplying two $n\times n$ matrices.

Let A, B and C be three vector spaces and let $T \in A \otimes B \otimes C$ be a tensor. As described in Section 9.2, we flatten T to obtain the linear map

$$\hat{T} : A^* \to B \otimes C. \tag{10.9}$$

Given bases, we can think of $\hat{T}$ as a matrix. If the tensor T has rank 1, i.e. $T = a \otimes b \otimes c$, then the matrix $\hat{T}$ has rank 1. If T is a sum of r tensors of rank 1, then $\hat{T}$ is a sum of r matrices of rank 1. Thus, the rank and border rank of the tensor T are bounded below by the rank of the matrix $\hat{T}$.

The best possible lower bound one can possibly obtain by flattening is the maximum of $\dim(A)$, $\dim(B)$ and $\dim(C)$. Our aim is to give larger lower bounds on the border rank of some tensors T. This requires new ideas. One of the tools—Strassen's invariant—was presented in Section 9.2. We now show how representation theory can lead us to a more general result.

The groups $\mathrm{GL}(A)$, $\mathrm{GL}(B)$ and $\mathrm{GL}(C)$ act on our tensors, and this action preserves the rank. In what follows, all spaces are viewed as representations of these groups, and linear maps are morphisms of representations. We first tensor the map (10.9) with the identity on $\bigwedge^k C$. This gives the linear map

$$\begin{array}{rccc} \hat{T} \otimes \mathrm{Id}_{\bigwedge^k C}: & A^* \otimes \bigwedge^k C & \to & B \otimes C \otimes \bigwedge^k C, \\ & f \otimes (c_1 \wedge \cdots \wedge c_k) & \mapsto & \hat{T}(f) \otimes (c_1 \wedge \cdots \wedge c_k). \end{array}$$

We have $\operatorname{brk} T \geq \operatorname{rk}(\hat{T}) = \operatorname{rk}(\hat{T} \otimes \mathrm{Id}_{\bigwedge^k C})/(\dim \bigwedge^k C)$. The tensor product $C \otimes \bigwedge^k C$ is a reducible representation of $\mathrm{GL}(C)$. For instance, the map

$$C \otimes \bigwedge^{k} C \to \bigwedge^{k+1} C,\ c_0 \otimes (c_1 \wedge \cdots \wedge c_k) \mapsto c_0 \wedge c_1 \wedge \cdots \wedge c_k$$

has nonzero kernel. By tensoring this map with Id_B we obtain a linear map

$$B \otimes C \otimes \bigwedge^{k} C \longrightarrow B \otimes \bigwedge^{k+1} C.$$

We now compose the last map with $\hat{T} \otimes \mathrm{Id}_{\bigwedge^k C}$ to obtain our final map

$$T_{k,C} \;:\; A^* \otimes \bigwedge^{k} C \;\longrightarrow\; B \otimes \bigwedge^{k+1} C.$$

It is instructive to fix bases for A, B and C and explicitly write the matrix that represents the linear map $T_{k,C}$. Each entry in that matrix either is zero or equals an entry of the given tensor $T = (t_{ijk})$, up to sign.

Example 10.28. Suppose that $A = K^2$, $B = K^3$, $C = K^4$ and $k = 2$. Then $T = (t_{ijk})$ is a $2 \times 3 \times 4$ tensor. The domain and range of our linear map $T_{2,C}$ are the 12-dimensional vector spaces $K^2 \otimes \wedge^2 K^4$ and $K^3 \otimes \wedge^3 K^4$. With respect to their standard bases, $T_{2,C}$ is given by the 12×12 matrix

	1|12	1|13	1|14	1|23	1|24	1|34	2|12	2|13	2|14	2|23	2|24	2|34
1|123	t_{113}	$-t_{112}$	0	t_{111}	0	0	t_{213}	$-t_{212}$	0	t_{211}	0	0
1|124	t_{114}	0	$-t_{112}$	0	t_{111}	0	t_{214}	0	$-t_{212}$	0	t_{211}	0
1|134	0	t_{114}	$-t_{113}$	0	0	t_{111}	0	t_{214}	$-t_{213}$	0	0	t_{211}
1|234	0	0	0	t_{114}	$-t_{113}$	t_{112}	0	0	0	t_{214}	$-t_{213}$	t_{212}
2|123	t_{123}	$-t_{122}$	0	t_{121}	0	0	t_{223}	$-t_{222}$	0	t_{221}	0	0
2|124	t_{124}	0	$-t_{122}$	0	t_{121}	0	t_{224}	0	$-t_{222}$	0	t_{221}	0
2|134	0	t_{124}	$-t_{123}$	0	0	t_{121}	0	t_{224}	$-t_{223}$	0	0	t_{221}
2|234	0	0	0	t_{124}	$-t_{123}$	t_{122}	0	0	0	t_{224}	$-t_{223}$	t_{222}
3|123	t_{133}	$-t_{132}$	0	t_{131}	0	0	t_{233}	$-t_{232}$	0	t_{231}	0	0
3|124	t_{134}	0	$-t_{132}$	0	t_{131}	0	t_{234}	0	$-t_{232}$	0	t_{231}	0
3|134	0	t_{134}	$-t_{133}$	0	0	t_{131}	0	t_{234}	$-t_{233}$	0	0	t_{231}
3|234	0	0	0	t_{134}	$-t_{133}$	t_{132}	0	0	0	t_{234}	$-t_{233}$	t_{232}

If the tensor T has rank 1, so that its entries factor as $t_{ijk} = x_i y_j z_k$, then the matrix $T_{2,C}$ has rank 3. Hence, if T has rank 2 then $T_{2,C}$ has rank ≤ 6, if T has rank 3 then $T_{2,C}$ has rank ≤ 9, and if T is general then $T_{2,C}$ is invertible.

Returning to the general case, suppose that $T = a \otimes b \otimes c$ is a rank-1 tensor in $A \otimes B \otimes C$. The image of the linear map $T_{k,C}$ equals the subspace

$$Kb \otimes \big(Kc \wedge \bigwedge^{k}(C/Kc)\big) \;\simeq\; K^1 \otimes \bigwedge^{k} K^{\dim(C)-1}.$$

In particular, the rank of $T_{k,C}$ is at most $\binom{\dim C-1}{k}$ when T has rank 1. Further, all our constructions are linear in the entries t_{ijk} of the tensor T, i.e. $(T_1 + \cdots + T_r)_{k,C} = (T_1)_{k,C} + \cdots + (T_r)_{k,C}$. Thus, if T has border rank at most r, then the rank of the matrix $T_{k,C}$ is at most $r \cdot \binom{\dim C-1}{k}$. This reasoning implies the following lower bound on the border rank of a tensor.

Proposition 10.29. *The border rank of any tensor T is bounded below by*

$$\mathrm{brk}(T) \;\geq\; \mathrm{rk}(T_{k,C})/\tbinom{\dim C-1}{k}.$$

This result implies that the $s \times s$ minors of the matrix $T_{k,C}$ vanish on all tensors whose border rank is less than $s/\binom{\dim C-1}{k}$. This gives a method for identifying explicit polynomials that vanish on tensors of border rank up to

$$\min\left((\dim B)\binom{\dim C}{k+1},\ (\dim A)\binom{\dim C}{k} \right)/\binom{\dim C-1}{k}. \tag{10.10}$$

These explicit polynomials are the appropriate minors of the matrix $T_{k,C}$.

Proposition 10.29 can now be used to derive lower bounds on the border rank of a specific tensor T we are interested in. This is done by showing that the matrix $T_{k,C}$ has high rank for some k. Here is a concrete numerical example. Fix an even integer $m > 0$, set $k = m/2$, and assume $\dim(A) = \dim(B) = \dim(C) = m$. Thus we consider $m \times m \times m$ tensors. The matrix $T_{k,C}$ has format $m\binom{m}{k} \times m\binom{m}{k+1}$. Therefore, its rank can be as large as $m \cdot \binom{m}{k+1}$. If this happens then we get the following lower bound:

$$\mathrm{brk}(T) \ \geq\ m \cdot \binom{m}{k+1}/\binom{m-1}{k} \ =\ \frac{m^2}{k+1}.$$

This should be compared with what one can hope to get from the usual flattening of T. The $m \times m^2$ matrix $\hat{T}$ can at best give the lower bound $\mathrm{brk}(T) \geq \mathrm{rk}(\hat{T}) = m$. Thus the bound from Proposition 10.29 is almost twice as good. This technique can be generalized by replacing the exterior powers in our construction by other irreducible representations $S^\lambda(C)$. The resulting generalizations of our matrices $T_{k,C}$ are known as the *Young flattenings* of the tensor T; see [**35**, §8.2].

Young flattenings have been a key tool recently for proving lower bounds on the complexity of matrix multiplication. This is done by finding a lower bound on the border rank of the matrix multiplication tensor $T = M_n$, which has format $m \times m \times m$ for $m = n^2$. To estimate the rank of the matrices $T_{k,C}$, one also uses representation theory. This is beyond the scope of the book. Just as an example, we state an inequality due to Landsberg and Ottaviani [**36**]. They proved that the border rank of the tensor M_n satisfies

$$\mathrm{brk}(M_n) \ \geq\ n^2\binom{2n-1}{n-1}/\binom{2n-2}{n-1} = 2n^2 - n.$$

Further improvements on the lower bounds can be found in [**34**].

Representation theory can be exploited to reduce the complexity of high-dimensional objects with symmetry. For instance, it is very useful for describing the ideals of varieties with a group action. Suppose that a group G acts on a projective space $\mathbb{P}^n$ and $X \subset \mathbb{P}^n$ is a variety which is mapped to itself by G. We will illustrate our approach with two particular examples:

(1) $X = G(2,5) \subset \mathbb{P}(\bigwedge^2 \mathbb{C}^5)$ is the Grassmannian and $G = \mathrm{GL}(5)$;

(2) $X = \sigma_4(\mathbb{P}^3 \times \mathbb{P}^3 \times \mathbb{P}^3) \subset \mathbb{P}(\mathbb{C}^4 \otimes \mathbb{C}^4 \otimes \mathbb{C}^4)$ is the secant variety of all tensors of border rank at most 4 and $G = \mathrm{GL}(4) \times \mathrm{GL}(4) \times \mathrm{GL}(4)$.

As G acts on $\mathbb{P}^n$, it also acts on the polynomial ring $K[\mathbf{x}] = K[x_0, \ldots, x_n]$, and this restricts to an action on the radical ideal $I(X)$. The vector space $I(X)_k$ of polynomials of degree k in this ideal is a representation of the group G. This has several advantages. Given just one polynomial $f \in I(X)$, we immediately obtain many, as $\langle Gf \rangle \subseteq I(X)$. Further, we can decompose $I(X)_k$ into irreducibles. Given just one polynomial from each irreducible piece, usually the highest weight vector, we can reconstruct the whole $I(X)_k$. This approach is extremely useful when combined with numerical methods. Indeed, $I(X)_k$ is often of huge dimension and it would be impossible to write down a basis. Yet it could happen that $I(X)_k$ has just a few irreducible representations and we just need to remember G and a few polynomials.

To make this explicit, we first note that there are no linear functions vanishing on either the Grassmannian or the secant variety $\sigma_4(\mathbb{P}^3 \times \mathbb{P}^3 \times \mathbb{P}^3)$. In both cases $I_1 = 0$. The space of linear functions on the ambient space of $G(2,5)$ equals $\bigwedge^2((\mathbb{C}^5)^*)$. This is the irreducible representation of $\mathrm{GL}(5)$ given by the Young diagram $\ydiagram{1,1}$, with highest weight vector $e_1^* \wedge e_2^*$.

The irreducible representations for a product of groups are the tensor products of irreducible representations of each factor. Hence, irreducible representations of $\mathrm{GL}(4) \times \mathrm{GL}(4) \times \mathrm{GL}(4)$ will be represented by triples of Young diagrams, each with at most four rows. The highest weight vector is just the tensor product of the individual highest weight vectors.

The linear functions on the ambient space of $\sigma_4(\mathbb{P}^3 \times \mathbb{P}^3 \times \mathbb{P}^3)$ are given by $\ydiagram{1} \otimes \ydiagram{1} \otimes \ydiagram{1}$. We see that a coordinate in the ambient space is a product of three coordinates, one for each $\mathbb{P}^3$. The space of quadrics is more complicated, but the ideal is still trivial. Indeed, for $X = \sigma_4(\mathbb{P}^3 \times \mathbb{P}^3 \times \mathbb{P}^3)$ we have $I(X)_2 = I(X)_3 = I(X)_4 = 0$. This is either a challenging exercise or a (less challenging) literature check [**32**]. Hence, the first interesting case is $I(X)_5$. This is a subspace of the space of degree-5 polynomials in 64 variables. The dimension of this subspace equals 1728. This sounds scary, but everything becomes manageable when we apply representation theory.

Let us begin by describing the quadrics in the ideal of $G(2,5)$. The space of all quadrics equals $S^2(\bigwedge^2(\mathbb{C}^5)^*)$, so its dimension is $\binom{\binom{5}{2}+1}{2} = 55$. The composition of symmetric and wedge products is an example of *plethysm*. No general formulas are known for plethysm of irreducible representations. However, each special case can be dealt with. The above representation is a sum of two irreducibles, namely $\ydiagram{2,2}$ and $\ydiagram{1,1,1,1}$, of dimensions 50 and 5.

From Example 10.24, we know that the second Veronese embedding of $G(2,5)$ spans the representation $\ydiagram{2,2}$. As the linear functions on the Veronese embedding are the quadratic functions before the embedding, we deduce that

$$\mathbb{C}[\mathbf{x}]_2/I(G(2,5))_2 \;\simeq\; \begin{array}{|c|c|}\hline \; & \; \\ \hline \; & \; \\ \hline \end{array} \qquad \text{and hence} \qquad I(G(2,5))_2 \;\simeq\; \begin{array}{|c|}\hline \; \\ \hline \; \\ \hline \; \\ \hline \; \\ \hline \end{array}\,.$$

Writing p_{ij} for $e_i^* \wedge e_j^*$, the highest weight vector of the 5-dimensional space $I(G(2,5))_2$ is the Plücker quadric $p_{23}p_{14} - p_{13}p_{24} + p_{12}p_{34}$. This is familiar from (5.3). Indeed, it has the correct weight and belongs to the ideal.

We now pass to the secant variety $X = \sigma_4(\mathbb{P}^3 \times \mathbb{P}^3 \times \mathbb{P}^3)$. We seek quintics that vanish on $4 \times 4 \times 4$ tensors of border rank 4. For this we apply the method described at the beginning of this section. We first focus on tensors $T \in \mathbb{C}^4 \otimes \mathbb{C}^4 \otimes \mathbb{C}^3$. Such a tensor may be represented as three 4×4 matrices A, B and C. The map $T_{1,\mathbb{C}^3}$ for $e_i \in \mathbb{C}^3$ and $f \in \mathbb{C}^4$ is defined by

$$T_{1,\mathbb{C}^3}(f \otimes e_i) \;=\; (Af) \otimes e_1 \wedge e_i \,+\, (Bf) \otimes e_2 \wedge e_i \,+\, (Cf) \otimes e_3 \wedge e_i.$$

Note that one of the three terms on the right is zero. In matrix form,

$$T_{1,\mathbb{C}^3} \;=\; \begin{pmatrix} 0 & -C & B \\ -C & 0 & A \\ -B & A & 0 \end{pmatrix}.$$

If the matrices A, B and C are symmetric, then negating the second column block gives a skew-symmetric matrix. The principal minors of a skew-symmetric matrix with unknown entries are squares of polynomials known as Pfaffians. The vanishing of Pfaffians of size $2s$, hence of degree s, ensures that a skew-symmetric matrix is of rank at most $2s - 2$. In our case, by Proposition 10.29, the 10×10 Pfaffians are the degree-5 polynomials vanishing on X, as they provide the correct rank condition for $T_{1,\mathbb{C}^3}$.

For the general case, let us first assume that B is invertible. Using row and column operations, we can transform the above matrix to the form

$$\begin{pmatrix} 0 & 0 & B \\ 0 & AB^{-1}C - CB^{-1}A & A \\ -B & A & 0 \end{pmatrix}.$$

The rank of this matrix equals 8 plus the rank of $AB^{-1}C - CB^{-1}A$. In particular, the rank is 8 if and only if $AB^{-1}C - CB^{-1}A = 0$. This is not a polynomial equation, but we can transform it by multiplying by $\det(B)$:

$$A \cdot \operatorname{adj}(B) \cdot C \;-\; C \cdot \operatorname{adj}(B) \cdot A \;=\; 0. \tag{10.11}$$

Here $\operatorname{adj}(B)$ is the adjugate matrix, whose entries are the signed 3×3 minors of B. Hence the entries of (10.11) have degree $5 = 1 + 3 + 1$ in the entries of A, B and C. They are known as *Strassen equations.* We note that they are related to the Aronhold invariant in (9.15). The latter is obtained by replacing the 4×4 matrices A, B and C in (10.11) with symmetric 3×3 matrices.

The upper left entry of (10.11) is a quintic with 180 terms. We regard it as a polynomial on the space $\mathbb{C}^4 \otimes \mathbb{C}^4 \otimes \mathbb{C}^4$, by restricting the last $\mathbb{C}^4$ to $\mathbb{C}^3$. In fact, this polynomial uses only 30 of the 64 variables. It vanishes on tensors of border rank 4. It is the highest weight vector of a 576-dimensional irreducible representation of $\mathrm{GL}(4) \times \mathrm{GL}(4) \times \mathrm{GL}(4)$, given combinatorially by

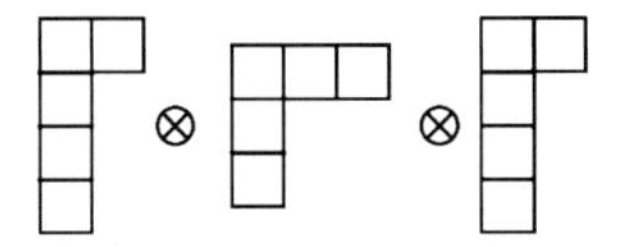

By permuting the three factors, we get two further irreducibles that vanish on the secant variety X. Together they span $I(X)_5$, which has dimension $1728 = 3 \cdot 576$. We construct all of these quintics by applying the symmetric group S_3 and $\mathrm{GL}(4) \times \mathrm{GL}(4) \times \mathrm{GL}(4)$ to one small quintic with 180 terms.

There is one big difference between our Grassmannian and our secant variety. A theorem of Kostant states that the ideal of a homogeneous variety, such as a Grassmannian, is generated by quadrics. By contrast, ideals of higher secant varieties are known in only very few cases. The above ideal $I(X)$ is not generated in degree 5. Describing the generators of $I(X)$ is known as the Salmon problem. Salmon is not a mathematician, but a fish. The problem was posed by Elizabeth Allman, to highlight the importance of the secant variety X for phylogenetics. It represents the general Markov model for a tree with three leaves attached to the root, each with four states.

We stress that there are many practical applications of representation theory. For instance, the book of Serre [**46**], to which we referred several times, grew out of lectures for quantum chemists and physicists. The book by Diaconis [**16**] highlights applications in probability and statistics.

Exercises

(1) (a) Prove that over an algebraically closed field, every irreducible representation of an abelian group is 1-dimensional.
    (b) Explain the correspondence between characters of a torus $T = (\mathbb{C}^*)^n$, as defined in Chapter 8, and irreducible representations of T.

(2) Derive the character table of the symmetric group S_4. Hint:

$$1^2 + 1^2 + 2^2 + 3^2 + 3^2 \;=\; 24.$$

What is the geometric meaning of the 3-dimensional representations?

(3) Let $f : V_1 \to V_2$ be a morphism between representations of a group G. Show that the kernel, image and cokernel of f are also representations of G. Prove also that morphisms between two representations are closed under taking scalar multiples and sums, i.e. they form a vector space.

(4) Derive the character table of the symmetric group S_5. Hint:

$$1^2 + 1^2 + 4^2 + 4^2 + 5^2 + 5^2 + 6^2 = 120.$$

Write matrices $\rho(g)$ for the 6-dimensional irreducible representation.

(5) Let V_1 and V_2 be two representations of a group G.
   (a) Prove that $\mathrm{Hom}(V_1, V_2)$ is also a representation of G. How can you characterize morphisms of representations inside $\mathrm{Hom}(V_1, V_2)$?
   (b) In terms of multiplicities of isotypic components, what is the dimension of the space of morphisms of representations $V_1 \to V_2$?
   (c) Conclude that the multiplicity of an irreducible representation W in V_1 equals the dimension of morphisms of representations $W \to V_1$.

(6) Let V be a representation of $\mathrm{GL}(n)$. Its character χ_V is a Laurent polynomial in $t_1, \ldots, t_n$. Show that $S^2(V)$ and $\bigwedge^2 V$ are also representations of $\mathrm{GL}(V)$, and compute the characters $\chi_{S^2(V)}$ and $\chi_{\bigwedge^2 V}$ in terms of χ_V.

(7) Describe the 2-dimensional irreducible representation from Example 10.15 explicitly. Map each of the six permutations of $\{1, 2, 3\}$ to a 2×2 matrix.

(8) Let ρ be the action of $\mathrm{GL}(6)$ on $\bigwedge^3 K^6$. What is the highest weight? What is the Young diagram? Find the entries of the 20×20 matrix $\rho(g)$.

(9) Is every $2 \times 2 \times 2$ tensor a symmetric tensor plus a skew-symmetric tensor?

(10) If $U = K^n$, what is the dimension of $S^{(2,1)}(U)$? Give a polynomial in n.

(11) What is the dimension of the vector space $S^3(S^3(K^3))$? Find a weight basis. Write down the character of this representation of $\mathrm{GL}(3)$. Can you decompose it into Schur polynomials? What does *plethysm* mean?

(12) What are the orbits of points in the adjoint representation? Are they closed? What is the dimension of a general orbit? What is the vanishing ideal of such an orbit, e.g. for $n = 3$?

(13) Show that the representation K^2 of the additive group $(K, +)$ in Example 10.11 is not a (nontrivial) sum of irreducible representations.

(14) What is the determinant of the "large matrix" from Example 10.3?

(15) What is the determinant of the 12×12 matrix $T_{2,C}$ in Example 10.28?

(16) Let $A = K^3$, $B = C = K^4$ and $k = 2$. What ranks are possible for the matrix $T_{2,C}$? Give an explicit tensor T such that $T_{2,C}$ has maximal rank.

(17) Write the space $I(G(3,6)_2)$ in Example 5.9 as a representation of $\mathrm{GL}(6)$.

Chapter 11

Invariant Theory

"Like the Arabian phoenix rising out of its ashes, the theory of invariants ... is once again at the forefront of mathematics",
Gian-Carlo Rota

What is geometry? An answer to this question was proposed by Felix Klein's *Erlanger Programm*. According to Klein, a quantity is geometric if it is invariant under the action of an underlying group of transformations. Thus, in short, geometry is invariant theory. For example, Euclidean geometry is the study of quantities, expressed in terms of the coordinates of points, that are invariant under the Euclidean group. From the modern point of view, invariant theory can be seen as a branch of representation theory. However, that view does not do justice to the tremendous utility of invariant theory for dealing with geometric objects. In particular, in algebraic geometry, invariants are used to construct quotients of algebraic varieties modulo groups that act on them. This results in a concise description of orbit spaces. The study of such spaces is *geometric invariant theory*. Our aim in this chapter is to give a first introduction to this theory, starting with actions by finite groups.

11.1. Finite Groups

We fix the polynomial ring $K[\mathbf{x}] = K[x_1, \ldots, x_n]$ over a field K of characteristic zero. The group $\mathrm{GL}(n, K)$ of invertible $n \times n$ matrices acts on K^n. This induces an action on the ring of polynomial functions on K^n. Namely, if $\sigma = (\sigma_{ij})$ is a matrix in $\mathrm{GL}(n, K)$ and f is a polynomial in $K[\mathbf{x}]$, then σf is the polynomial that is obtained from f by replacing the variable x_i with the linear form $\sum_{j=1}^n \sigma_{ji} x_j$ for $i = 1, \dots, n$. Thus the variable x_i is mapped to the linear form whose coefficients appear in the ith column of σ. This convention ensures compatibility with (10.1). In this book, all matrices in a representation of a group G act by left multiplication on column vectors.

Let G be a subgroup of $\mathrm{GL}(n, K)$. A polynomial $f \in K[\mathbf{x}]$ is an *invariant* of the group G if $\sigma f = f$ for all $\sigma \in G$. We write $K[\mathbf{x}]^G$ for the set of all such invariants. This set is a subring of the polynomial ring $K[\mathbf{x}]$ because the sum of two invariants is again an invariant, and the same for the product.

In this chapter we discuss two scenarios. In this section we consider finite groups G, and in Section 11.2 we consider representations of nice (i.e. reductive) infinite groups such as $\mathrm{SL}(d, K)$ and $\mathrm{SO}(d, K)$. A celebrated theorem of Hilbert shows that the invariant ring is finitely generated in this case. After two initial examples, we prove this result for finite groups G.

Example 11.1. Let G be the group of $n \times n$ permutation matrices. The invariant ring $K[\mathbf{x}]^G$ consists of all polynomials f that are invariant under permutation of the coordinates. This means that f satisfies

$$f(x_{\pi_1}, x_{\pi_2}, \dots, x_{\pi_n}) = f(x_1, x_2, \dots, x_n) \quad \text{for all permutations } \pi \text{ of } \{1, 2, \dots, n\}.$$

Such polynomials are called *symmetric*. The invariant ring $K[\mathbf{x}]^G$ is generated by the n elementary symmetric polynomials $E_1, \dots, E_n$ defined below. See [**51**, Theorem 1.1.1] for a proof. The elementary symmetric polynomials E_i are the coefficients of the following auxiliary polynomial in one variable z:

$$(z + x_1)(z + x_2) \cdots (z + x_n) = z^n + \sum_{i=1}^n E_i(\mathbf{x}) z^{n-i}. \tag{11.1}$$

We also set $E_0 = 1$. Alternatively, since $\mathrm{char}(K) = 0$, the invariant ring $K[\mathbf{x}]^G$ can also be generated by the power sums

$$P_j(\mathbf{x}) = x_1^j + x_2^j + \cdots + x_n^j \qquad \text{for } j = 1, 2, \dots, n.$$

The formulas that express the elementary symmetric polynomials E_i in terms of the power sums P_j, and vice versa, are known as *Newton's identities*:

$$\begin{aligned} kE_k &= \textstyle\sum_{i=1}^k (-1)^{i-1} E_{k-i} P_i \qquad \text{and} \\ P_k &= (-1)^{k-1} k E_k + \textstyle\sum_{i=1}^{k-1} (-1)^{k-1-i} E_{k-i} P_i \quad \text{for } 1 \le k \le n. \end{aligned} \tag{11.2}$$

Next, we provide geometric motivation for studying the ring of invariants. Suppose a group G acts on an n-dimensional vector space K^n. Our aim is to describe the space of orbits Q, i.e. a geometric object whose points correspond to orbits. We are not claiming that such a space always has the structure of a variety, but let us assume this for the moment. Following the approach presented in Chapters 1 and 2, we try to describe Q through polynomial functions on it. By mapping each point to the orbit it belongs to, we expect to get a quotient map $K^n \to Q$. This map of varieties is dual to a ring map, namely the inclusion from the ring of polynomial functions on Q into $K[\mathbf{x}]$, the coordinate ring of K^n. Under this inclusion, a polynomial function on Q gives rise to a polynomial that is constant on the orbits of G.

As invariants are exactly the polynomial functions that are constant along G-orbits, we see that $K[Q]$ maps to $K[\mathbf{x}]^G$. Thus invariants offer an algebraic view on the space of orbits. The quotient space $Q = K^n//G$ is the spectrum of $K[\mathbf{x}]^G$. Its (closed) points correspond to the orbits. This interpretation is only informal, as the details are subtle. In particular, we have not proved that there is indeed a bijection between (closed) points of Q and orbits. This is the case when G is finite, for example, but not in general. Making it all precise is the aim of *geometric invariant theory.*

Example 11.2. For $n = 2$, consider the following representation of the *cyclic group of order* 4. This gives the rotational symmetries of the square:

$$G = \left\{ \begin{pmatrix} 1 & 0 \\ 0 & 1 \end{pmatrix}, \begin{pmatrix} -1 & 0 \\ 0 & -1 \end{pmatrix}, \begin{pmatrix} 0 & 1 \\ -1 & 0 \end{pmatrix}, \begin{pmatrix} 0 & -1 \\ 1 & 0 \end{pmatrix} \right\}. \tag{11.3}$$

As we will prove in Example 11.7, its invariant ring is generated by

$$I_1 = x_1^2 + x_2^2, \quad I_2 = x_1^2 x_2^2, \quad I_3 = x_1^3 x_2 - x_1 x_2^3.$$

These three invariants are algebraically dependent. Using implicitization, we compute the cubic relation they satisfy. With this, we can write

$$K[x_1, x_2]^G = K[I_1, I_2, I_3] \simeq K[y_1, y_2, y_3]/\langle y_1^2 y_2 - 4y_2^2 - y_3^2 \rangle. \tag{11.4}$$

The spectrum of the ring (11.4) corresponds to the cubic surface in K^3 defined by the equation $y_1^2 y_2 = 4y_2^2 + y_3^2$. The points on this surface are in one-to-one correspondence with the G-orbits on K^2. This bijection is given by the map $K^2 \to K^3$, $(x_1, x_2) \mapsto (I_1, I_2, I_3)$, which is constant on orbits.

In what follows, let G be a finite subgroup of $\mathrm{GL}(n, K)$. One can create invariants by averaging polynomials. The *Reynolds operator*, denoted by a star, is the following map from arbitrary polynomials to invariants:

$$* : K[\mathbf{x}] \to K[\mathbf{x}]^G, \quad f \mapsto f^* := \frac{1}{|G|} \sum_{\sigma \in G} \sigma f. \tag{11.5}$$

Each of the following properties of the Reynolds operator is easily verified:

Lemma 11.3. *The Reynolds operator $*$ has the following three properties:*

(a) *The map $*$ is K-linear, i.e. $(\lambda f + \nu g)^* = \lambda f^* + \nu g^*$ for all $f, g \in K[\mathbf{x}]$ and $\lambda, \nu \in K$.*

(b) *The map $*$ restricts to the identity on $K[\mathbf{x}]^G$, i.e. $I^* = I$ for all invariant polynomials I.*

(c) *The map $*$ is a $K[\mathbf{x}]^G$-module homomorphism, i.e. $(fI)^* = f^*I$ for all $f \in K[\mathbf{x}]$ and $I \in K[\mathbf{x}]^G$.*

The next result marks the beginning of commutative algebra in 1890.

Theorem 11.4 (Hilbert's Finiteness Theorem). *The invariant ring $K[\mathbf{x}]^G$ of any finite matrix group $G \subset \mathrm{GL}(n, K)$ is finitely generated as a K-algebra.*

We present the proof under our standing assumption that K has characteristic zero. However, the result holds for every field K. For a proof in characteristic p see [**15**]. That setting is known as *modular invariant theory*.

Proof. Let $\mathcal{I}_G = \langle K[\mathbf{x}]^G_+ \rangle$ be the ideal in $K[\mathbf{x}]$ that is generated by all homogeneous invariants of positive degree. By Lemma 11.3(a), every invariant is a K-linear combination of the special invariants $(\mathbf{x}^{\mathbf{a}})^*$. These special invariants are homogeneous polynomials, obtained by averaging the images of the monomial functions $\mathbf{x}^{\mathbf{a}}$ over the group G. Thus the ideal $\mathcal{I}_G$ is generated by the infinite set $\{ (\mathbf{x}^{\mathbf{a}})^* : \mathbf{a} \in \mathbb{N}^n \backslash \{0\} \}$. By Hilbert's Basis Theorem (Corollary 1.14), the ideal $\mathcal{I}_G$ is finitely generated, so that a finite subset of integer vectors $\mathbf{a}$ in $\mathbb{N}^n$ suffices for generation. In conclusion, there exist homogeneous invariants $I_1, I_2, \ldots, I_m$ such that $\mathcal{I}_G = \langle I_1, I_2, \ldots, I_m \rangle$.

We claim that these m invariants generate the invariant ring $K[\mathbf{x}]^G$ as a K-algebra. Suppose the contrary, and let I be a homogeneous element of minimal degree in the difference $K[\mathbf{x}]^G \backslash K[I_1, I_2, \ldots, I_m]$. Since $I \in \mathcal{I}_G$, we have $I = \sum_{j=1}^m f_j I_j$ for some polynomials $f_j \in K[\mathbf{x}]$. As the invariants I_j are homogeneous, so are the polynomials f_j. The degree of each f_j is strictly less than the degree of the invariant I that is not in the subring.

We now apply the Reynolds operator to both sides of the equation $I = \sum_{j=1}^m f_j I_j$. This yields

$$I = I^* = \Big(\sum_{j=1}^m f_j I_j\Big)^* = \sum_{j=1}^m f_j^* I_j.$$

Here we are using the properties (b) and (c) in Lemma 11.3. The new coefficients f_j^* are homogeneous invariants whose degrees are less than $\deg(I)$. From the minimality assumption on the degree of I, we get $f_j^* \in K[I_1, \ldots, I_m]$

for $j = 1, \ldots, m$. This implies $I \in K[I_1, \ldots, I_m]$, which is a contradiction to our assumption. This completes the proof of Theorem 11.4. $\square$

Theorem 11.5 (Noether's degree bound). *If G is finite and $\mathrm{char}(K) = 0$, then the invariant ring $K[\mathbf{x}]^G$ is generated by homogeneous invariants whose degree is at most the group order $|G|$.*

Proof. Let $\mathbf{u} = (u_1, \ldots, u_n)$ be new variables. We extend the action of G to $K[\mathbf{u}, \mathbf{x}]$ by $gu_i = u_i$ for all i. For any $d \in \mathbb{N}$, we consider the expression

$$\begin{aligned} S_d(\mathbf{u}, \mathbf{x}) &= \left[(u_1x_1 + \cdots + u_nx_n)^d\right]^* \\ &= \frac{1}{|G|} \sum_{\sigma \in G} \left[u_1(\sigma x_1) + \cdots + u_n(\sigma x_n)\right]^d. \end{aligned}$$

This is a polynomial in $\mathbf{u}$ whose coefficients are polynomials in $\mathbf{x}$. Up to a multiplicative constant, these coefficients are the invariants $(\mathbf{x}^{\mathbf{a}})^*$ where $|\mathbf{a}| = d$. By definition, all polynomials in $\mathbf{u}$ are fixed under the action of $*$.

Consider the $|G|$ expressions $u_1(\sigma x_1) + \cdots + u_n(\sigma x_n)$, one for each element σ of the group G. The polynomial $S_d(\mathbf{u}, \mathbf{x})$ is the dth power sum of these $|G|$ expressions, up to a multiplicative constant. It follows from Newton's identities (11.2) that this dth power sum for $d > |G|$ is a polynomial with rational coefficients in the kth power sums for $k \leq |G|$. Hence $S_d(\mathbf{u}, \mathbf{x})$ is a polynomial with rational coefficients in $S_k(\mathbf{u}, \mathbf{x})$ for $k \leq |G|$. All of these are polynomials in $\mathbf{u}$ whose coefficients are polynomials in $\mathbf{x}$. It follows that these coefficients for $d > |G|$ are polynomials in the coefficients for $k \leq |G|$. Therefore, all invariants $(\mathbf{x}^{\mathbf{a}})^*$ with $|\mathbf{a}| > |G|$ are polynomial functions (over $\mathbb{Q} \subseteq K$) in the invariants $(\mathbf{x}^{\mathbf{b}})^*$ with $|\mathbf{b}| \leq |G|$. This proves the claim. $\square$

We note that Example 11.2 attains Noether's degree bound. Here, the group has order 4, and the invariant ring requires a generator of degree 4.

Our next theorem is a useful tool for constructing the invariant ring. It says that we can count invariants by averaging the reciprocals of the characteristic polynomials of all matrices in the group.

Theorem 11.6 (Molien). *Let G be a finite group of $n \times n$ matrices. The Hilbert series of the invariant ring $K[\mathbf{x}]^G$ is the rational generating function*

$$\sum_{d=0}^{\infty} \dim_K\big(K[\mathbf{x}]_d^G\big) z^d \quad = \quad \frac{1}{|G|} \sum_{\sigma \in G} \frac{1}{\det(\mathrm{Id} - z\sigma)}. \tag{11.6}$$

The coefficient of z^d in this formal generating function is the number of linearly independent homogeneous invariants of degree d.

Proof. See [**51**, Theorem 2.2.1]. $\square$

Example 11.7. Consider the cyclic group $G = \mathbb{Z}_4$ in Example 11.2. For the four matrices σ in Example 11.2, the quadratic polynomials $\det(\mathrm{Id} - z\sigma)$ are $(1-z)^2$, $(1+z)^2$ and twice $1+z^2$. Adding up their reciprocals and dividing by $|G| = 4$, we see that the Hilbert series of $K[\mathbf{x}]^G$ is

$$(11.7) \qquad \frac{1+z^4}{(1-z^2)(1-z^4)} \quad = \quad 1 + z^2 + 3z^4 + 3z^6 + 5z^8 + \cdots .$$

This agrees with the Hilbert series of the ring on the right in (11.4), where $\deg(y_1) = 2$ and $\deg(y_2) = \deg(y_3) = 4$. Indeed, every element of that ring can be uniquely represented as a sum of a polynomial in $K[y_1, y_2]$ and y_3 times a polynomial in $K[y_1, y_2]$. One says that the ring is a free module with basis $\{1, y_3\}$ over $K[y_1, y_2]$. This explains the numerator and denominator on the left of (11.7). It proves that the invariants I_1, I_2, I_3 generate $K[\mathbf{x}]^G$.

11.2. Classical Invariant Theory

Hilbert's Finiteness Theorem also holds for an infinite group $G \subset \mathrm{GL}(n, K)$ that has a Reynolds operator $*$ which satisfies (a), (b) and (c) in Lemma 11.3. This property is equivalent to G being *reductive*. Recall from Section 10.1 that a group G is reductive if every finite-dimensional representation is completely reducible. We apply this result to the space of homogeneous polynomials of degree d in n variables. These form a representation $K[\mathbf{x}]_d$, and $K[\mathbf{x}]_d^G$ is a subrepresentation. If there exists a complementary subrepresentation H such that $H \oplus K[\mathbf{x}]_d^G = K[\mathbf{x}]_d$, then we may define $*$ as the projection with kernel H. Summing over all degrees $d \in \mathbb{N}$ and extending the projection $*$ linearly to $K[\mathbf{x}]$, we get the Reynolds operator for G.

Corollary 11.8. *Fix a reductive group G of $n \times n$ matrices, and let $\mathcal{I}_G$ be the ideal in $K[\mathbf{x}]$ generated by all homogeneous invariants of positive degree. If $\{g_1, g_2, \ldots, g_m\}$ is any set of homogeneous polynomials that generates $\mathcal{I}_G$, then its image $\{g_1^*, g_2^*, \ldots, g_m^*\}$ under the Reynolds operator generates the invariant ring $K[\mathbf{x}]^G$ as a K-algebra.*

Proof. Let $M = \langle x_1, \ldots, x_n \rangle$ be the homogeneous maximal ideal in $K[\mathbf{x}]$, and consider the finite-dimensional vector space $\mathcal{I}_G / M\mathcal{I}_G$. It has a basis of invariants since $\mathcal{I}_G$ is generated by invariants. This means that the Reynolds operator acts as the identity on $\mathcal{I}_G / M\mathcal{I}_G$. The residue classes of $g_1, g_2, \ldots, g_m$ modulo $M\mathcal{I}_G$ also span $\mathcal{I}_G / M\mathcal{I}_G$ as a vector space, and hence so do the invariants $g_1^*, g_2^*, \ldots, g_m^*$. By the graded Nakayama's Lemma, we find that $g_1^*, g_2^*, \ldots, g_m^*$ generate the ideal $\mathcal{I}_G$. As in the proof of Theorem 11.4, we conclude that $g_1^*, g_2^*, \ldots, g_m^*$ generate the K-algebra $K[\mathbf{x}]^G$. $\square$

Classical invariant theory was primarily concerned with the case where G is a representation of the group $\mathrm{SL}(d, K)$ of $d \times d$ matrices with determinant 1. Here d is an integer that is usually much smaller than n. We continue to assume that K is a field of characteristic zero. We have assumed that G is the image of a group homomorphism $\mathrm{SL}(d, K) \to \mathrm{GL}(n, K)$. It is known that $\mathrm{SL}(d, K)$ is a reductive group, i.e. there also exists an averaging operator $* : K[\mathbf{x}] \to K[\mathbf{x}]^G$ which has the same formal properties as the averaging operator of a finite group, stated in Lemma 11.3. See also Remark 10.9.

That Reynolds operator $*$ can be realized either by integration or by differentiation. In the first realization, one replaces the sum in (11.5) by an integral. Namely, one takes $K = \mathbb{C}$ and integrates over the compact subgroup $\mathrm{SU}(d, \mathbb{C})$ with respect to Haar measure. The same kind of integral also works in Theorem 11.6. If $G = \mathrm{SL}(d, \mathbb{C})$ then one can compute the Hilbert series of the invariant ring by averaging reciprocal characteristic polynomials. In other words, Molien's Theorem generalizes to $G = \mathrm{SL}(d, \mathbb{C})$.

An alternative to integrating with respect to Haar measure on $\mathrm{SU}(d, \mathbb{C})$ is to apply a certain differential operator known as *Cayley's Ω-process.* This process, which is explained in [**51**, §4.3], can also be used to transform arbitrary polynomials into invariants.

A third method for computing invariants is to use plain old linear algebra. Indeed, suppose we fix an integer $d \in \mathbb{N}$ and seek a basis for the space $K[\mathbf{x}]_d^G$ of homogeneous invariants of degree d. We then pick a general polynomial f of degree d with unknown coefficients, and we examine the equations $\sigma f = f$ for $\sigma \in G$. Each of these translates into a linear system of equations in the unknown coefficients of f. By taking enough matrices σ, we obtain a linear system of equations whose solutions are precisely the invariants of degree d. In the case where G is a connected Lie group, such as $\mathrm{SL}(d, \mathbb{C})$, one can replace the condition $\sigma f = f$ with the requirement that f be annihilated by the associated *Lie algebra.* Setting up these linear equations and solving them is usually quite efficient on small examples. See [**51**, §4.5].

In what follows we take the matrix group to be an n-dimensional polynomial representation of $G = \mathrm{SL}(d, K)$ for some $d, n \in \mathbb{N}$. This is a direct sum of irreducible representations, indexed by partitions, as in Chapter 10.

Example 11.9. Let $U = (K^d)^m$ be the space of $d \times m$ matrices. The group $G = \mathrm{SL}(d, K)$ acts naturally on column vectors of length d, so we view U as the direct sum of m copies of the defining representation, one copy for each column in our matrix. Equivalently, G acts on U by matrix multiplication on the left. This induces an action on the ring $K[U]$ of polynomials in the entries of a $d \times m$ matrix of variables. If $m < d$ then this action has no nonconstant invariants. If $m \geq d$ then the $\binom{m}{d}$ maximal minors of the $d \times m$

matrix are invariants. This invariance holds because the determinant of the product of two $d \times d$ matrices is the product of the determinants. It is known that the invariant ring $K[U]^G$ is generated by these $\binom{m}{d}$ determinants. This is the First Fundamental Theorem of Invariant Theory; cf. [**51**, §3.2].

Note that we already encountered the invariant ring $K[U]^G$ in Chapter 5. This ring is the coordinate ring of the Grassmannian of d-dimensional subspaces in K^m. Thus, $K[U]^G$ is isomorphic to a polynomial ring in $\binom{m}{d}$ variables modulo the ideal of quadratic Plücker relations.

Arguably, the most important irreducible representations of the group $G = \mathrm{SL}(d, K)$ are the pth symmetric powers of the defining representation K^d. Here p can be any positive integer. We denote such a symmetric power by $V = K[u_1, \ldots, u_d]_p = \mathrm{Sym}_p(K^d)$. Its elements are homogeneous polynomials of degree p in d variables. The G-module V has dimension $n = \binom{p+d-1}{p}$. A basis consists of the monomials of degree p in d variables. The action of G on V is simply by linear change of coordinates.

Example 11.10 ($d = 2, p = 3$)**.** Fix the space $V = \mathrm{Sym}_3(K^2)$ of binary cubics

$$f(u_1, u_2) \quad = \quad x_1 u_1^3 \,+\, x_2 u_1^2 u_2 \,+\, x_3 u_1 u_2^2 \,+\, x_4 u_2^3. \tag{11.8}$$

The coefficients x_i are the coordinates on $V \simeq K^4$. The way we set things up, the group $\mathrm{SL}(2, K)$ acts on this space by left multiplication, in its guise as the group G of 4×4 matrices of the form

$$\phi(\sigma) \;=\; \begin{bmatrix} \sigma_{11}^3 & \sigma_{11}^2\sigma_{12} & \sigma_{11}\sigma_{12}^2 & \sigma_{12}^3 \\ 3\sigma_{11}^2\sigma_{21} & \sigma_{11}^2\sigma_{22} + 2\sigma_{11}\sigma_{12}\sigma_{21} & \sigma_{12}^2\sigma_{21} + 2\sigma_{11}\sigma_{12}\sigma_{22} & 3\sigma_{12}^2\sigma_{22} \\ 3\sigma_{11}\sigma_{21}^2 & \sigma_{12}\sigma_{21}^2 + 2\sigma_{11}\sigma_{21}\sigma_{22} & \sigma_{11}\sigma_{22}^2 + 2\sigma_{12}\sigma_{21}\sigma_{22} & 3\sigma_{12}\sigma_{22}^2 \\ \sigma_{21}^3 & \sigma_{21}^2\sigma_{22} & \sigma_{21}\sigma_{22}^2 & \sigma_{22}^3 \end{bmatrix}.$$

For a given 2×2 matrix $\sigma \in G = \mathrm{SL}(2, K)$, the determinant of the 4×4 matrix $\phi(\sigma)$ is equal to $(\sigma_{11}\sigma_{22} - \sigma_{12}\sigma_{21})^6 = 1$. The G-action on V is given by $x \mapsto \phi(\sigma)x$ where x is the column vector $(x_1, x_2, x_3, x_4)^T$. One invariant under this action is the discriminant of the binary cubic $f(u_1, u_2)$, which is

$$\Delta \;=\; 27x_1^2x_4^2 \;-\; 18x_1x_2x_3x_4 \;+\; 4x_1x_3^3 \;+\; 4x_2^3x_4 \;-\; x_2^2x_3^2. \tag{11.9}$$

It turns out that the discriminant generates the invariant ring:

$$K[\mathbf{x}]^G \;=\; K[\Delta].$$

Every G-invariant can be written uniquely as a polynomial in Δ.

Invariants of binary forms ($d = 2$) are a well-studied subject in invariant theory. Complete lists of generators for the invariant ring are known up to degree $p = 10$. For $p = 2$, there is also only the discriminant $\Delta = x_2^2 - 4x_1x_3$. For $p = 4$, we have two generating invariants of degrees 2 and 3. However, for $p = 10$, the invariant ring has 106 minimal generators.

We close with an example seen twice in the previous chapter.

Example 11.11. Following Example 10.4, consider the action of $G = \mathrm{SO}(2, K)$ on the space V of binary cubics. The invariant ring $K[\mathbf{x}]^G$ is generated by two quadrics and two quartics. These are $z_1 z_2, z_3 z_4$ and $z_1 z_3^3, z_2 z_4^3$ for the corresponding action via the diagonal matrix $D(\theta)$ that was derived for $K = \mathbb{C}$ in Example 10.10. In the notation of (11.8), this translates into

$$
\begin{aligned}
I_1 &= x_1^2 - 2x_1x_3 + x_2^2 - 2x_2x_4 + x_3^2 + x_4^2, \\
I_2 &= 9x_1^2 + 6x_1x_3 + x_2^2 + 6x_2x_4 + x_3^2 + 9x_4^2, \\
I_3 &= 27x_1^3x_4 - 9x_1^2x_2x_3 + 2x_1x_2^3 + 9x_1x_2^2x_4 - 6x_1x_2x_3^2 - 9x_1x_3^2x_4 \\
&\quad - 27x_1x_4^3 + x_2^3x_3 + 6x_2^2x_3x_4 - x_2x_3^3 + 9x_2x_3x_4^2 - 2x_3^3x_4, \\
I_4 &= 27x_1^4 + 18x_1^2x_2^2 - 18x_1^2x_3^2 - 162x_1^2x_4^2 + 24x_1x_2^2x_3 + 72x_1x_2x_3x_4 - x_2^4 \\
&\quad - 8x_1x_3^3 - 8x_2^3x_4 + 6x_2^2x_3^2 - 18x_2^2x_4^2 + 24x_2x_3^2x_4 - x_3^4 + 18x_3^2x_4^2 + 27x_4^4.
\end{aligned}
$$

These four invariants can be found by Derksen's algorithm (Theorem 11.13).

11.3. Geometric Invariant Theory

According to Felix Klein, invariant theory plays a fundamental role in geometry. A polynomial in the coordinates of a space is invariant under the group of interest if and only if that polynomial expresses a geometric property. For instance, consider the space V of binary cubics f in Example 11.10. The hypersurface defined by f in $\mathbb{P}^1$ consists of three points. The vanishing of the invariant Δ means that these three points are not all distinct.

In geometric invariant theory, one considers the variety $\mathcal{V}(\mathcal{I}_G)$ defined by all homogeneous invariants of positive degree. This variety is known as the *nullcone*. Its points are known as *unstable points*. For a finite group G, the nullcone consists of just the origin. In symbols, $\mathcal{V}(\mathcal{I}_G) = \{0\}$.

For $G = \mathrm{SL}(d, K)$ the situation is more interesting, and the geometry of the nullcone is very important for understanding the invariant ring $K[\mathbf{x}]^G$. Corollary 11.8 says, more or less, that computing $K[\mathbf{x}]^G$ is equivalent to finding polynomial equations that define the nullcone.

Example 11.12 ($d = p = 3$). We consider the 10-dimensional vector space $V = \mathrm{Sym}_3(K^3)$ whose elements are the *ternary cubics*

$$
\begin{aligned}
f(\mathbf{u}) = {}& x_1u_1^3 + x_2u_2^3 + x_3u_3^3 + x_4u_1^2u_2 + x_5u_1^2u_3 \\
&+ x_6u_2^2u_1 + x_7u_2^2u_3 + x_8u_3^2u_1 + x_9u_3^2u_2 + x_0u_1u_2u_3.
\end{aligned}
$$

The group $G = \mathrm{SL}(3, K)$ acts on V by linear change of coordinates. The corresponding invariant ring is generated by two invariants I_4 and I_6 of degrees 4 and 6 respectively. In symbols, $K[\mathbf{x}]^G = K[I_4, I_6]$. The degree-4 invariant is unique up to scaling. It is the following sum of 25 monomials:

$$\begin{aligned} I_4 = x_0^4 &- 8x_0^2x_4x_9 - 8x_0^2x_5x_7 - 8x_0^2x_6x_8 - 216x_0x_1x_2x_3 + 24x_0x_1x_7x_9 \\ &+ 24x_0x_2x_5x_8 + 24x_0x_3x_4x_6 + 24x_0x_4x_7x_8 + 144x_1x_2x_8x_9 \\ &+ 24x_0x_5x_6x_9 + 144x_1x_3x_6x_7 - 48x_1x_6x_9^2 - 48x_1x_7^2x_8 - 48x_2x_4x_8^2 \\ &+ 144x_2x_3x_4x_5 - 48x_2x_5^2x_9 - 48x_3x_4^2x_7 - 48x_3x_5x_6^2 + 16x_4^2x_9^2 \\ &- 16x_4x_5x_7x_9 - 16x_4x_6x_8x_9 + 16x_5^2x_7^2 - 16x_5x_6x_7x_8 + 16x_6^2x_8^2. \end{aligned}$$

The degree-6 invariant is similarly unique. It is a sum of 103 monomials:

$$\begin{aligned} I_6 = x_0^6 &- 12x_0^4x_4x_9 - 12x_0^4x_5x_7 - 12x_0^4x_6x_8 \\ &+ 540x_0^3x_1x_2x_3 + \cdots + 96x_5x_6^2x_7x_8^2 - 64x_6^3x_8^3. \end{aligned}$$

The invariant I_4 is the *Aronhold invariant.* This plays an important role in the theory of tensor decomposition. Indeed, we can regard f as a symmetric $3 \times 3 \times 3$ tensor. A random tensor f has rank 4. The Aronhold invariant f vanishes for those tensors of border rank ≤ 3. In other words, $I_4 = 0$ holds if and only if f is a sum of three cubes of linear forms, or can be approximated by a sequence of such. See the discussion of ranks of tensors in Chapter 9.

On the geometric side, we can identify f with the cubic curve $\mathcal{V}(f)$ it defines in the projective plane $\mathbb{P}^2$. To a number theorist, this is an *elliptic curve.* An important invariant of this curve is the *discriminant* Δ.

The discriminant of a ternary cubic has degree 12. An explicit formula is

$$\Delta \;=\; I_4^3 - I_6^2. \tag{11.10}$$

This expression vanishes if and only if the cubic curve $\mathcal{V}(f)$ has a singular point. Typically, this singularity is a *node.* In the special case where both I_4 and I_6 vanish, the singular point is a *cusp.* Thus, for ternary cubics, the nullcone $\mathcal{V}(\mathcal{I}_G)$ is given by plane cubics that have a cusp. The moduli space of elliptic curves is parametrized by the *j-invariant*, which equals I_4^3/Δ.

We now present a general purpose algorithm, due to Harm Derksen, for computing the invariant ring of a reductive algebraic group G that acts polynomially on a vector space $V = K^n$. The group G can be represented as an algebraic variety inside $\mathrm{GL}(n, K)$, that is, by polynomial equations in the entries of an unknown $n \times n$ matrix. This works for both finite groups and polynomial representations of $\mathrm{SL}(d, K)$, such as the ones discussed above. As before, we use the notation $\sigma \mapsto \phi(\sigma)$ to write the representation of G on $V = K^n$ explicitly. The ring $K[\sigma]$ denotes the coordinate ring of G, which is typically represented as a polynomial ring modulo an ideal.

The product $G \times V \times V$ is an algebraic variety, with coordinates $(\sigma, \mathbf{x}, \mathbf{y})$. Inside this coordinate ring $K[\sigma, \mathbf{x}, \mathbf{y}]$, let $\mathcal{J}_G$ be the ideal generated by the

n entries of the vector $\mathbf{y} - \phi(\sigma)\mathbf{x}$. This ideal is radical, and it is prime when G is a connected group such as $\mathrm{SL}(d, K)$. Its variety describes the action of the group. The elimination ideal $\mathcal{J}_G \cap K[\mathbf{x}, \mathbf{y}]$ is also radical (resp. prime). Its variety contains pairs of points in V that lie in the same G-orbit.

Theorem 11.13 (Derksen's algorithm). *The ideal $\mathcal{I}_G$ that defines the null-cone is the image in $K[\mathbf{x}]$ of the elimination ideal $\mathcal{J}_G \cap K[\mathbf{x}, \mathbf{y}]$ under the substitution $\mathbf{y} = 0$. From any finite list of generators for the ideal $\mathcal{I}_G$, we can construct algebra generators for the invariant ring $K[\mathbf{x}]^G$ via Corollary 11.8.*

Proof. Let I be any homogeneous invariant of positive degree. Then $I(\mathbf{x}) \equiv I(\phi(\sigma)\mathbf{x}) \equiv I(\mathbf{y})$ modulo the ideal $\mathcal{J}_G$ that defines the group action. Therefore, $I(\mathbf{x}) - I(\mathbf{y})$ lies in the elimination ideal $\mathcal{J}_G \cap K[\mathbf{x}, \mathbf{y}]$. We find $I(\mathbf{x})$ in the ideal obtained by substituting $\mathbf{y} = 0$. This proves that $\mathcal{I}_G$ is contained in the ideal that is output by Derksen's algorithm. For the converse direction, we refer to the argument given in the proof of [**14**, Theorem 3.1]. □

Remark 11.14. The ideal $\mathcal{I}_G$ is generally not the radical ideal of the null-cone. It is generated by homogeneous invariants of positive degree. For instance, if $n = 2$ and $G = \{\pm \mathrm{Id}_2\}$, then $\mathcal{I}_G = \langle x_1^2, x_1x_2, x_2^2 \rangle$ is not radical.

Example 11.15 ($p = d = 2$). Consider the 3-dimensional space $V = \mathrm{Sym}_2(K^2)$ of binary quadrics

$$f(u_1, u_2) \quad = \quad x_1u_1^2 \,+\, x_2u_1u_2 \,+\, x_3u_2^2.$$

The coordinate ring of the variety $\mathrm{SL}(2, K) \times V \times V$ is the polynomial ring

$$K[\sigma, \mathbf{x}, \mathbf{y}] \quad = \quad K\big[\,\sigma_{11}, \sigma_{12}, \sigma_{21}, \sigma_{22},\, x_1, x_2, x_3,\, y_1, y_2, y_3\,\big]$$

modulo the principal ideal $\langle \sigma_{11}\sigma_{22} - \sigma_{12}\sigma_{21} - 1 \rangle$. Note that this ring has 10 generators. The ideal that encodes our action is

$$\begin{aligned} \mathcal{J}_G \quad = \quad \big\langle\, & \sigma_{11}^2x_1 + \sigma_{11}\sigma_{21}x_2 + \sigma_{21}^2x_3 \,-\, y_1,\ \sigma_{12}^2x_1 + \sigma_{12}\sigma_{22}x_2 + \sigma_{22}^2x_3 - y_3, \\ & 2\sigma_{11}\sigma_{12}x_1 + (\sigma_{11}\sigma_{22} + \sigma_{12}\sigma_{21})x_2 + 2\sigma_{21}\sigma_{22}x_3 \,-\, y_2 \,\big\rangle. \end{aligned}$$

We eliminate the four variables σ_{ij} that represent the group elements. This yields the principal ideal

$$\mathcal{J}_G \cap K[\mathbf{x}, \mathbf{y}] \quad = \quad \langle\, 4x_1x_3 - x_2^2 \,-\, 4y_1y_3 + y_2^2 \,\rangle.$$

We now set $y_1 = y_2 = y_3 = 0$. The result is the familiar discriminant $\Delta = 4x_1x_3 - x_2^2$. In this manner, Derksen's algorithm finds the invariant ring for binary quadrics. It confirms that $K[\mathbf{x}]^G = K[\Delta]$.

In Example 11.10, we determined the invariant ring for $\mathrm{SL}(2, K)$ acting on $2 \times 2 \times 2$ tensors that are symmetric. In what follows, we extend this computation to nonsymmetric tensors. Thus, we present a case study in invariant theory for $d = 2$ and $n = 8$. We identify K^8 with the space $(K^2)^{\otimes 3}$ of $2 \times 2 \times 2$ tensors. The corresponding polynomial ring is denoted by

$$K[\mathbf{x}] \;=\; K[x_{111}, x_{112}, x_{121}, x_{122}, x_{211}, x_{212}, x_{221}, x_{222}].$$

The group $G = \mathrm{SL}(2, K)$ acts on K^2 by matrix-vector multiplication. This action extends naturally to the triple tensor product of K^2. Explicitly, if $\sigma = \begin{pmatrix} \sigma_{11} & \sigma_{12} \\ \sigma_{21} & \sigma_{22} \end{pmatrix}$ is a 2×2 matrix in G, then σ acts by performing the following substitution in each polynomial on $K[\mathbf{x}]$:

$$x_{ijk} \;\mapsto\; \sum_{r=1}^{2}\sum_{s=1}^{2}\sum_{t=1}^{2} x_{rst}\sigma_{ri}\sigma_{sj}\sigma_{tk}. \tag{11.11}$$

Here are two nice polynomials that are invariant under this action:

Example 11.16. Up to scaling, there is a unique polynomial of degree 2 that is invariant under $G = \mathrm{SL}(2, K)$. That invariant is the following quadric, which we call the *hexagon invariant*:

$$\mathrm{Hex}(\mathbf{x}) \;=\; x_{112}x_{122} - x_{122}x_{121} + x_{121}x_{221} - x_{221}x_{211} + x_{211}x_{212} - x_{212}x_{112}.$$

Another nice invariant is the *hyperdeterminant*, which has degree 4:

$$\begin{aligned} \mathrm{Det}(\mathbf{x}) = {} & x_{221}^2x_{112}^2 + x_{211}^2x_{122}^2 + x_{121}^2x_{212}^2 + x_{111}^2x_{222}^2 \\ & + 4x_{111}x_{221}x_{122}x_{212} + 4x_{121}x_{211}x_{112}x_{222} \\ & - 2x_{211}x_{221}x_{112}x_{122} - 2x_{121}x_{221}x_{112}x_{212} - 2x_{121}x_{211}x_{122}x_{212} \\ & - 2x_{111}x_{221}x_{112}x_{222} - 2x_{111}x_{211}x_{122}x_{222} - 2x_{111}x_{121}x_{212}x_{222}. \end{aligned}$$

We saw this in Example 4.10. One checks by computation that the substitution (11.11) maps the hexagon invariant $\mathrm{Hex}(\mathbf{x})$ to itself times the third power of $\det(\sigma) = \sigma_{11}\sigma_{22} - \sigma_{12}\sigma_{21}$. Similarly, the hyperdeterminant $\mathrm{Det}(\mathbf{x})$ transforms to itself times $\det(\sigma)^6$. Both are left invariant when $\det(\sigma) = 1$.

Invariants can be used to test whether two tensors lie in the same G-orbit. Here is a concrete example. We write our $2 \times 2 \times 2$ tensors as vectors in $\mathbb{R}^8$ as follows: $\mathbf{c} = (c_{111}, c_{112}, c_{121}, c_{122}, c_{211}, c_{212}, c_{221}, c_{222})$. The following two tensors appear in the theory of signatures of paths, seen briefly in Section 8.3. It is of interest to know whether their G-orbits agree up to scaling:

$$\begin{aligned} \mathbf{c}_{\text{axis}} &= \left(\tfrac{1}{6}, \tfrac{1}{2}, 0, \tfrac{1}{2}, 0, 0, 0, \tfrac{1}{6}\right) \\ \text{and} \quad \mathbf{c}_{\text{mono}} &= \left(\tfrac{1}{6}, \tfrac{1}{4}, \tfrac{1}{6}, \tfrac{4}{15}, \tfrac{1}{12}, \tfrac{2}{15}, \tfrac{1}{10}, \tfrac{1}{6}\right). \end{aligned}$$

The two polynomials in Example 11.16 are invariants of $\mathrm{SL}(2, K)$ acting on the tensor space $\mathbb{R}^8$. The following rational function is invariant under $\mathrm{GL}(2, K)$. It is homogeneous of degree 0,so it represents a rational function on the projective space $\mathbb{P}^7$, invariant under projective transformations:

$$\frac{\mathrm{Hex}(\mathbf{x})^2}{\mathrm{Det}(\mathbf{x})}. \tag{11.12}$$

We find that the rational invariant (11.12) evaluates to the number 81 on the tensor $\mathbf{c}_{\text{axis}}$, and it evaluates to 45 on the tensor $\mathbf{c}_{\text{mono}}$. Hence the orbit closures of our two special core tensors of format $2 \times 2 \times 2$ are disjoint in $\mathbb{P}^7$.

We now come to determination of the ring of invariants for the G-action on the space K^8 of $2 \times 2 \times 2$ tensors. Using Derksen's algorithm, we derive the following result:

Theorem 11.17. *The invariant ring $K[\mathbf{x}]^{\mathrm{SL}(2)}$ of $2 \times 2 \times 2$ tensors has Krull dimension 5. It is minimally generated by 13 invariants, namely the hexagon invariant of degree 2, eight invariants of degree 4 (including the hyperdeterminant), and four invariants of degree 6.*

In addition to the hyperdeterminant, there are three additional invariants of degree 4 that deserve special attention. Each has 17 terms when expanded. One of these invariants is

$$\begin{aligned}(x_{111}x_{222} - x_{212}x_{121})^2 \; + \; & x_{121}x_{222}x_{112}^2 + x_{111}x_{212}x_{122}^2 + x_{121}x_{222}x_{211}^2 \\ + \; x_{111}x_{212}x_{221}^2 - & (x_{122} + x_{221})(x_{112} + x_{211})(x_{111}x_{222} + x_{212}x_{121}) \\ & + 2x_{111}x_{122}x_{212}x_{221} + 2x_{112}x_{121}x_{211}x_{222}.\end{aligned} \tag{11.13}$$

The other two invariants in this family are obtained by permuting indices.

Corollary 11.18. *The three quartics in* (11.13) *together with* Hex *and* Det *are algebraically independent. All other invariants in $K[\mathbf{x}]^{\mathrm{SL}(2)}$ are integral over the polynomial subring generated by these five. The five invariants cut out the nullcone $\mathcal{V}(K[\mathbf{x}]_+^{\mathrm{SL}(2)})$, which has dimension 4 and degree 12 in $\mathbb{P}^7$.*

It is instructive to restrict the 13 generating invariants in Theorem 11.17 to the 4-dimensional subspace $\mathrm{Sym}_3(K^2)$ of symmetric $2 \times 2 \times 2$ tensors. We discussed this in Example 11.10. The restriction is carried out by setting

$$x_{111} = x_1, \; x_{112} = x_{121} = x_{211} = \frac{1}{3}x_2, \; x_{122} = x_{212} = x_{221} = \frac{1}{3}x_3, \; x_{222} = x_4.$$

The resulting symmetric tensors correspond to binary cubics (11.8). The hyperdeterminant and the five other generators of degree 4 all specialize to the *discriminant* Δ of the binary cubic. The other eight generators of $K[\mathbf{x}]^{\mathrm{SL}(2)}$, including the hexagon invariant, specialize to zero. In this manner, the invariant ring in Theorem 11.17 maps onto the invariant ring of binary cubics. What about the other irreducible SL(2) modules in $(K^2)^{\otimes 3}$?

Exercises

(1) Let G be the symmetry group of the square $[-1,1]^2$ in the plane $\mathbb{R}^2$. This is an order-8 subgroup in $\mathrm{GL}(2,\mathbb{R})$. List all eight matrices. Determine the invariant ring $\mathbb{R}[x_1,x_2]^G$.

(2) Let G be the symmetry group of the regular 3-cube, as a subgroup of $\mathrm{GL}(3,\mathbb{R})$. How many matrices are in G, and what are their characteristic polynomials? Determine the Molien series (11.7) of this group. What does it tell you about the invariant ring?

(3) Fix $n=5$. Let $\psi(j)$ denote the number of monomials in the expansion of the power sum P_j in terms of the elementary symmetric functions E_1,E_2,E_3,E_4,E_5. Compute $\psi(j)$ for some small values, say $j\leq 20$. Guess a formula for $\psi(j)$. Can you prove it?

(4) Show that Noether's degree bound is always tight for finite cyclic groups.

(5) Find a subgroup of $\mathrm{GL}(4,K)$ that has order 15. Compute the invariant ring of your matrix group.

(6) Let T be the group of 3×3 diagonal matrices with determinant 1, acting on the space $V=\mathrm{Sym}_3(K^3)$ of ternary cubics. This group is the torus $T\simeq(K^*)^2$. Determine the invariant ring $K[V]^T$. Do you see any relationship to the invariants in Example 11.12?

(7) Let $G=A_n$ be the *alternating group* of order $n!/2$. Its elements are the even permutation matrices. Determine the invariant ring $K[\mathbf{x}]^G$.

(8) List all 103 monomials of the invariant I_6 of ternary cubics in Example 11.12. Compute the explicit formulas, in terms of the 10 coefficients $x_1,x_2,\ldots,x_9,x_0$, for the discriminant and the j-invariant.

(9) Consider the action of $\mathrm{SL}(3,K)$ on the space $\mathrm{Sym}_2(K^3)\simeq K^6$ of symmetric 3×3 matrices. The entries of the 6×6 matrix $\phi(\sigma)$ are quadratic forms in $\sigma_{11},\sigma_{12},\ldots,\sigma_{33}$. Write this matrix explicitly, similarly to the 4×4 matrix $\phi(\sigma)$ shown in Example 11.10. What is the invariant ring?

(10) Using an internet source, find generators for the ring of invariants of the action of $\mathrm{SL}(2,K)$ on the space $\mathrm{Sym}_5(K^2)$ of binary quintics. Write them explicitly as polynomials in the coefficients x_1,x_2,x_3,x_4,x_5,x_6.

(11) Using Derksen's algorithm, compute the invariant ring for binary quartics ($p=4,d=2$). How many minimal generators does this ring have?

(12) The rotation group $\mathrm{SO}(2,\mathbb{R})$ acts by left multiplication on the space of 2×2 matrices. Describe the orbits and determine the invariant ring.

(13) Consider the action of $\mathrm{SO}(2, \mathbb{R})$ on the 5-dimensional space of binary quartics. Describe the orbits and determine the invariant ring.

(14) Hilbert's 14th problem asked: Is the invariant ring of every matrix group $G \subset \mathrm{GL}(n, K)$ finitely generated? Find the answer and explain it.

Chapter 12

Semidefinite Programming

"*Premature optimization is the root of all evil*", Donald Knuth

The transition from linear algebra to nonlinear algebra has a natural counterpart in convex optimization, namely the passage from linear programming to semidefinite programming. This transition is what we now embark on. The term "program" or "programming", as it is used in this chapter, is simply an old-fashioned way of saying "optimization problem".

Linear programming concerns the solution of linear systems of inequalities and the optimization of linear functions subject to linear constraints. The feasible region is a *convex polyhedron*, and the optimal solutions form a face of that polyhedron. In *semidefinite programming* we work in the space of symmetric $n \times n$ matrices. The inequality constraints now stipulate that some linear combination of matrices be positive semidefinite. The feasible region given by such constraints is a closed convex set, known as a *spectrahedron*. We again seek to optimize a linear function, but over a spectrahedron instead of a polyhedron. The condition for a polynomial to be a sum of squares may be regarded as a semidefinite program. This gives a connection to the real Nullstellensatz (Chapter 6), and it establishes semidefinite programming as a key tool for computing in real algebraic geometry.

12.1. Spectrahedra

In this chapter we work over the field $\mathbb{R}$ of real numbers. This field comes with an order. The Spectral Theorem of linear algebra states that all eigenvalues of a symmetric matrix $A \in \mathrm{Sym}_2(\mathbb{R}^n)$ are real. Moreover, there is an orthonormal basis of $\mathbb{R}^n$ consisting of eigenvectors of A. We say that the matrix A is *positive definite* if it satisfies the following conditions. It is a basic fact about quadratic forms that these three conditions are equivalent:

(1) All n eigenvalues of A are positive real numbers.

(2) All 2^n principal minors of A are positive real numbers.

(3) Every nonzero column vector $\mathbf{u} \in \mathbb{R}^n$ satisfies $\mathbf{u}^T A \mathbf{u} > 0$.

Here, by a *principal minor* we mean the determinant of any square submatrix of A whose set of column indices agrees with its set of row indices. For the empty set, we get the 0×0 minor of A, which equals 1. Next there are the n diagonal entries of A, which are the 1×1 principal minors. We continue with principal minors of size 2×2, then 3×3, etc. At the end of this list, we get to the determinant of A, which is the unique $n \times n$ principal minor.

Each of the three conditions (1), (2) and (3) behaves as expected when we pass to the closure. This is not obvious because the closure of an open semialgebraic set $\{f > 0\}$, where $f \in \mathbb{R}[\mathbf{x}]$, is generally smaller than the closed semialgebraic set $\{f \geq 0\}$ that is defined by the same polynomial f.

Example 12.1. Let $f = x^3 + x^2y + xy^2 + y^3 - x^2 - y^2$. The set $\{f > 0\}$ is the open halfplane above the line $x + y = 1$ in $\mathbb{R}^2$. The closure of the set $\{f > 0\}$ is the corresponding closed halfplane. It is strictly contained in the set $\{f \geq 0\}$. Namely, $\{f \geq 0\}$ is the union of $\{f > 0\}$ and the origin $(0, 0)$. To see this, it helps to factor the polynomial f into irreducible factors.

Luckily, no such thing happens with condition (2) for positive definite matrices. In fact, the closure of the set of positive definite matrices is obtained by allowing the case of equality in each of the three conditions.

Theorem 12.2. *For a symmetric $n \times n$ matrix A, the following three conditions are equivalent. If these hold then A is called* positive semidefinite*:*

(1') *All n eigenvalues of A are nonnegative real numbers.*

(2') *All 2^n principal minors of A are nonnegative real numbers.*

(3') *Every nonzero column vector $\mathbf{u} \in \mathbb{R}^n$ satisfies $\mathbf{u}^T A \mathbf{u} \geq 0$.*

The semialgebraic set PSD_n of all positive semidefinite symmetric $n \times n$ matrices is a full-dimensional closed convex cone in $\mathrm{Sym}_2(\mathbb{R}^n)$.

We use the notation $X \succeq 0$ to express that a symmetric matrix X is positive semidefinite. A *spectrahedron* $\mathcal{S}$ is the intersection of the cone PSD_n with an affine-linear subspace $\mathcal{L}$ of the ambient space $\mathrm{Sym}_2(\mathbb{R}^n)$. Hence, spectrahedra are closed convex semialgebraic sets.

An affine-linear space $\mathcal{L}$ of symmetric matrices is given either by a linear parametrization or as the solution set to a system of (usually inhomogeneous) linear equations. In the latter representation,

$$\mathcal{L} = \{ X \in \mathrm{Sym}_2(\mathbb{R}^n) : \langle A_1, X \rangle = b_1, \ldots, \langle A_s, X \rangle = b_s \}. \tag{12.1}$$

Here $A_1, A_2, \ldots, A_s \in \mathrm{Sym}_2(\mathbb{R}^n)$ are fixed matrices and $b_1, b_2, \ldots, b_s \in \mathbb{R}$ are fixed scalars. We employ the standard inner product in the space of square matrices, which is given by the trace of the matrix product:

$$\langle A, X \rangle := \mathrm{trace}(AX) = \sum_{i=1}^{n} \sum_{j=1}^{n} a_{ij} x_{ij}. \tag{12.2}$$

The associated spectrahedron $\mathcal{S} = \mathcal{L} \cap \mathrm{PSD}_n$ consists of all positive semidefinite matrices that lie in the subspace $\mathcal{L}$. The same matrices can be used to describe a different subspace in its parametric representation, namely

$$\{ A_0 + x_1 A_1 + \cdots + x_s A_s : (x_1, \ldots, x_s) \in \mathbb{R}^s \} \subset \mathrm{Sym}_2(\mathbb{R}^n). \tag{12.3}$$

Here A_0 is an additional matrix. If this formulation is used, then it is customary to identify the spectrahedron with its preimage in $\mathbb{R}^s$. In symbols,

$$\mathcal{S} = \{ (x_1, \ldots, x_s) \in \mathbb{R}^s : A_0 + x_1 A_1 + \cdots + x_s A_s \succeq 0 \}. \tag{12.4}$$

Proposition 12.3. *Every convex polyhedron is a spectrahedron. Convex polyhedra are precisely the spectrahedra that arise when the ambient affine-linear subspace of* $\mathrm{Sym}_2(\mathbb{R}^n)$ *consists only of diagonal* $n \times n$ *matrices.*

Proof. Suppose that the matrices $A_0, A_1, \ldots, A_s$ are diagonal matrices. Then (12.4) is the solution set in $\mathbb{R}^s$ of a system of n inhomogeneous linear inequalities. Such a set is a convex polyhedron. Every convex polyhedron in $\mathbb{R}^s$ has such a representation. We simply write its defining linear inequalities as the diagonal entries of the $n \times n$ matrix $A_0 + x_1 A_1 + \cdots + x_s A_s$. □

In the previous proposition, the subspace $\mathcal{L}$ is an affine-linear subspace of $\mathrm{Sym}_2(\mathbb{R}^n)$ that consists only of diagonal matrices. The formula $\mathcal{S} = \mathcal{L} \cap \mathrm{PSD}_n$ with $\mathcal{L}$ as in (12.1) corresponds to the standard representation of a convex polyhedron as the set of nonnegative points in an affine-linear space. Here the equations for X in (12.1) include those that require the off-diagonal entries of each X to be zero:

$$\langle X, E_{ij} \rangle = x_{ij} = 0 \quad \text{for } i \neq j.$$

The other matrices A_i are diagonal and their b_i are typically nonzero.

Example 12.4. Let $\mathcal{L}$ be the space of symmetric 3×3 matrices whose three diagonal entries are all equal to 1. This is an affine-linear subspace of dimension $s = 3$ in $\mathrm{Sym}_2(\mathbb{R}^3) \simeq \mathbb{R}^6$. The spectrahedron $\mathcal{S} = \mathcal{L} \cap \mathrm{SDP}_3$ is the yellow convex body seen in Figure 1.1. To draw this spectrahedron in $\mathbb{R}^3$, one uses the representation (12.4) with $s = 3$, namely

$$\mathcal{S} \quad = \quad \left\{ (x, y, z) \in \mathbb{R}^3 : \begin{pmatrix} 1 & x & y \\ x & 1 & z \\ y & z & 1 \end{pmatrix} \succeq 0 \right\}.$$

The boundary of $\mathcal{S}$ consists of all points (x, y, z) where the matrix has determinant zero and its nonzero eigenvalues are positive. The determinant is a polynomial of degree 3 in x, y, z, so the boundary lies in a cubic surface in $\mathbb{R}^3$. This cubic surface also contains points where the three eigenvalues are positive, zero and negative. Such points are drawn in red in Figure 1.1. They lie in the Zariski closure of the yellow boundary points.

We next slice our 3-dimensional spectrahedron to get a planar picture.

Example 12.5. Suppose that $\mathcal{L} \subset \mathrm{Sym}_2(\mathbb{R}^3)$ is a general 2-dimensional plane that intersects the 6-dimensional cone PSD_3. The spectrahedron $\mathcal{S}$ is a convex body in $\mathbb{R}^2$ whose boundary is a smooth cubic curve, drawn in red in Figure 12.1 on the left. On that boundary, the 3×3 determinant vanishes and the other two eigenvalues are positive. For points $(x, y) \in \mathbb{R}^2 \backslash \mathcal{S}$, the matrix has at least one negative eigenvalue. The black curve lies in the Zariski closure of the red curve. It separates points in $\mathbb{R}^2 \backslash \mathcal{S}$ whose remaining two eigenvalues are positive from those with two negative eigenvalues.

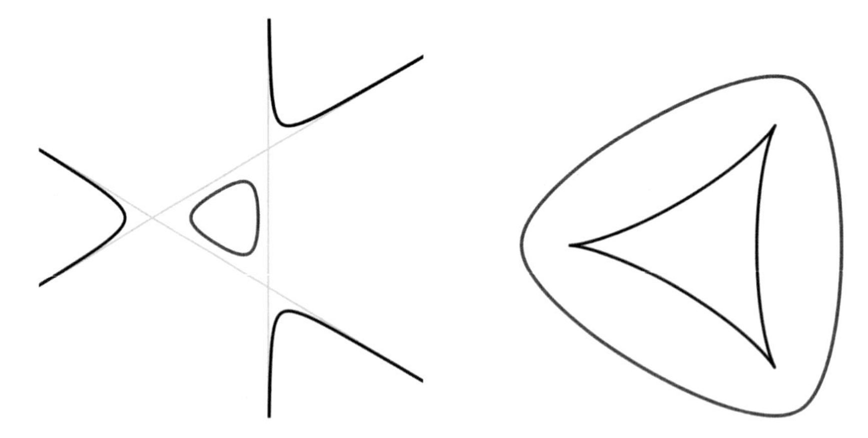

Figure 12.1. A plane curve of degree 3 (left) and its dual curve of degree 6 (right). The red part on the left bounds a spectrahedron while that on the right bounds its convex dual. The points on the dual curve are lines that are tangent to the primal curve. Biduality holds.

To be explicit, suppose that our planar cubic spectrahedron is written as

$$\mathcal{S} \; = \; \left\{ (x, y) \in \mathbb{R}^2 : \begin{pmatrix} 1 & x & x+y \\ x & 1 & y \\ x+y & y & 1 \end{pmatrix} \succeq 0 \right\}. \tag{12.5}$$

The cubic curve is the locus where the 3×3 matrix in (12.5) is singular. Its determinant equals

$$f \quad = \quad 2x^2y + 2xy^2 - 2x^2 - 2xy - 2y^2 + 1. \tag{12.6}$$

The curve $\{f = 0\}$ has four connected components in $\mathbb{R}^2$, one in red and three in black, as shown in Figure 12.1 on the left. The boundary of the cubic spectrahedron $\mathcal{S}$ is the convex part of the curve that is shown in red.

The picture on the right in Figure 12.1 shows the *dual curve*. This lives in the dual plane whose points (u, v) represent the lines $\ell = \{(x, y) : ux + vy = 1\}$ in the given plane $\mathbb{R}^2$. The points in the dual curve correspond to lines ℓ that are tangent to the original curve. For an introduction to duality in algebraic geometry we refer to [**5**, §5.2.4]. Chapter 5 in [**5**] argues that this algebraic duality agrees with duality in convexity or optimization.

The curve that is dual to (12.6) has degree 6, and its equation can be found by the following computation of an elimination ideal in $\mathbb{R}[x, y, u, v]$:

$$\begin{aligned} &\langle\, f(x,y)\,,\; u\cdot x + v\cdot y - 1\,,\; \partial f/\partial x \cdot v - \partial f/\partial y \cdot u \,\rangle \,\cap\, \mathbb{R}[u,v] \\ = \quad &\langle\, 8u^6 - 24u^5v + 21u^4v^2 - 2u^3v^3 + 21u^2v^4 - 24uv^5 \\ &+ 8v^6 - 24u^5 + 60u^4v - 24u^3v^2 - 24u^2v^3 + 60uv^4 \\ &- 24v^5 + 12u^4 - 24u^3v + 36u^2v^2 - 24uv^3 + 12v^4 + 24u^3 \\ &- 36u^2v - 36uv^2 + 24v^3 - 24u^2 + 24uv - 24v^2 + 4\,\rangle. \end{aligned} \tag{12.7}$$

Do try this out! The black points on the sextic in Figure 12.1 correspond to lines that are tangent at black points of the cubic, and similarly for the red points. Moreover, the convex set enclosed by the red sextic on the right in Figure 12.1 is dual as a convex set to the spectahedron on the left.

The polynomials in (12.6) and (12.7) have degrees 3 and 6 respectively, confirming what was asserted in the caption to Figure 12.1. A random line L will meet the curve in three (left) or six (right) complex points. Consider the point p on the other side of duality that corresponds to the line L. The points of L correspond to the lines through p. There are three (right) or six (left) complex lines through p that are tangent to the curve.

12.2. Optimization and Duality

We now finally come to *semidefinite programming* (SDP). This refers to the problem of maximizing or minimizing a linear function over a spectrahedron. Linear programming is the special case where the spectrahedron consists of diagonal matrices. If the spectrahedron is given in its standard form representation (12.1), then we get the SDP in its primal form:

$$\begin{aligned} &\text{Minimize } \langle C, X\rangle \text{ subject to } \langle A_1, X\rangle = b_1, \\ &\langle A_2, X\rangle = b_2, \ldots, \langle A_s, X\rangle = b_s \text{ and } X \succeq 0. \end{aligned} \tag{12.8}$$

Here $C = (c_{ij})$ is a $n \times n$ matrix that represents the cost function. Every convex optimization problem has a dual problem. At first glance, it is not so easy to relate SDP duality to duality in algebraic geometry, as seen for plane curves in Figure 12.1. Making that connection is the point of [**5**, Chapter 5].

We abbreviate $b^T x = \sum_{i=1}^{s} b_i x_i$. The semidefinite programming problem dual to (12.8) takes the following form:

(12.9) Maximize $b^T x$ subject to $C - x_1 A_1 - x_2 A_2 - \cdots - x_s A_s \succeq 0$.

In this formulation, the spectrahedron of feasible points lives in $\mathbb{R}^s$, similarly to (12.4). We refer to either of the formulations (12.8) and (12.9) as a *semidefinite program*, also abbreviated SDP. The relationship between the primal SDP and the dual SDP is given by the following theorem:

Theorem 12.6 (Weak Duality Theorem). *If x is any feasible solution to (12.9) and X is any feasible solution to (12.8), then $b^T x \leq \langle C, X \rangle$. If the equality $b^T x = \langle C, X \rangle$ holds, then both x and X are optimal.*

The term *feasible* means only that the point x (resp. X) satisfies the equations and inequalities that are required in (12.9) (resp. (12.8)). The point is *optimal* if it is feasible and it solves the program, i.e. it attains the maximum (resp. minimum) value for that optimization problem.

Proof. The inner product of two positive semidefinite matrices is a nonnegative real number:

$$0 \leq \langle C - \sum_{i=1}^{s} x_i A_i, X \rangle \;=\; \langle C, X \rangle - \sum_{i=1}^{s} x_i \cdot \langle A_i, X \rangle \;=\; \langle C, X \rangle - b^T x.$$

This shows that the optimal value of the minimization problem (12.8) is an upper bound for the optimal value of the maximization problem (12.9). If the equality is attained by a pair (X, x) of feasible solutions, then X must be optimal for (12.8) and x must be optimal for (12.9). □

There is also a Strong Duality Theorem which states that, under suitable hypotheses, the *duality gap* $\langle C, X \rangle - b^T x$ must attain the value zero for some feasible pair (X, x). These hypotheses are always satisfied for diagonal matrices, and we recover the Duality Theorem of Linear Programming as a special case. Note that this is equivalent to Farkas' Lemma in Theorem 6.13.

Interior point methods for linear programming are numerical algorithms that start at an interior point of the feasible polyhedron and create a path from that point towards an optimal vertex. The same class of algorithms works for semidefinite programming. These algorithms run in polynomial time and are well behaved in practice. Semidefinite programming has a much greater expressive power than linear programming. Many more problems can be phrased as an SDP. We illustrate this with an example from linear algebra.

Example 12.7 (The largest eigenvalue). Let A be a positive definite symmetric $n \times n$ matrix. Consider the problem of computing its largest eigenvalue $\lambda_{\max}(A)$. We can solve this without writing down the characteristic polynomial and extracting its roots. Let $C = \mathrm{Id}$ be the identity matrix and consider the SDP problems (12.8) and (12.9) with $s = 1$ and $b = 1$. They are

(12.8)′ Minimize $\mathrm{trace}(X)$ subject to $X \succeq 0$ and $\langle A, X\rangle = 1$.

(12.9)′ Maximize x subject to $\mathrm{Id} - xA \succeq 0$.

If x^* is the common optimal value of (12.8)′ and (12.9)′, then $\lambda_{\max}(A) = 1/x^*$.

The inner product $\langle A, X\rangle = \mathrm{trace}(A \cdot X)$ of two positive semidefinite matrices A and X can only be zero when their matrix product $A \cdot X$ is zero. We record this for our situation:

Lemma 12.8. *Let X be a feasible solution of* (12.8) *and let x be a feasible solution of* (12.9). *Then the duality gap $\langle C, X\rangle - b^T x$ is zero if and only if the product of symmetric matrices $(C - \sum_{i=1}^s x_i A_i) \cdot X$ is the zero matrix.*

This lemma implies the following algebraic reformulation of duality:

Corollary 12.9. *Consider the following system of s linear equations and n^2 bilinear equations in the $\binom{n+1}{2} + s$ coordinates of the pair (X, x):*

$$\langle A_1, X\rangle = b_1\,, \;\ldots\;, \langle A_s, X\rangle = b_s \quad \text{and} \quad (C - \sum_{i=1}^{s} x_i A_i) \cdot X = 0. \tag{12.10}$$

Let (X, x) be as in Lemma 12.8 *and assume $X \succeq 0$, $C - \sum_{i=1}^s x_i A_i \succeq 0$ and that* (12.10) *holds. Then X is optimal for* (12.8) *and x is optimal for* (12.9).

The equations (12.10) are the *Karush-Kuhn-Tucker (KKT) equations.* These play a major role when one is exploring semidefinite programming from an algebraic perspective. In particular, they allow one to study the nature of the optimal solution as a function of the data. A key feature of the KKT system is that the two optimal matrices have complementary ranks. This follows from the *complementary slackness* condition on the right of (12.10):

$$\mathrm{rank}\big(C - \sum_{i=1}^{s} x_i A_i\big) \,+\, \mathrm{rank}(X) \,\leq\, n.$$

In particular, if X is known to be nonzero, then the determinant of the matrix $C - \sum_{i=1}^s x_i A_i$ vanishes. For instance, for the eigenvalue problem in Example 12.7, we have $(\mathrm{Id} - xA) \cdot X = 0$ and $\langle A, X\rangle = 1$. This implies that $\det(\mathrm{Id} - xA) = 0$, so $1/x$ is a root of the characteristic polynomial.

Example 12.10. Consider the problem of maximizing a linear function $\ell(x,y) = ux + vy$ over the spectrahedron $\mathcal{S}$ in (12.5). This is an SDP in the dual formulation (12.9) with $s = 2$ and $b = (u, v)$ and with

$$A_1 = -\begin{pmatrix} 0 & 1 & 1 \\ 1 & 0 & 0 \\ 1 & 0 & 0 \end{pmatrix} \quad \text{and} \quad A_2 = -\begin{pmatrix} 0 & 0 & 1 \\ 0 & 0 & 1 \\ 1 & 1 & 0 \end{pmatrix}.$$

The KKT system (12.10) is a system of equations in eight unknowns:

$$2x_{12} + 2x_{13} + u = 2x_{13} + 2x_{23} + v = 0 \quad \text{and}$$

$$\begin{pmatrix} 1 & x & x{+}y \\ x & 1 & y \\ x{+}y & y & 1 \end{pmatrix} \cdot \begin{pmatrix} x_{11} & x_{12} & x_{13} \\ x_{12} & x_{22} & x_{23} \\ x_{13} & x_{23} & x_{33} \end{pmatrix} = \begin{pmatrix} 0 & 0 & 0 \\ 0 & 0 & 0 \\ 0 & 0 & 0 \end{pmatrix}.$$

Here u and v are parameters. By eliminating the variables x_{ij} from the equations above, we obtain an ideal I in $\mathbb{Q}[u, v, x, y]$ that characterizes the optimal solution (x^*, y^*) to our SDP as an algebraic function of (u, v). Let ℓ^* now be a new unknown, and consider the elimination ideal $\big(I + \langle ux + vy - \ell^* \rangle\big) \cap \mathbb{Q}[u, v, \ell^*]$. Its generator is a ternary sextic in u, v, ℓ^*. This is precisely the homogenization of the dual sextic in (12.7). It expresses the optimal value ℓ^* as an algebraic function of degree 6 in the cost (u, v).

This relationship between the dual hypersurface and the optimal value function generalizes to arbitrary polynomial optimization problems, including semidefinite programs. This is the content of [**5**, Theorem 5.23]. We refer to the book [**5**], and especially Chapter 5, for further reading on spectrahedra, semidefinite programming, and the relevant duality theory.

A fundamental task in convex algebraic geometry [**5**] is computing the convex hull of a given algebraic variety or semialgebraic set. Recall that the *convex hull* of a set is the smallest convex set containing the given set. Spectrahedra or their linear projections, known as *spectrahedral shadows*, can be used for this task. This matters for optimization since minimizing a linear function over a set is equivalent to minimizing over its convex hull.

Example 12.11 (Toeplitz spectrahedron)**.** Consider the convex body

$$K = \left\{ (x, y, z) \in \mathbb{R}^3 \, : \, \begin{bmatrix} 1 & x & y & z \\ x & 1 & x & y \\ y & x & 1 & x \\ z & y & x & 1 \end{bmatrix} \succeq 0 \right\}. \tag{12.11}$$

The determinant of the given *Toeplitz matrix* of size 4×4 factors as

$$(x^2 + 2xy + y^2 - xz - x - z - 1)(x^2 - 2xy + y^2 - xz + x + z - 1).$$

The spectrahedron (12.11) is the convex hull of the *cosine moment curve*

$$\big\{ \big(\cos(\theta), \cos(2\theta), \cos(3\theta)\big) \, : \, \theta \in [0, 2\pi] \big\}. \tag{12.12}$$

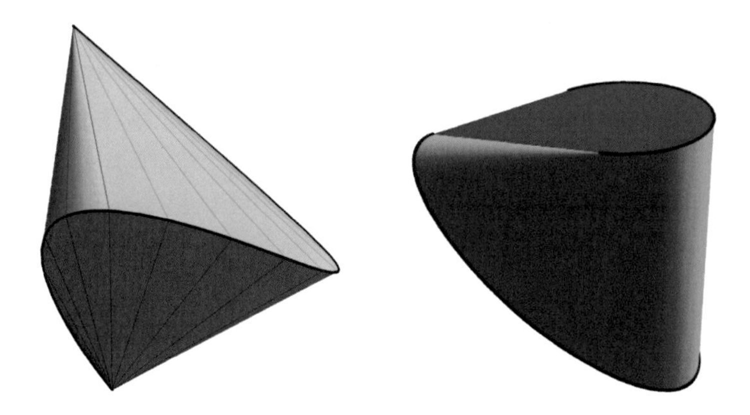

Figure 12.2. The Toeplitz spectrahedron and its dual convex body.

The curve and its convex hull are shown on the left in Figure 12.2. The two endpoints, $(x, y, z) = (1, 1, 1)$ and $(x, y, z) = (-1, 1, -1)$, correspond to matrices of rank 1. All other points on the curve have rank 2.

We call (12.11) the *Toeplitz spectrahedron*. To construct this convex body geometrically, we form the cone from each endpoint over the cosine curve, and we intersect these two quadratic cones. The two cones intersect along this curve and the line through the endpoints of the cosine curve.

Shown on the right in Figure 12.2 is the convex body K^* dual to the Toeplitz spectrahedron K. Its points (a_1, a_2, a_3) correspond to the trigonometric polynomials $1+a_1\cos(\theta)+a_2\cos(2\theta)+a_3\cos(3\theta)$ that are nonnegative on the interval $[0, 2\pi]$. This convex body K^* is not a spectrahedron because it has a nonexposed edge, that is, a 1-dimensional face which is not the intersection of K^* with a supporting hyperplane (cf. [**5**, Exercise 6.13]).

We close this section with a reformulation of SDP that will appeal to algebraists. It expresses the primal (12.8) and the dual (12.9) in a form that looks symmetric. After replacing the constraint matrices $A_1, A_2, \ldots, A_s$ by linear combinations, we may assume that $b_1 = 1$ and $b_2 = \cdots = b_s = 0$.

Let $\mathcal{V}$ be the subspace of $\mathrm{Sym}_2(\mathbb{R}^n)$ spanned by $\{A_2, \ldots, A_s\}$, and let $\mathcal{U}$ be the subspace of $\mathrm{Sym}_2(\mathbb{R}^n)$ spanned by $\{C, A_1\}$ and $\mathcal{V}$. We assume $\dim(\mathcal{V}) = s - 1$ and $\dim(\mathcal{U}) = s + 1$. Our instance defines a pair of flags

$$(12.13) \qquad \mathcal{V} \subset \mathcal{U} \subset \mathrm{Sym}_2(\mathbb{R}^n) \quad \text{and} \quad \mathcal{U}^\perp \subset \mathcal{V}^\perp \subset \mathrm{Sym}_2(\mathbb{R}^n)^*.$$

In what follows we assume that $\mathcal{U}$ and $\mathcal{V}$ are generic subject to the inclusion (12.13). This ensures that strong duality holds, i.e. the duality gap is zero.

The KKT equations (12.10) can now be written as

$$(12.14) \qquad X \in \mathcal{V}^\perp, \quad Y \in \mathcal{U} \quad \text{and} \quad X \cdot Y = 0.$$

The matrices X and Y are considered up to scaling, that is, they are unknown points in the projective space $\mathbb{P}(\mathrm{Sym}_2(\mathbb{R}^n)) = \mathbb{P}^{\binom{n+1}{2}-1}$. The flag $\mathcal{V} \subset \mathcal{U}$ constitutes our data, specifying a variety in $\mathbb{P}^{\binom{n+1}{2}-1} \times \mathbb{P}^{\binom{n+1}{2}-1}$ via (12.14).

Corollary 12.12. *The semidefinite programming problem* (12.8) *is equivalent to solving the bilinear system of equations* (12.14) *subject to* $X, Y \succeq 0$.

Proof. Using strong duality, this is a reformulation of Corollary 12.9. □

12.3. Sums of Squares

Semidefinite programming can be used to model and solve arbitrary polynomial optimization problems. The key to doing so is the representation of nonnegative polynomials in terms of sums of squares, or, more generally, the real Nullstellensatz (see Chapter 6). We explain this for the simplest scenario, namely the problem of unconstrained polynomial optimization.

Let $f(x_1, \ldots, x_n)$ be a polynomial of even degree $d = 2p$, and suppose that f attains a minimal real value f^* on $\mathbb{R}^n$. Our goal is to compute f^* and a point $\mathbf{u}^* \in \mathbb{R}^n$ such that $f(\mathbf{u}^*) = f^*$. Minimizing a function is equivalent to finding the best possible lower bound λ for that function. Our goal is therefore equivalent to solving the following optimization problem:

$$\text{Maximize } \lambda \text{ such that } \quad f(\mathbf{x}) - \lambda \geq 0 \quad \text{for all } \mathbf{x} \in \mathbb{R}^n. \tag{12.15}$$

This is a difficult problem. Instead, we consider the following relaxation:

$$\text{Maximize } \lambda \text{ such that } \quad f(\mathbf{x}) - \lambda \quad \text{is a sum of squares in } \mathbb{R}[\mathbf{x}]. \tag{12.16}$$

Here *relaxation* means that we have modified the set of feasible solutions. Indeed, every sum of squares is nonnegative, but not every nonnegative polynomial is a sum of squares of polynomials. For instance, the Motzkin polynomial $x^4y^2 + x^2y^4 + 1 - 3x^2y^2$ is nonnegative, but it is not a sum of squares of polynomials, by Example 6.11. For that reason, the optimal value of (12.16) is always a lower bound for the optimal value of (12.15), but the two values can be different in some cases. However, here is the good news:

Proposition 12.13. *The optimization problem* (12.16) *is a semidefinite program.*

Proof. Let $\mathbf{x}^{[p]}$ be the column vector whose entries are all monomials in $x_1, \ldots, x_n$ of degree $\leq p$. Thus $\mathbf{x}^{[p]}$ has length $\binom{n+p}{n}$. Let $G = (g_{ij})$ be a symmetric $\binom{n+p}{n} \times \binom{n+p}{n}$ matrix with unknown entries. Then the scalar $(\mathbf{x}^{[p]})^T \cdot G \cdot \mathbf{x}^{[p]}$ is a polynomial of degree $d = 2p$ in $x_1, \ldots, x_n$. We set

$$f(\mathbf{x}) - \lambda \quad = \quad (\mathbf{x}^{[p]})^T \cdot G \cdot \mathbf{x}^{[p]}. \tag{12.17}$$

By collecting coefficients of the $\mathbf{x}$-monomials, we get a system of $\binom{2p+n}{n}$ linear equations in $1 + \binom{\binom{n+p}{n}+1}{2}$ unknowns, namely λ and the matrix entries g_{ij}.

Suppose the linear system (12.17) has a solution (G, λ) such that G is positive semidefinite. Then we can write $G = H^T H$ where H is a real matrix with r rows and $\binom{p+n}{n}$ columns. (This is known as a *Cholesky factorization* of G; see Exercise 2.) The polynomial in (12.17) then equals

$$f(\mathbf{x}) - \lambda \quad = \quad (H\mathbf{x}^{[p]})^T \cdot (H\mathbf{x}^{[p]}). \tag{12.18}$$

This is the scalar product of a vector of length r with itself. Hence $f(\mathbf{x}) - \lambda$ is a sum of squares of polynomials of degree $\leq p$. Conversely, every representation of $f(\mathbf{x}) - \lambda$ as a sum of squares of polynomials uses polynomials of degree $\leq p$ and can hence be written in the form (12.18).

Our argument shows that the problem (12.16) is equivalent to

$$\begin{array}{c}\text{Maximize } \lambda \text{ subject to } (G, \lambda) \text{ satisfying} \\ \text{the linear equations (12.17) and } G \succeq 0.\end{array} \tag{12.19}$$

This is a semidefinite programming problem, so the proof is complete. □

If $n = 1$ or $d = 2$ or ($n = 2$ and $d = 4$), then every nonnegative polynomial of degree d in n real variables is a sum of squares. These are precisely the cases where (12.15) and (12.19) are equivalent. See [**5**, §3.1.2].

Example 12.14 ($n = 1, p = 2, d = 4$)**.** Suppose we seek to find the minimum of the degree-4 polynomial $f(x) = 3x^4 + 4x^3 - 12x^2$. Of course, we know how to do this using calculus. Here we explain the SDP approach.

The linear equations (12.17) have a 1-dimensional space of solutions. Introducing a parameter μ for that line, the solutions can be written as

$$f(x) - \lambda \quad = \quad \begin{pmatrix} x^2 & x & 1 \end{pmatrix} \begin{pmatrix} 3 & 2 & \mu - 6 \\ 2 & -2\mu & 0 \\ \mu - 6 & 0 & -\lambda \end{pmatrix} \begin{pmatrix} x^2 \\ x \\ 1 \end{pmatrix}. \tag{12.20}$$

Consider the set of all pairs (λ, μ) such that the 3×3 matrix in (12.20) is positive semidefinite. This set is a cubic spectrahedron in the plane $\mathbb{R}^2$, just like that shown on the left in Figure 12.1. We seek to maximize λ over all points in that spectrahedron. The optimal point is $(\lambda^*, \mu^*) = (-32, -2)$. Substituting this into the matrix in (12.20), we obtain a positive semidefinite matrix of rank 2. This can be factored as $G = H^T H$, where H has format 2×3. The resulting representation (12.18) as a sum of two squares is

$$f(x) - \lambda^* \; = \; f(x) + 32 \; = \; \left((\sqrt{3}x - \frac{4}{\sqrt{3}}) \cdot (x+2)\right)^2 \; + \; \frac{8}{3}(x+2)^2.$$

The right-hand side is nonnegative for all x. It takes the value 0 at $x = -2$.

Any polynomial optimization problem can be translated into a relaxation that is a semidefinite programming problem. For unconstrained optimization we saw this in Proposition 12.13. Remarkably, the same approach works for arbitrary optimization problems where both the objective function and the constraints are given by polynomials. Indeed, suppose we wish to minimize $f(\mathbf{x})$ subject to some polynomial constraints. We seek a certificate for $f(\mathbf{x}) - \lambda < 0$ to have no solution. This certificate is promised by the real Nullstellensatz or the Postitivstellensatz. We discussed these in Chapter 6.

If we fix a degree bound, then the existence of a certificate translates into a semidefinite program, and so does the additional requirement for λ to be minimal. The translation is explained in [**5**, §3.4.3]. This relaxation may or may not give the correct solution for some fixed degree bound.

If one increases the degree bound, then the SDP formulation is more likely to succeed, albeit at the expense of having to solve a much larger problem. This is a powerful and widely used approach to polynomial optimization, known as *SOS programming*. The term *Lasserre hierarchy* refers to varying the degree bounds. It is often found in the computer science literature.

Every spectrahedron $\mathcal{S} = \mathcal{L} \cap \mathrm{PSD}_n$ has a special point in its relative interior. This point, defined as the unique matrix in $\mathcal{S}$ whose determinant is maximal, is known as the *analytic center*. Finding the analytic center of $\mathcal{S}$ is a convex optimization problem, since the function $X \mapsto \log\det(X)$ is strictly concave on the cone of positive definite matrices X. The analytic center is important for semidefinite programming because it serves as the starting point for interior point methods. Indeed, the *central path* of an SDP starts at the analytic center and runs to the optimal face. It is computed by a sequence of numerical approximations.

Example 12.15. The determinant function takes on all values between 0 and 1 on the spectrahedron $\mathcal{S}$ in (12.5). The value 1 is attained only by the identity matrix, for $(x, y) = (0, 0)$. This point is the analytic center of $\mathcal{S}$.

We close this section by relating spectrahedra and their analytic centers to statistics. For further reading on this connection we refer to the article [**55**]. Every positive definite $n \times n$ matrix $\Sigma = (\sigma_{ij})$ is the *covariance matrix* of a multivariate normal distribution. Its inverse Σ^{-1} is also symmetric and positive definite. The matrix Σ^{-1} is known as the *concentration matrix*.

A *Gaussian graphical model* is specified by requiring that some off-diagonal entries of Σ^{-1} be zero. These entries correspond to the nonedges of the graph. Maximum likelihood estimation for this graphical model translates into a matrix completion problem. Suppose that S is the sample covariance matrix of a given data set. We regard S as a partial matrix, with visible entries only on the diagonal and on the edges of the graph. One considers the set of all completions of the nonedge entries that make the matrix S positive definite. The set of all these completions is a spectrahedron. Maximum likelihood estimation for the data S in the graphical model amounts to maximizing the logarithm of the determinant. We hence seek to compute the analytic center of the spectrahedron of all matrix completions.

Example 12.16 (Positive definite matrix completion)**.** Suppose that the eight entries σ_{ij} in the following symmetric 4×4 matrix are visible, but the two entries x and y are unknown:

$$\Sigma \quad = \quad \begin{pmatrix} \sigma_{11} & \sigma_{12} & x & \sigma_{14} \\ \sigma_{12} & \sigma_{22} & \sigma_{23} & y \\ x & \sigma_{23} & \sigma_{33} & \sigma_{34} \\ \sigma_{14} & y & \sigma_{34} & \sigma_{44} \end{pmatrix}. \tag{12.21}$$

This corresponds to the graphical model of the 4-cycle whose edges are $12, 23, 34, 41$. Given visible entries σ_{ij}, we consider the set of pairs (x, y) that make Σ positive definite. This set is the interior of a planar spectrahedron $\mathcal{S}_\sigma$ bounded by a quartic curve. The maximum likelihood estimator is the analytic center of $\mathcal{S}_\sigma$.

One is also interested in conditions on the σ_{ij} such that $\mathrm{int}(\mathcal{S}_\sigma)$ is nonempty. When can we find (x, y) that make Σ positive definite? One necessary condition is that the diagonal entries σ_{ii} are positive. A further condition is that the four visible principal 2×2 minors are positive:

$$\sigma_{11}\sigma_{22} > \sigma_{12}^2, \quad \sigma_{22}\sigma_{33} > \sigma_{23}^2, \quad \sigma_{33}\sigma_{44} > \sigma_{34}^2, \quad \sigma_{11}\sigma_{44} > \sigma_{14}^2. \tag{12.22}$$

But these necessary conditions are not sufficient. The answer is as follows.

The region of all matrices σ with $\mathrm{int}(S_\sigma) \neq \emptyset$ is a semialgebraic convex cone. Its boundary is a hypersurface of degree 8. The polynomial defining that hypersurface is

$$\begin{gathered} \sigma_{11}^2\sigma_{44}^2\sigma_{23}^4 - 2\sigma_{22}\sigma_{33}^2\sigma_{44}\sigma_{12}^2\sigma_{14}^2 - 2\sigma_{11}\sigma_{33}\sigma_{44}^2\sigma_{12}^2\sigma_{23}^2 - 2\sigma_{11}\sigma_{22}\sigma_{33}\sigma_{44}\sigma_{14}^2\sigma_{23}^2 \\ + 4\sigma_{33}\sigma_{44}\sigma_{12}^2\sigma_{14}^2\sigma_{23}^2 + 8\sigma_{11}\sigma_{22}\sigma_{33}\sigma_{44}\sigma_{12}\sigma_{14}\sigma_{23}\sigma_{34} - 4\sigma_{33}\sigma_{44}\sigma_{12}^3\sigma_{14}\sigma_{23}\sigma_{34} \\ - 4\sigma_{22}\sigma_{33}\sigma_{12}\sigma_{14}^3\sigma_{23}\sigma_{34} + \sigma_{22}^2\sigma_{33}^2\sigma_{14}^4 - 4\sigma_{11}\sigma_{44}\sigma_{12}\sigma_{14}\sigma_{23}^3\sigma_{34} - 2\sigma_{11}\sigma_{22}\sigma_{33}\sigma_{44}\sigma_{12}^2\sigma_{34}^2 \\ - 2\sigma_{11}\sigma_{22}^2\sigma_{33}\sigma_{14}^2\sigma_{34}^2 + 4\sigma_{22}\sigma_{33}\sigma_{12}^2\sigma_{14}^2\sigma_{34}^2 - 2\sigma_{11}^2\sigma_{22}\sigma_{44}\sigma_{23}^2\sigma_{34}^2 + 4\sigma_{11}\sigma_{44}\sigma_{12}^2\sigma_{23}^2\sigma_{34}^2 \\ + 4\sigma_{11}\sigma_{22}\sigma_{14}^2\sigma_{23}^2\sigma_{34}^2 - 4\sigma_{11}\sigma_{22}\sigma_{12}\sigma_{14}\sigma_{23}\sigma_{34}^3 + \sigma_{11}^2\sigma_{22}^2\sigma_{34}^4 + \sigma_{33}^2\sigma_{44}^2\sigma_{12}^4. \end{gathered}$$

This polynomial is found by eliminating x and y from the determinant and its partial derivatives with respect to x and y, after saturating by the ideal of 3×3 minors. For more details on this example see [**55**, §4.2].

Exercises

(1) Prove Theorem 12.2.

(2) Show that a real symmetric matrix G is positive semidefinite if and only if it admits a Cholesky factorization $G = H^T H$ over the real numbers, with H upper triangular.

(3) Consider the problem of maximizing the smallest eigenvalue among all matrices in an affine-linear subspace of $\mathrm{Sym}_2(\mathbb{R}^n)$. Express this problem as a semidefinite program. For $n = 3$, solve it for the subspace in (12.5).

(4) Maximize and minimize the linear function $13x + 17y + 23z$ over the spectrahedron $\mathcal{S}$ in Example 12.4. Use SDP software if you can.

(5) Maximize and minimize the linear function $13x + 17y + 23z$ over the Toeplitz spectrahedron in Example 12.11. Use SDP software if you can.

(6) Write the dual SDP and solve the KKT system for the previous two problems. Also write down the symmetric formulation in Corollary 12.12.

(7) Compute the convex body dual to the spectrahedron $\mathcal{S}$ in Example 12.4.

(8) Consider the problem of minimizing the univariate polynomial $x^6+5x^3+7x^2+x$. Express this problem as a semidefinite program and solve it.

(9) True or false? A spectrahedron $\mathcal{S} = \mathcal{L} \cap \mathrm{PSD}_n$ is a polyhedron if and only if the linear space $\mathcal{L}$ consists only of diagonal matrices.

(10) In the partial matrix (12.21) set $\sigma_{11} = \sigma_{22} = \sigma_{33} = \sigma_{44} = 5$, $\sigma_{12} = \sigma_{23} = \sigma_{34} = 1$ and $\sigma_{14} = 2$. Compute the spectrahedron $\mathcal{S}_\sigma$, draw a picture, and find the analytic center of $\mathcal{S}_\sigma$. What is the statistical interpretation?

(11) Find numerical values for the eight entries σ_{ij} of the partial 4×4 matrix in (12.21) such that (12.22) holds but $\mathrm{int}(\mathcal{S}_\sigma) = \emptyset$.

(12) Let $s = 3$ and $n = 4$, pick a general flag $\mathcal{V} \subset \mathcal{U}$ as in (12.13), and consider the subvariety of $\mathbb{P}^9 \times \mathbb{P}^9$ defined by the KKT equations (12.14). Working over $\mathbb{C}$, show that this variety is finite and determine its cardinality.

(13) Study the convex hull of following curve which seems similar to (12.12):
$$\big\{ \big(\cos(\theta), \cos(2\theta), \sin(3\theta)\big) \;:\; \theta \in [0, 2\pi] \big\}.$$
Can you draw a picture? Is this convex body a spectrahedron?

Chapter 13

Combinatorics

"Combinatorics is the nanotechnology of mathematics",
Sara Billey

Combinatorics interacts in many fruitful ways with algebra and geometry, e.g. in the interplay between polytopes and toric varieties in Chapter 8. We here discuss combinatorial themes that are important for nonlinear algebra. The first such theme is matroid theory. Matroids encode independence, just like groups encode symmetry. The theory of matroids has many connections to toric geometry, and we will present a few of them. One of our aims is to highlight connections between toric varieties, matroids, and Grassmannians (Chapter 5). In these connections, *lattice polytopes* play a prominent role. We conclude by presenting a snapshot of *generating functions.* Their role as *Hilbert series* brings us back to the two key invariants of an algebraic variety: dimension and degree. The emphasis of this chapter is not on combinatorics per se, but rather on the role it plays within nonlinear algebra. In short, what this chapter offers is a pinch of combinatorics in a vast sea of algebra.

13.1. Matroids

In this section we give an introduction to the theory of *matroids.* The name matroids reveals that these objects can be seen as generalizations of matrices. As we will see, every matrix over a field defines a matroid.

We fix a finite set E, which will be the ground set for our matroids. We shall distinguish a family of subsets of E that are called *independent.*

Thus a matroid M is a family $\mathfrak{I} \subset 2^E$ of subsets of E that we refer to as independent sets. These are assumed to satisfy certain axioms, which reflect the familiar setting where E is a set of vectors in a vector space V. In that setting, being independent simply means being linearly independent.

A first observation is that whenever we have an independent set $I \subset E$, it is reasonable to assume that every subset of I is also independent. We thus obtain the first axiom of a matroid for the family $\mathfrak{I}$:

1. If $I \in \mathfrak{I}$ and $J \subset I$, then $J \in \mathfrak{I}$.

What we have defined so far is a *simplicial complex*. Another requirement for $\mathfrak{I}$ is to be nonempty. Equivalently, we want the empty set to be independent:

2. We have $\emptyset \in \mathfrak{I}$.

To obtain a matroid, we need one more axiom. To motivate it we make the following observation. Consider two finite subsets I and J of a vector space V. Suppose that each of these sets consists of elements that are linearly independent and that $|I| < |J|$. Then we can extend I by an element $j \in J$ in such a way that $I \cup \{j\}$ is still linearly independent. This fact from linear algebra is precisely what we need to get the last axiom for the family $\mathfrak{I}$:

3. If $I, J \in \mathfrak{I}$ and $|I| < |J|$, then $I \cup \{j\} \in \mathfrak{I}$ holds for some $j \in J \backslash I$.

Definition 13.1. A *matroid* is a family of subsets $\mathfrak{I}$ satisfying Axioms 1, 2 and 3.

Exercise 1 asks the reader to prove the assertion in the next sentence.

Example 13.2. The following families $\mathfrak{I}$ are matroids:

- (Representable matroid) Let V be a vector space over an arbitrary field F. Let $E \subset V$ be a nonempty, finite subset. We define $\mathfrak{I}$ to be the family of subsets of E that are linearly independent. We say that this matroid is *representable* over F. In coordinates, $V \simeq F^n$, and we may write the set E as the rows of an $|E| \times n$ matrix.
- (Graphic matroid) Let G be a graph with edge set E. Let $\mathfrak{I}$ be the family of those subsets of E that do not contain a cycle. Equivalently, $\mathfrak{I}$ is the family of forests in G.
- (Algebraic matroid) Let $F \subset K$ be an arbitrary field extension. Let E be a finite subset of K. Let $\mathfrak{I}$ be the family of subsets of E that are algebraically independent over F.
- (Uniform matroid) Let E be a finite set and $k \leq |E|$. Let $\mathfrak{I}$ be the family of subsets of cardinality at most k in E. This matroid is denoted by $U_{k,E}$ or $U_{k,|E|}$.

Matroids are known for having many equivalent definitions, depending on one's point of view. For example, thanks to the first axiom, to determine a matroid we do not have to know all independent sets, just those that are inclusion-maximal. By analogy to linear algebra, the inclusion-maximal independent sets are called *bases*. It turns out (as the reader is asked to prove in Exercise 2) that a nonempty family $\mathfrak{B} \subset 2^E$ of subsets of E is the set of bases of some matroid if and only if the following axiom is satisfied:

- For all bases $B_1, B_2 \in \mathfrak{B}$ and each element $b_2 \in B_2 \backslash B_1$ there exists $b_1 \in B_1 \backslash B_2$ such that $(B_1 \backslash \{b_1\}) \cup \{b_2\} \in \mathfrak{B}$.

The seemingly weak axiom on $\mathfrak{B}$ in fact implies the following two statements:

- For all $B_1, B_2 \in \mathfrak{B}$ and $b_2 \in B_2 \backslash B_1$ there exists $b_1 \in B_1 \backslash B_2$ such that both $(B_1 \backslash \{b_1\}) \cup \{b_2\}$ and $(B_2 \backslash \{b_2\}) \cup \{b_1\}$ are bases in $\mathfrak{B}$.
- For all $B_1, B_2 \in \mathfrak{B}$ and any subset $A_2 \subset B_2 \backslash B_1$ there exists a subset $A_1 \subset B_1 \backslash B_2$ such that both $(B_1 \backslash A_1) \cup A_2$ and $(B_2 \backslash A_2) \cup A_1$ are in $\mathfrak{B}$.

The first point is known as the *symmetric exchange property* and the second as the *multiple symmetric exchange property*. The fact that both exchange properties hold is nontrivial; we refer to the proofs in [**37**, **59**]. We will soon see the algebraic meaning of the exchange properties.

Exercise 3 states that all bases of a matroid have the same cardinality. The cardinality of a basis is known as the *rank* of a matroid. More generally, for a matroid on a ground set E, we may define the rank of any subset $A \subset E$.

Definition 13.3. For a matroid on the ground set E, specified by its collection of independent sets $\mathfrak{I} \subset 2^E$, we define the *rank* function

$$r : 2^E \to \mathbb{Z}, \quad A \mapsto \max_{I \in \mathfrak{I}} \{|I \cap A|\}.$$

Thus, the rank of a set is the largest cardinality of any independent subset.

We note that for a representable matroid, the rank is simply the dimension of the vector subspace spanned by the given vectors. For any matroid, representable or not, the rank function r satisfies the following:

- $0 \leq r(A)$ for all $A \subset E$, and $r(\emptyset) = 0$.
- $r(A) \leq r(A \cup \{b\}) \leq r(A) + 1$ for all $A \subset E$ and $b \in E$.

Further, the rank function has one more property, known as *submodularity*:

- For all $A, B \subset E$ we have $r(A \cup B) + r(A \cap B) \leq r(A) + r(B)$.

In Exercise 7 the reader is asked to prove that any function $r : 2^E \to \mathbb{Z}$ satisfying the three axioms above is a rank function of a matroid. The independent sets can be reconstructed as those $I \subset E$ for which $r(I) = |I|$. Thus, the above rank axioms give another possible definition of a matroid.

Matroid theory has numerous connections to topics seen earlier in this book. Two of these connections will be explored in the next two sections. We close this section with two other connections, namely those to solving polynomial equations (Chapter 3) and to tropical algebra (Chapter 7).

Suppose we are given a matroid M and a field K. Two basic questions arise: Is M representable over K? What is the set of all matrices that represent M? This leads to a system of polynomial equations. Namely, let $E = \{1, 2, \ldots, m\}$ and let $n = r(E)$ be the rank of M. Let $\mathcal{R}(M)$ denote the set of all matrices $X \in K^{m\times n}$ whose rows represent M. This is known as the *realization space* of the matroid M. Thus a subset I of the rows of X is linearly independent in K^n if and only if it is independent in M. For any subset $B \subset E$ with $|B| = n$, we write $\det_B(X)$ for the $n \times n$ determinant given by the rows of X that are indexed by B. By definition, $\mathcal{R}(M)$ equals

$$\big\{X \in K^{m\times n} : \det_B(X) \neq 0 \text{ for } B \in \mathfrak{B} \text{ and } \det_B(X) = 0 \text{ for } B \notin \mathfrak{B}\big\}.$$

This is an affine variety. The conditions can be written as one equation

$$z \cdot \prod_{B\in\mathfrak{B}} \det_B(X) \;=\; 1, \qquad \text{where } z \text{ is a new variable.}$$

With this, the set $\mathcal{R}(M)$ is a closed subvariety in the affine space K^{mn+1}.

Example 13.4 (Fano plane). The Fano plane is a matroid M of rank $n = 3$ on $m = 7$ elements. The nonbases are $124, 235, 346, 457, 156, 267$ and 137. These are the triples in the ground set that are not bases. Note that scaling the rows and columns of a matrix by a nonzero scalar does not change the matroid. Every element $X \in \mathcal{R}(M)$ satisfies $\det_{123}(X) \neq 0$. To check whether M is representable, we may assume that X has an identity matrix in rows 1, 2 and 3. After further scaling of the rows and columns, the transpose of X equals

$$X^T \;=\; \begin{pmatrix} 1 & 0 & 0 & 1 & 0 & 1 & z \\ 0 & 1 & 0 & x & 1 & 1 & 0 \\ 0 & 0 & 1 & 0 & y & 1 & 1 \end{pmatrix}.$$

The zero entries in columns 4, 5 and 7 come from the nonbases 124, 235 and 713. Similarly, the nonzero entries are specified by the bases. Whenever possible, these are scaled to 1. The bases $134, 125$ and 237 imply $x, y, z \neq 0$.

We now consider the constraints $\det_B(X) = 0$ that are imposed by the four nonbases 346, 457, 561 and 672. These give the generators of the ideal

$$\langle\, x-1,\, xyz+1,\, y-1,\, z-1\,\rangle \quad \subset \; K[x,y,z].$$

If the characteristic of our field K is different from 2, then this is the unit ideal and hence M is not representable. However, if $\mathrm{char}(K) = 2$, then the point $(x,y,z) = (1,1,1)$ is a solution and gives a representation of M.

The Fano plane is depicted in Figure 13.1. Here, lines represent linear dependencies. As the matroid is not representable over $\mathbb{R}$, one encodes the fact that 457 is not a basis by putting these three points on a circle.

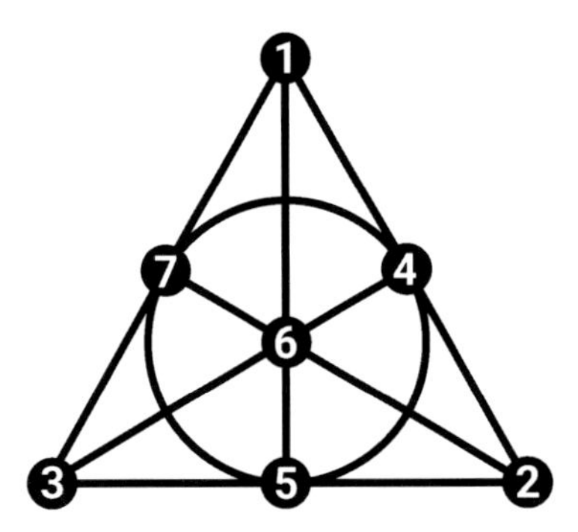

Figure 13.1. The Fano plane.

We have seen that representability of matroids leads to varieties. Remarkably, also the converse holds. Given any affine variety V that is defined over $\mathbb{Z}$, one can construct a matroid M such that $\mathcal{R}(M)$ is essentially equal to V. This is the content of the celebrated *Mnëv's Universality Theorem.*

Matroids play a fundamental role also in tropical geometry. Following Chapter 7, for every ideal I in a Laurent polynomial ring $K[\mathbf{x}^{\pm}]$ we have an associated tropical variety $\mathrm{trop}(\mathcal{V}(I))$. This tropical variety is a subset of $\mathbb{R}^n$. It is called a *tropical linear space* if I is generated by linear polynomials.

Example 13.5. Let $n = 3$ and $I = \langle x_1 + x_2 + x_3 + 1, x_1 + 2x_2 + 3x_3 + 4\rangle$. The classical variety $\mathcal{V}(I)$ is a line in the affine 3-space K^3. Its tropicalization $\mathrm{trop}(\mathcal{V}(I))$ is a balanced graph in $\mathbb{R}^3$ with four unbounded rays in the coordinate directions e_1, e_2, e_3 and $-e_1 - e_2 - e_3$. This *tropical line* captures the structure of the associated matroid, i.e. the uniform matroid $U_{2,4}$. It is represented by the matrix seen in the ideal generators: $X = \begin{bmatrix} 1 & 1 & 1 & 1 \\ 1 & 2 & 3 & 4 \end{bmatrix}^T$.

In general one defines a tropical linear space $\mathrm{trop}(M)$ for any matroid M. This construction is explained in [**38**, §4.2]. Tropical linear spaces are the linear spaces in tropical geometry. They are fundamental to the concept of a tropical manifold. In classical mathematics, a manifold is a space that looks locally like a linear space. A *tropical manifold* is a space that looks locally like a matroid M. Each of its charts is a tropical linear space $\text{trop}(M)$.

13.2. Lattice Polytopes

In this section we study the interplay between toric geometry and matroids. Starting from the basics, we recall the definition of a lattice polytope that we know from Chapter 8. We work in the real vector space $\mathbb{R}^n$. A polytope P is the convex hull of a finite set of points $p_1, \ldots, p_k \in \mathbb{R}^n$:

$$P := \left\{x \in \mathbb{R}^n : x = \sum_{i=1}^{k} \lambda_i p_i, \text{ where } \lambda_1, \ldots, \lambda_k \geq 0 \text{ and } \sum_{i=1}^{k} \lambda_i = 1\right\}.$$

We call P a *lattice polytope* if $p_1, \ldots, p_k \in \mathbb{Z}^n$. If none of the p_i is redundant in the representation of P, then $p_1, \ldots, p_k$ are the *vertices* of P.

To pass from a combinatorial object, such as a matroid, to its polytope, we proceed as follows. Elements $e \in E$ correspond to standard basis vectors b_e of $\mathbb{R}^{|E|}$. Any subset $A \subset E$ can be identified with a point $p_A := \sum_{e \in A} b_e \in \mathbb{R}^{|E|}$. In this way a family of subsets may be identified with a set of points.

Definition 13.6 (Matroid basis polytope)**.** Let M be a matroid on the ground set E, with the set of bases $\mathfrak{B}$. We use the notation introduced above. We define the *matroid basis polytope* $P_M \subset \mathbb{R}^{|E|}$ as the convex hull of the points $p_B := \sum_{e \in B} b_e \in \mathbb{R}^{|E|}$, where B ranges over the set $\mathfrak{B}$ of bases.

By construction, P_M is a lattice polytope. We will study the associated toric variety X_{P_M}, as defined in Section 8.2. From now on we write X_M since P_M is determined by the matroid M. By definition, X_M is a toric variety in the projective space $\mathbb{P}^{|\mathfrak{B}|-1}$, parametrized by the monomials $\prod_{i \in B} t_i$ where B runs over $\mathfrak{B}$. We have $\dim(X_M) \leq |E| - 1$ and equality usually holds.

Example 13.7. Fix the uniform matroid $M = U_{2,4}$. Then P_M is the regular octahedron. The toric variety $X_M \subset \mathbb{P}^5$ has the parametric representation

$$(t_1, t_2, t_3, t_4) \mapsto (t_1t_2, t_1t_3, t_1t_4, t_2t_3, t_2t_4, t_3t_4). \tag{13.1}$$

The variety X_M has dimension 3 and degree 4. Using implicitization as in Section 4.2, we find that its defining prime ideal is $\langle y_1y_6 - y_2y_5, y_2y_5 - y_3y_4\rangle$.

One important property a lattice polytope and its toric variety can have is *normality*. This was introduced in Definitions 8.7 and 8.19. Neil White [**58**] showed that toric varieties arising from matroids are always normal.

Theorem 13.8. *A matroid basis polytope is normal in the lattice it spans.*

We shall present a proof of this theorem. Our approach rests on the following important combinatorial result. For a proof we refer to [**42**, 12.3.1].

Theorem 13.9 (Matroid Union Theorem). *Let $M_1, \dots, M_k$ be matroids on the same ground set E with respective families of independent sets $\mathfrak{I}_1, \dots, \mathfrak{I}_k$ and rank functions $r_1, \dots, r_k$. Let*

$$\mathfrak{I} \; := \; \{I \subset E \, : \, I = \bigcup_{i=1}^{k} I_i \;\; \textit{for} \;\; I_i \in \mathfrak{I}_i\}.$$

Then $\mathfrak{I}$ is also the family of independent sets for a matroid, known as the join of $M_1, \dots, M_k$. The rank function for the join matroid satisfies

$$r(A) \; = \; \min_{S \subset A}\{|A \setminus S| + \sum_{i=1}^{k} r_i(S)\}.$$

As a corollary, we obtain the following theorem due to Jack Edmonds.

Theorem 13.10. *Let M be a matroid with rank function r_M. Then its ground set E can be partitioned into k independent sets if and only if*

$$|A| \; \leq \; k \cdot r_M(A) \quad \textit{for all subsets } A \subset E.$$

Proof. The $\Rightarrow$ implication is straightforward.

For the opposite implication consider the join U of M with itself k times. We apply the matroid union theorem to compute the rank of E:

$$r_U(E) \; = \; \min_{A \subset E}\{|E| - |A| + k \cdot r_M(A)\}.$$

By the assumption, we have $|E| - |A| + k \cdot r_M(A) \geq |E|$, and equality holds for $A = \emptyset$. Hence, $r_U(E) = |E|$. This means that E is an independent set in U, and hence by definition it is a union of k independent sets of M. □

Let M be a matroid with the family $\mathfrak{I}$ of independent sets. We take $E = \{1, 2, \dots, n\}$ and consider any subset $E' \subset E$. The *restriction* of M to E' is the matroid where $A \subset E'$ is independent if and only if $A \in \mathfrak{I}$.

Proof of Theorem 13.8. Let $p \in (kP_M) \cap \mathbb{Z}^n$. We know that

$$p \; = \; \sum_{B \in \mathfrak{B}} \lambda_B \cdot p_B \quad \text{with} \quad \sum_{B \in \mathfrak{B}} \lambda_B = k \;\; \text{and} \;\; 0 \leq \lambda_B \in \mathbb{Q}.$$

By clearing denominators, we find a positive integer d such that

$$d \cdot p \; = \; \sum \lambda'_B \cdot p_B,$$

where $\sum \lambda'_B = d \cdot k$ and $0 \leq \lambda'_B \in \mathbb{Z}$. By restricting the matroid M to a subset, we may assume that all coordinates of $p = (p_1, \dots, p_n)$ are nonzero.

We next define two matroids. The first matroid N has the ground set $E_N := \{(i,j) : 1 \le i \le n, 1 \le j \le p_i\}$. In other words, we replace a point i in the original matroid by p_i new points. A subset $\{(i_1,j_1),\dots,(i_s,j_s)\} \subset E_N$ is independent in the matroid N if and only if

- all i_q's are distinct, and
- $\{i_1,\dots,i_s\}$ is an independent set in M.

Bases of N map naturally to bases of M. Also, the rank function for N is the same as the one for M if we forget the second coordinates. The point p has a decomposition as a sum of k points corresponding to bases of M if and only if the matroid N is covered by k bases (i.e. the ground set is a union of k bases). Hence, by Theorem 13.10, our aim is to prove the following statement: *For any* $A \subset E_N$ *we have* $|A| \le k \cdot r_N(A)$.

For the second matroid N', we replace any point of N by d new points. The ground set of N' is $E_{N'} := \{(i,j,l) : 1 \le i \le n, 1 \le j \le p_i, 1 \le l \le d\}$. A subset $\{(i_1,j_1,l_1),\dots,(i_s,j_s,l_s)\} \subset E_{N'}$ is independent if and only if

- all i_q's are distinct, and
- $\{i_1,\dots,i_s\}$ is an independent set in M.

We have a natural projection $\pi : E_{N'} \to E_N$ by forgetting the last coordinate. We note that $r_{N'}(\pi^{-1}(A)) = r_N(A)$. As the point $d \cdot p$ is decomposable, we know that the matroid N' can be covered by kd bases. Hence, $|B| \le dk \cdot r_{N'}(B)$ for any $B \subset E_{N'}$. Applying this to $\pi^{-1}(A)$, we obtain

$$d|A| \;=\; |\pi^{-1}(A)| \;\le\; dk \cdot r_{N'}(\pi^{-1}(A)) \;=\; dk \cdot r_N(A).$$

This is equivalent to the statement we wanted to prove. □

Our next aim is to relate matroids to the geometry of special subvarieties of Grassmannians. We recall that one definition of the Grassmannian is that $G(k,n)$ is the orbit of $[e_1 \wedge \cdots \wedge e_k] \in \mathbb{P}(\bigwedge^k K^n)$ under the action of the group $\mathrm{GL}(n)$ of invertible $n \times n$ matrices. While the Grassmannian is an orbit of the big group $\mathrm{GL}(n)$, we may ask how smaller groups act on $G(k,n)$. In particular, consider the torus $T := (K^*)^n$ of diagonal matrices. This acts on $\mathbb{P}(\bigwedge^k K^n)$ and on $G(k,n)$. However, $G(k,n)$ is generally not an orbit or orbit closure of T, since $\dim(G(k,n)) = k(n-k)$ exceeds $\dim(T) = n$.

We fix a point $p \in G(k,n)$. The following questions motivate us:

- What is the T-orbit of p?
- What is the closure of this orbit?
- How can we describe this variety?

A beautiful answer is provided by the characterization of matroid polytopes due to Gelfand, Goresky, MacPherson and Serganova [**38**, Theorem 4.2.12].

A point $p = [v_1 \wedge \cdots \wedge v_k] \in G(k,n)$ represents the subspace $V = \langle v_1, \ldots, v_k \rangle$ in K^n. We write the vectors $v_1, \ldots, v_k$ as a $k \times n$ matrix N_p. From Chapter 5 we know that the coordinates of $p \in \mathbb{P}(\bigwedge^k K^n)$ are the maximal minors of N_p. How does a torus element $t = (t_1, \ldots, t_n) \in T$ act on p? In general, t acts on a standard basis vector $e_{i_1} \wedge \cdots \wedge e_{i_k}$ by rescaling it with $t_{i_1} \cdots t_{i_k}$. Hence, the orbit of p is the image of the monomial map

$$T \ni (t_1, \ldots, t_n) \mapsto \big(t_{i_1} \cdots t_{i_k} \det_{i_1 \ldots i_k}(N_p) \big)_{1 \leq i_1 < \cdots < i_k \leq n} \in \mathbb{P}^{\binom{n}{k}-1}.$$

Example 13.11. Fix $k = 2$ and $n = 4$. Let $p \in G(2,4)$ be the row space of

$$N_p = \begin{bmatrix} 1 & 1 & 1 & 1 \\ 1 & 2 & 3 & 4 \end{bmatrix}.$$

This is the same subspace as that seen in Examples 13.5 and 13.7. In Plücker coordinates on the Grassmannian, the corresponding point in $\mathbb{P}^5$ is

$$\begin{aligned} & (e_1 + e_2 + e_3 + e_4) \wedge (e_1 + 2e_2 + 3e_3 + 4e_4) \\ = \ & e_1 \wedge e_2 + 2e_1 \wedge e_3 + 3e_1 \wedge e_4 + e_2 \wedge e_3 + 2e_2 \wedge e_4 + e_3 \wedge e_4. \end{aligned}$$

Hence, the T-orbit of p in $G(2,4) \subset \mathbb{P}^5$ is parametrized as follows:

$$(t_1, t_2, t_3, t_4) \mapsto (t_1t_2, 2t_1t_3, 3t_1t_4, t_2t_3, 2t_2t_4, t_3t_4). \tag{13.2}$$

This is almost the same as the monomial map in (13.1). The only difference is the constants given by minors of the matrix N_p. However, these constants do not depend on $t \in T$. We can define an automorphism of $\mathbb{P}(\bigwedge^k K^n)$ that turns the orbit into an image of a monomial map, by rescaling the coordinates. The prime ideal of this orbit is $\langle 4y_1y_6 - y_2y_5, 3y_2y_5 - 4y_3y_4 \rangle$.

In general, if all $k \times k$ minors of N_p are nonzero, in which case the matroid of N_p is uniform, then the T-orbit of p is isomorphic to the image of the map given by all square-free monomials of degree k. The associated basis polytope is the *hypersimplex*. If the matroid is not uniform, then we delete all zero coordinates of p and consider the torus orbit in $\mathbb{P}^{|\mathfrak{B}|-1}$.

For instance, if the lower right matrix entry "4" in Example 13.11 is replaced by "3", then (13.2) becomes $(t_1, t_2, t_3, t_4) \mapsto (t_1t_2, 2t_1t_3, 2t_1t_4, t_2t_3, t_2t_4)$ and the closure of the image of this map is the hypersurface in $\mathbb{P}^4$ defined by $\langle y_2y_5 - y_3y_4 \rangle$. Our discussion establishes the following general result.

Proposition 13.12. *The closure of the T-orbit of any point $p = [v_1 \wedge \cdots \wedge v_k]$ in a Grassmannian $G(k,n)$ is isomorphic to the toric variety represented by the matroid basis polytope, for the representable matroid defined by columns of the $k \times n$ matrix N_p with ith row equal to v_i.*

The results of Chapter 8 combined with Theorem 13.8 imply:

Corollary 13.13. *Any torus orbit closure in a Grassmannian in its Plücker embedding is projectively normal.*

We conclude that the toric varieties associated with matroids include all torus orbit closures in Grassmannians. But they are more general because there exist matroids that are not representable. We now present a toric variety that arises as a torus orbit closure in $G(3,7)$ only when $\mathrm{char}(K) = 2$.

Example 13.14 (Fano matroid). Fix $n = 7$ and let M be the Fano matroid from Example 13.4. It has seven nonbases and hence $|\mathfrak{B}| = 28 = \binom{7}{3} - 7$. Thus, the toric variety X_M lives in $\mathbb{P}^{27}$. It has dimension 6 and degree 232. The prime ideal of X_M is minimally generated by 140 quadrics $y_i y_j - y_k y_l$.

We now turn to the interpretation of basis exchange properties in terms of toric ideals. Consider a matroid with basis polytope P. We recall that

- the ideal of the associated toric variety is generated by binomials;
- every binomial in the ideal corresponds to an integral relation among lattice points of P.

How do these statements specialize in the case of matroids? The vertices of P are characteristic vectors of bases. A sum of vertices is a sum of characteristic vectors. This corresponds to taking a union of bases as *multisets*.

Example 13.15. Consider the uniform matroid $U_{2,4}$. Its basis polytope is a regular octahedron. A typical integral relation between its vertices is

$$(1,1,0,0) + (0,0,1,1) \;=\; (1,0,1,0) + (0,1,0,1).$$

As a union of bases, this corresponds to the identity

$$\{1,2\} \cup \{3,4\} \;=\; \{1,3\} \cup \{2,4\}.$$

With appropriate indexing of the variables, this translates into the quadratic binomial $y_{12}y_{34} - y_{13}y_{24}$ in the ideal of the associated toric threefold in $\mathbb{P}^5$.

Two multisets of bases are *compatible* if their unions (as multisets) are the same. Equivalently, every element of E belongs to the same number of bases in the first and second multisets of bases. We conclude that the binomials in the ideal of the toric variety represented by a matroid basis polytope are in bijection with pairs of compatible multisets of bases. The quadrics in a toric ideal correspond to pairs of basis pairs $(\{B_1, B_2\}, \{B_3, B_4\})$ satisfying $B_1 \cup B_2 = B_3 \cup B_4$. In transitioning between these basis pairs, we change

- B_1 by subtracting a set $A_1 \subset B_1 \backslash B_2$ and adding $A_2 \subset B_2 \backslash B_1$, and
- B_2 by adding A_1 to it and subtracting A_2.

We see that quadrics in the ideal of X_M correspond to multiple symmetric exchanges. It follows that symmetric basis exchanges, i.e. the case where $|A_1| = |A_2| = 1$, form a distinguished set of quadrics in the ideal. The following four conjectures are due to Neil White.

Conjecture 13.16. *The following is true for an arbitrary matroid M.*

- *Representable case: The ideal of any T-orbit in a Grassmannian is*
  (1) *generated by quadrics, and*
  (2) *generated by quadrics coming from symmetric basis exchanges.*
- *General case: Any two finite multisets of bases (B_i) and (B_j) such that $\bigcup B_i = \bigcup B_j$ can be transformed to one another through a finite sequence of the following steps:*
  (1) *replace two bases B and B' in one multiset by two bases $\tilde{B}$ and $\tilde{B}'$ obtained by multiple symmetric exchanges (i.e. $B \cup B' = \tilde{B} \cup \tilde{B}'$);*
  (2) *replace two bases B and B' in one multiset by two bases $\tilde{B}$ and $\tilde{B}'$ obtained by a symmetric exchange (i.e. $\tilde{B} = B \cup \{b_1\} \backslash \{b_2\}$ and $\tilde{B}' = B' \cup \{b_2\} \backslash \{b_1\}$).*

In these conjectures, the general case implies the representable case.

13.3. Generating Functions

In this section we introduce multivariate generating functions that are given by rational functions. The key example is multigraded Hilbert series. We discuss methods for computing them, and we explore connections to regular triangulations. In particular, we discuss Ehrhart series of lattice polytopes.

We start with the familiar example of the polynomial ring $K[\mathbf{x}]$. In Chapter 1 we introduced its Hilbert function $d \mapsto h(d)$. The value $h(d) = \dim_K K[\mathbf{x}]_d$ is the number of monomials of degree d or, equivalently, the number of lattice points u satisfying $u \geq 0$ and $\sum_{i=1}^n u_i = d$. Let Δ be the standard simplex, i.e. the convex hull of the standard basis vectors. The Hilbert function counts the number of lattice points in dilations of Δ, i.e. $h(d) = |d\Delta \cap \mathbb{Z}^n|$. Further, by Example 1.22, the Hilbert series equals

$$\mathrm{HS}(z) \;=\; \sum_{q=0}^{\infty} |q\Delta \cap \mathbb{Z}^n| \cdot z^q \;=\; \frac{1}{(1-z)^n}.$$

Our next aim is to refine our counting. So far we have treated all monomials of the same degree on an equal footing. What happens if we try to remember each monomial, not only its degree? Then we obtain the generating function

$$\mathrm{MHS}(x) \;=\; \frac{1}{\prod_{i=1}^n (1-x_i)} \;=\; \sum_{u \in \mathbb{N}^n} x_1^{u_1} x_2^{u_2} \cdots x_n^{u_n}. \tag{13.3}$$

Note that the polytopes $q\Delta$ are slices of the nonnegative orthant $C = \mathbb{R}^n_{\geq 0}$. The sum in (13.3) may be regarded as a sum over all lattice points of C. We next replace C by more general cones.

Definition 13.17. Let $C \subset \mathbb{R}^n$ be a rational pointed polyhedral cone. We define the associated *multigraded Hilbert series* to be the formal power series

$$\mathrm{MHS}_C(\mathbf{x}) \quad = \sum_{c \in C \cap \mathbb{Z}^n} \mathbf{x}^c.$$

If $C \subset \mathbb{R}^n_{\geq 0}$ then we reconstruct the Hilbert series of $K[C]$ from the multigraded Hilbert series by setting $x_1 = \cdots = x_n = z$. However, MHS_C remembers much more information: all lattice points of the cone. Our next aim is to represent MHS_C as a rational function, just as we did for $C = \mathbb{R}^n_{\geq 0}$. A cone generated by linearly independent vectors is called *simplicial.*

Lemma 13.18. *Let C be a simplicial cone with generators $c_1, \ldots, c_d \in \mathbb{Z}^n$. There is a Laurent polynomial $\kappa_C(\mathbf{x})$ with nonnegative coefficients such that*

$$\mathrm{MHS}_C(\mathbf{x}) \quad = \quad \frac{\kappa_C(\mathbf{x})}{\prod_{i=1}^d (1 - \mathbf{x}^{\mathbf{c}_i})}.$$

Proof. Consider the following half-open parallelepiped:

$$P \; := \; \{x \in C \, : \, x = \sum_{i=1}^{d} \lambda_i c_i, \; 0 \leq \lambda_1, \ldots, \lambda_d < 1\}.$$

Since the c_i are linearly independent and generate C as a cone, every lattice point $c \in C$ has a unique representation $c = p + \sum_{i=1}^d s_i c_i$ where $p \in P$ is a lattice point and the s_i are nonnegative integers. We obtain

$$\begin{aligned}
\mathrm{MHS}_C(\mathbf{x}) \; &= \sum_{c \in C \cap \mathbb{Z}^n} \mathbf{x}^c \; = \sum_{p \in P \cap \mathbb{Z}^n} \mathbf{x}^p \left(\prod_{i=1}^{d} \left(\sum_{s_i = 0}^{\infty} \mathbf{x}^{s_i c_i} \right) \right) \\
&= \sum_{p \in P \cap \mathbb{Z}^n} \mathbf{x}^p \left(\prod_{i=1}^{d} \frac{1}{1 - \mathbf{x}^{c_i}} \right) \; = \; \frac{\sum_{p \in P \cap \mathbb{Z}^n} \mathbf{x}^p}{\prod_{i=1}^d (1 - \mathbf{x}^{c_i})}.
\end{aligned}$$

Hence $\kappa_C(\mathbf{x})$ is the sum of all monomials representing lattice points in P. $\square$

Remark 13.19. We freely manipulated infinite summations in Lemma 13.18. This is justified by the hypothesis that C is pointed, which ensures that the series is absolutely convergent on some open set in $\mathbb{R}^n$.

Proposition 13.20. *Let C be a pointed, rational polyhedral cone with generators $c_1, \ldots, c_d \in \mathbb{Z}^n$. Here C is not necessarily simplicial. Then we have*

$$\mathrm{MHS}_C(\mathbf{x}) \quad = \quad \frac{\kappa_C(\mathbf{x})}{\prod_{i=1}^d (1 - \mathbf{x}^{\mathbf{c}_i})},$$

where $\kappa_C(\mathbf{x})$ is a Laurent polynomial with integral coefficients.

Proof. We *triangulate* the cone C, i.e. we write it as a union of simplicial cones with rays c_i that intersect in lower-dimensional simplicial cones. This may be done e.g. by induction on the number d. Using inclusion-exclusion over all cones, we write $\mathrm{MHS}_C(\mathbf{x})$ as an alternating sum of multivariate Hilbert series for simplicial cones. Lemma 13.18 now implies the claim. □

Remark 13.21. The proofs of Proposition 13.20 and Lemma 13.18 suggest an algorithm for computing the multigraded Hilbert series. This procedure also gives a combinatorial interpretation of the numerator $\kappa_C(\mathbf{x})$.

The case where C is a cone over a lattice polytope $P \subset \mathbb{R}^n$ is particularly nice. We identify P with the polytope $P \times \{1\} \subset \mathbb{R}^n \times \mathbb{R}$. Let $C \subset \mathbb{R}^{n+1}$ be the cone over P, i.e. the smallest cone that contains $P \times \{1\}$.

Proposition 13.22. *Let C and P be as defined above. The Hilbert function for C with respect to the grading induced by the last variable is given by*

$$h(q) \;=\; |qP \cap \mathbb{Z}^{n+1}|.$$

The function h coincides with a polynomial for all $q \in \mathbb{Z}_{\geq 0}$. This is the Ehrhart polynomial *of P, which counts the lattice points in dilations of P.*

Proof. The only nontrivial statement is that $h(q)$ is equal to a polynomial for *all* positive integers q. By induction on $d := \dim P$ and by triangulating P, it is enough to prove the statement in the case where P is a simplex with vertices $v_1, \ldots, v_d$. Let $f(q) := \binom{d+q-1}{q} = \binom{d+q-1}{d-1}$ for $q \geq 0$ and $f(q) = 0$ for $q < 0$. Similiarly to the proof of Lemma 13.18, we have

$$h(q) \;=\; \sum_{i=0}^{d-1} a_i \cdot f(q-i),$$

where a_i is the number of lattice points with their last coordinate i in the set

$$\Big\{ x \in \mathbb{R}^{n+1} \;:\; x = \sum_{i=1}^{d} \lambda_i (v_i, 1) \text{ where } 0 \leq \lambda_1, \ldots, \lambda_d < 1 \Big\}.$$

We only have to consider a_i for $i < d$, as there are no lattice points in this parallelepiped with last coordinate greater than or equal to d.

We note that f is *not* a polynomial if we consider the negative arguments. The punchline is that the polynomial $g(q) := (d+q-1)(d+q-2)\cdots q/(d-1)!$ equals f also for negative integers q, as long as $q \geq -d+1$. Hence,

$$h(q) \;=\; \sum_{i=0}^{d-1} a_i \cdot g(q-i) \;\text{ for } q \in \mathbb{Z}_{\geq 0}.$$

This sum is a polynomial in q. It is the *Ehrhart polynomial* of P. □

Given a lattice polytope P with N lattice points, we associate to it a toric variety X_P in $\mathbb{P}^{N-1}$ as in Chapter 8. Hence, we obtain a binomial ideal $I_P \subset K[\mathbf{x}] = K[x_1, \ldots, x_N]$. Fix a term order $\prec$. The initial ideal $\mathrm{in}_\prec(I_P)$ is a monomial ideal. Its radical $\mathrm{rad}(\mathrm{in}_\prec(I_P))$ has the following property:

- If a product m of distinct variables does not belong to $\mathrm{rad}(\mathrm{in}_\prec(I_P))$, then neither does any monomial that divides m.

This may be restated as follows:

- The subsets S of the set of variables $\{x_1, \ldots, x_N\}$ that satisfy $\prod_{x \in S} x \notin \mathrm{rad}(\mathrm{in}_\prec(I_P))$ form a simplicial complex.

Our aim is to give a geometric description of this simplicial complex. The main idea is that the variables x_i are in bijection with lattice points of P. Let Δ be the collection of polytopes contained in P that are convex hulls of sets of points S such that the product of variables in S is not in $\mathrm{rad}(\mathrm{in}_\prec(I_P))$.

Example 13.23. Consider the square $P = \mathrm{conv}((0,0,1),(0,1,1),(1,0,1),(1,1,1))$. The associated variety is the surface in $\mathbb{P}^3$ defined by $x_1x_4 - x_2x_3$. We fix a term order for which x_1x_4 is the leading term. The subdivision Δ of P contains two triangles: $\mathrm{conv}((0,0,1),(0,1,1),(1,0,1))$ and $\mathrm{conv}((0,1,1),(1,0,1),(1,1,1))$. The minimal nonface is the pair of vertices $(0,0,1)$ and $(1,1,1)$. This pair corresponds to the unique generator x_1x_4 of the initial ideal. If we change the term order so that x_2x_3 becomes the leading term, then we obtain a different triangulation of P, given by the other diagonal.

Our next result relates Gröbner bases to triangulations.

Theorem 13.24. *Using the notation introduced above, Δ is a triangulation of P. The minimal nonfaces of Δ correspond to generators of* $\mathrm{rad}(\mathrm{in}_\prec(I_P))$.

Proof. The proof can be found in [**52**, Chapter 8]. □

Definition 13.25. The triangulations of the form Δ, induced by any term order, are called *regular*. There exist triangulations that are not regular.

The story that we are telling has three aspects: combinatorial, algebraic and geometric. From the combinatorial point of view, we are triangulating a lattice polytope P into simplices. The algebraic part is the finest, in that we *degenerate* a toric ideal I_P to a monomial ideal that shares with I_P all the most important invariants, such as dimension and degree. Let us now describe the geometry in this picture. Here we degenerate the variety $\mathcal{V}(I_P)$ to $\mathcal{V}(\mathrm{in}_\prec(I_P))$. One of the problems is that $\mathrm{in}_\prec(I_P)$ may be not radical, so we may lose some information, but let us ignore this for a moment.

What is $\mathcal{V}(\mathrm{in}_\prec(I_P))$? This variety equals $\mathcal{V}(\mathrm{rad}(\mathrm{in}_\prec(I_P)))$, so the question is: What is the set of zeros of a square-free monomial ideal? The answer is given by the solution to Exercise 12 in Chapter 2.

We find that the variety $\mathcal{V}(\mathrm{rad}(\mathrm{in}_\prec(I_P)))$ is a union of coordinate subspaces. Each subspace is spanned by basis vectors $(e_i)_{i\in S}$ with $\prod_{i\in S} x_i \notin \mathrm{rad}(\mathrm{in}_\prec(I_P))$. The simplices in the induced triangulation of P correspond naturally to components of $\mathcal{V}(\mathrm{in}_\prec(I_P))$; as the triangulation breaks the polytope into simple pieces, our variety is split into simple components.

We note that the idea of computing the dimension and degree of an ideal by passing to the initial ideal is equivalent to the idea of computing the Hilbert series of a cone by subdividing it into simplicial cones.

We know that the dimension of the (projective) toric variety associated to a polytope P equals the dimension of P. What about the degree?

Proposition 13.26. *Let $P \subset \mathbb{R}^d$ be a lattice polytope, where $P \cap \mathbb{Z}^d$ spans the lattice $\mathbb{Z}^d$. The degree of the ideal I_P equals the volume of P times $d!$.*

Sketch of the proof. The degree divided by the factorial of $\dim(X_P)$ is the leading coefficient of the Ehrhart polynomial h. Thus it is enough to show the following claim: For any $\epsilon > 0$ there exists a constant C such that

$$(\mathrm{vol}\, P - \epsilon)q^d - C \leq h_P(q) \leq (\mathrm{vol}\, P + \epsilon)q^d + C \text{ for all } q \in \mathbb{Z}_{>0}. \tag{13.4}$$

Here, as before, $h_P(q)$ is the number of lattice points in qP. It is easy to prove the inequality (13.4) for *rational* polytopes that are products of intervals. This fact implies the claim, by covering P with small products of intervals, according to the definition of the Lebesgue measure.

Thus, h_P is of degree d and has leading coefficient equal to $\mathrm{vol}\, P$. The proposition follows from the definition of the degree given in Section 1.3. □

Example 13.27. Let P be a d-dimensional simplex, given as the convex hull of 0 and d basis vectors. The Ehrhart polynomial is given by $h(q) = \binom{d+q}{d} = \frac{1}{d!}q^d +$ lower-order terms. Indeed, $\mathrm{vol}\, P = 1/d!$ and $\dim P = d$.

The usual Euclidean volume multiplied by $d!$ is known as the *normalized volume*. The simplex $\mathrm{conv}(0, e_1, e_2, \ldots, e_d)$ has normalized volume 1. The normalized volume of any lattice polytope is a positive integer. This integer equals the degree of the toric variety, if one works in the correct lattice.

How is the triangulation compatible with the degree computation? The sum of volumes of its (maximal) simplices is equal to the volume of P.

Theorem 13.28. *Let P be a d-dimensional lattice polytope whose lattice points generate the lattice $\mathbb{Z}^d$. Let I_P be the associated toric ideal. If $\prec$ is a term order and Δ the associated triangulation of P, then the following hold:*

(1) *The minimal primes of $\mathrm{in}_\prec(I_P)$ are in bijection with the maximal simplices in the triangulation Δ.*

(2) *The primary ideal corresponding to a minimal prime of $\mathrm{in}_\prec(I_P)$ has as degree the normalized volume of the associated simplex in Δ.*

Exercises

(1) Show that Example 13.2 presents matroids.

(2) (a) Fix the family of independent sets $\mathfrak{I}$ for a matroid M. Prove that the inclusion-maximal elements in $\mathfrak{I}$ satisfy the axioms for bases.
  (b) Fix a nonempty set $\mathfrak{B} \subset 2^E$ satisfying the axioms for bases of a matroid. Prove that $\mathfrak{I} := \{I \subset E \mid \exists B \in \mathfrak{B} : I \subset B\}$ satisfies the axioms for independent sets of a matroid.

(3) Prove that all bases of a matroid have the same cardinality.

(4) Prove that the points p_B in Definition 13.6 are vertices of the polytope P_M. Prove that these are the only lattice points in P_M.

(5) In this exercise we examine matroid duality.
  (a) Let $\mathfrak{B} \subset 2^E$ be a set of bases of a matroid M. Let $\mathfrak{B}^* := \{B \subset E : E \setminus B \in \mathfrak{B}\}$. Prove that $\mathfrak{B}^*$ is a set of bases of a matroid M^*. The matroid M^* is known as the dual matroid (of M).
  (b) Prove that a dual of a representable matroid is representable.

(6) Prove that for any matroid the rank function is submodular.

(7) Prove that any function $2^E \to \mathbb{Z}$ satisfying the three axioms of the rank function is indeed a rank function of some matroid.

(8) How many distinct torus orbit closures are there in $G(2,4)$? How many are there up to isomorphism (of algebraic varieties)?

(9) Prove White's conjectures for uniform matroids.

(10) Let M be the rank-3 matroid on $E = \{1, 2, \dots, 8\}$ with nonbases 124, 235, 346, 457, 568, 671, 782 and 813. Is M representable? Study $\mathcal{R}(M)$.

(11) Let I be an ideal generated by three linear polynomials in five variables, so that $\mathcal{V}(I)$ is a plane in K^5. Determine the tropical plane $\mathrm{trop}(\mathcal{V}(I))$.

(12) Use Theorem 13.28 to triangulate the matroid basis polytopes of $U_{3,5}$ and $U_{3,6}$.

(13) Carry out the computation reported in Example 13.14. Identify an initial monomial ideal of this toric ideal. Compute its Hilbert series MHS.

(14) Find a matroid M of rank 4 that has precisely 12 bases. Determine its rank function r and its polytope P_M. Verify White's conjectures for M.

(15) Let C be the cone over the regular 3-cube $[0,1]^3$. Compute MHS_C.

Bibliography

[1] H. Abo, A. Seigal, and B. Sturmfels, *Eigenconfigurations of tensors*, Algebraic and geometric methods in discrete mathematics, Contemp. Math., vol. 685, Amer. Math. Soc., Providence, RI, 2017, pp. 1–25. MR3625569

[2] B. Alexeev, M. A. Forbes, and J. Tsimerman, *Tensor rank: some lower and upper bounds*, 26th Annual IEEE Conference on Computational Complexity, IEEE Computer Soc., Los Alamitos, CA, 2011, pp. 283–291. MR3025382

[3] M. F. Atiyah and I. G. Macdonald, *Introduction to commutative algebra*, Addison-Wesley Publishing Co., Reading, Mass.-London-Don Mills, Ont., 1969. MR0242802

[4] S. Basu, R. Pollack, and M.-F. Roy, *Algorithms in real algebraic geometry*, 2nd ed., Algorithms and Computation in Mathematics, vol. 10, Springer-Verlag, Berlin, 2006. MR2248869

[5] G. Blekherman, P. A. Parrilo, and R. R. Thomas (eds.), *Semidefinite optimization and convex algebraic geometry*, MOS-SIAM Series on Optimization, vol. 13, Society for Industrial and Applied Mathematics (SIAM), Philadelphia, PA; Mathematical Optimization Society, Philadelphia, PA, 2013. MR3075433

[6] J. Bochnak, M. Coste, and M.-F. Roy, *Real algebraic geometry*, translated from the 1987 French original, revised by the authors, Ergebnisse der Mathematik und ihrer Grenzgebiete (3) [Results in Mathematics and Related Areas (3)], vol. 36, Springer-Verlag, Berlin, 1998. MR1659509

[7] P. Butkovič, *Max-linear systems: theory and algorithms*, Springer Monographs in Mathematics, Springer-Verlag London, Ltd., London, 2010. MR2681232

[8] D. Cartwright and B. Sturmfels, *The number of eigenvalues of a tensor*, Linear Algebra Appl. **438** (2013), no. 2, 942–952, DOI 10.1016/j.laa.2011.05.040. MR2996375

[9] L. Colmenarejo, F. Galuppi, and M. Michałek, *Toric geometry of path signature varieties*, Adv. in Appl. Math. **121** (2020), 102102, 35, DOI 10.1016/j.aam.2020.102102. MR4140555

[10] D. Cox, J. Little, and D. O'Shea, *Ideals, varieties, and algorithms*, An introduction to computational algebraic geometry and commutative algebra, 3rd ed., Undergraduate Texts in Mathematics, Springer, New York, 2007. MR2290010

[11] D. A. Cox, J. Little, and D. O'Shea, *Using algebraic geometry*, 2nd ed., Graduate Texts in Mathematics, vol. 185, Springer, New York, 2005. MR2122859

[12] D. A. Cox, J. B. Little, and H. K. Schenck, *Toric varieties*, Graduate Studies in Mathematics, vol. 124, American Mathematical Society, Providence, RI, 2011. MR2810322

[13] G. Craciun, A. Dickenstein, A. Shiu, and B. Sturmfels, *Toric dynamical systems*, J. Symbolic Comput. **44** (2009), no. 11, 1551–1565, DOI 10.1016/j.jsc.2008.08.006. MR2561288

[14] H. Derksen, *Computation of invariants for reductive groups*, Adv. Math. **141** (1999), no. 2, 366–384, DOI 10.1006/aima.1998.1787. MR1671758

[15] H. Derksen and G. Kemper, *Computational invariant theory*, Invariant Theory and Algebraic Transformation Groups, I, Encyclopaedia of Mathematical Sciences, 130, Springer-Verlag, Berlin, 2002. MR1918599

[16] P. Diaconis, *Group representations in probability and statistics*, Institute of Mathematical Statistics Lecture Notes—Monograph Series, vol. 11, Institute of Mathematical Statistics, Hayward, CA, 1988. MR964069

[17] A. Dickenstein and E. Feliu, *Algebraic Methods for Biochemical Reaction Networks*, textbook in preparation.

[18] V. Dolotin and A. Morozov, *Introduction to non-linear algebra*, World Scientific Publishing Co. Pte. Ltd., Hackensack, NJ, 2007. MR2361550

[19] M. Drton, B. Sturmfels, and S. Sullivant, *Lectures on algebraic statistics*, Oberwolfach Seminars, vol. 39, Birkhäuser Verlag, Basel, 2009. MR2723140

[20] S. Friedland and G. Ottaviani, *The number of singular vector tuples and uniqueness of best rank-one approximation of tensors*, Found. Comput. Math. **14** (2014), no. 6, 1209–1242, DOI 10.1007/s10208-014-9194-z. MR3273677

[21] W. Fulton and J. Harris, *Representation theory: A first course*, Readings in Mathematics, Graduate Texts in Mathematics, vol. 129, Springer-Verlag, New York, 1991. MR1153249

[22] I. M. Gel′fand, M. M. Kapranov, and A. V. Zelevinsky, *Discriminants, resultants, and multidimensional determinants*, Mathematics: Theory & Applications, Birkhäuser Boston, Inc., Boston, MA, 1994. MR1264417

[23] G.-M. Greuel and G. Pfister, *A* **Singular** *introduction to commutative algebra*, second, extended edition, with contributions by Olaf Bachmann, Christoph Lossen and Hans Schönemann; with 1 CD-ROM (Windows, Macintosh and UNIX), Springer, Berlin, 2008. MR2363237

[24] C. Harris, M. Michałek, and E. C. Sertöz, *Computing images of polynomial maps*, Adv. Comput. Math. **45** (2019), no. 5-6, 2845–2865, DOI 10.1007/s10444-019-09715-8. MR4047019

[25] R. Hartshorne, *Algebraic geometry*, Graduate Texts in Mathematics, No. 52, Springer-Verlag, New York-Heidelberg, 1977. MR0463157

[26] C. J. Hillar and L.-H. Lim, *Most tensor problems are NP-hard*, J. ACM **60** (2013), no. 6, Art. 45, 39, DOI 10.1145/2512329. MR3144915

[27] M. Joswig, *Essentials of tropical combinatorics*, to appear.

[28] F. Kirwan, *Complex algebraic curves*, London Mathematical Society Student Texts, vol. 23, Cambridge University Press, Cambridge, 1992. MR1159092

[29] K. Kozhasov, *On fully real eigenconfigurations of tensors*, SIAM J. Appl. Algebra Geom. **2** (2018), no. 2, 339–347, DOI 10.1137/17M1145902. MR3814009

[30] M. Kreuzer and L. Robbiano, *Computational linear and commutative algebra*, Springer, Cham, 2016. MR3559741

[31] K. Kubjas, P. A. Parrilo, and B. Sturmfels, *How to flatten a soccer ball*, Homological and computational methods in commutative algebra, Springer INdAM Ser., vol. 20, Springer, Cham, 2017, pp. 141–162. MR3751884

[32] J. M. Landsberg, *Tensors: geometry and applications*, Graduate Studies in Mathematics, vol. 128, American Mathematical Society, Providence, RI, 2012. MR2865915

[33] J. M. Landsberg and M. Michałek, *Towards finding hay in a haystack: explicit tensors of border rank greater than* $2.02m$ *in* $\mathbb{C}^m \otimes \mathbb{C}^m \otimes \mathbb{C}^m$, `arXiv:1912.11927`.

[34] J. M. Landsberg and M. Michałek, *A* $2n^2 - \log_2(n) - 1$ *lower bound for the border rank of matrix multiplication*, Int. Math. Res. Not. IMRN **15** (2018), 4722–4733, DOI 10.1093/imrn/rnx025. MR3842382

[35] J. M. Landsberg, *Geometry and complexity theory*, Cambridge Studies in Advanced Mathematics, vol. 169, Cambridge University Press, Cambridge, 2017. MR3729273

[36] J. M. Landsberg and G. Ottaviani, *New lower bounds for the border rank of matrix multiplication*, Theory Comput. **11** (2015), 285–298, DOI 10.4086/toc.2015.v011a011. MR3376667

[37] M. Lasoń, *List coloring of matroids and base exchange properties*, European J. Combin. **49** (2015), 265–268, DOI 10.1016/j.ejc.2015.04.004. MR3349540

[38] D. Maclagan and B. Sturmfels, *Introduction to tropical geometry*, Graduate Studies in Mathematics, vol. 161, American Mathematical Society, Providence, RI, 2015. MR3287221

[39] L. Manivel, *Symmetric functions, Schubert polynomials and degeneracy loci*, translated from the 1998 French original by John R. Swallow, SMF/AMS Texts and Monographs, vol. 6, Cours Spécialisés [Specialized Courses], 3, American Mathematical Society, Providence, RI; Société Mathématique de France, Paris, 2001. MR1852463

[40] M. Marshall, *Positive polynomials and sums of squares*, Mathematical Surveys and Monographs, vol. 146, American Mathematical Society, Providence, RI, 2008. MR2383959

[41] E. Miller and B. Sturmfels, *Combinatorial commutative algebra*, Graduate Texts in Mathematics, vol. 227, Springer-Verlag, New York, 2005. MR2110098

[42] J. G. Oxley, *Matroid theory*, Oxford Science Publications, The Clarendon Press, Oxford University Press, New York, 1992. MR1207587

[43] L. Pachter and B. Sturmfels, *Algebraic Statistics for Computational Biology*, Cambridge University Press, 2005, MR2205868.

[44] L. Qi, H. Chen, and Y. Chen, *Tensor eigenvalues and their applications*, Advances in Mechanics and Mathematics, vol. 39, Springer, Singapore, 2018. MR3791481

[45] A. Seigal, *Ranks and symmetric ranks of cubic surfaces*, J. Symbolic Comput. **101** (2020), 304–317, DOI 10.1016/j.jsc.2019.10.001. MR4109719

[46] J.-P. Serre, *Linear representations of finite groups*, translated from the second French edition by Leonard L. Scott, Graduate Texts in Mathematics, Vol. 42, Springer-Verlag, New York-Heidelberg, 1977. MR0450380

[47] I. R. Shafarevich, *Basic algebraic geometry. 1*, Varieties in projective space, 2nd ed., translated from the 1988 Russian edition and with notes by Miles Reid, Springer-Verlag, Berlin, 1994. MR1328833

[48] Y. Shitov, *A counterexample to Comon's conjecture*, SIAM J. Appl. Algebra Geom. **2** (2018), no. 3, 428–443, DOI 10.1137/17M1131970. MR3852707

[49] A. V. Smirnov, *The bilinear complexity and practical algorithms for matrix multiplication* (Russian, with Russian summary), Zh. Vychisl. Mat. Mat. Fiz. **53** (2013), no. 12, 1970–1984, DOI 10.1134/S0965542513120129; English transl., Comput. Math. Math. Phys. **53** (2013), no. 12, 1781–1795. MR3146566

[50] F. Sottile, *Toric ideals, real toric varieties, and the moment map*, Topics in algebraic geometry and geometric modeling, Contemp. Math., vol. 334, Amer. Math. Soc., Providence, RI, 2003, pp. 225–240, DOI 10.1090/conm/334/05984. MR2039975

[51] B. Sturmfels, *Algorithms in invariant theory*, Texts and Monographs in Symbolic Computation, Springer-Verlag, Vienna, 1993. MR1255980

[52] B. Sturmfels, *Gröbner bases and convex polytopes*, University Lecture Series, vol. 8, American Mathematical Society, Providence, RI, 1996. MR1363949

[53] B. Sturmfels, *Solving systems of polynomial equations*, CBMS Regional Conference Series in Mathematics, vol. 97, Published for the Conference Board of the Mathematical Sciences, Washington, DC; by the American Mathematical Society, Providence, RI, 2002. MR1925796

[54] B. Sturmfels and S. Sullivant, *Toric ideals of phylogenetic invariants*, Journal of Computational Biology **12** (2005) 204–228.

[55] B. Sturmfels and C. Uhler, *Multivariate Gaussian, semidefinite matrix completion, and convex algebraic geometry*, Ann. Inst. Statist. Math. **62** (2010), no. 4, 603–638, DOI 10.1007/s10463-010-0295-4. MR2652308

[56] B. Sturmfels, C. Uhler, and P. Zwiernik, *Brownian motion tree models are toric*, `arXiv:1902.09905`, to appear in Kybernetika.

[57] S. Sullivant, *Algebraic statistics*, Graduate Studies in Mathematics, vol. 194, American Mathematical Society, Providence, RI, 2018. MR3838364

[58] N. L. White, *The basis monomial ring of a matroid*, Advances in Math. **24** (1977), no. 3, 292–297, DOI 10.1016/0001-8708(77)90060-3. MR437366

[59] D. R. Woodall, *An exchange theorem for bases of matroids*, J. Combinatorial Theory Ser. B **16** (1974), 227–228, DOI 10.1016/0095-8956(74)90067-7. MR389631

[60] R. Vakil, *The Rising Sea: Foundations Of Algebraic Geometry Notes*, available at `http://math.stanford.edu/~vakil/216blog/index.html`.

Index

Selected Published Titles in This Series

211 **Mateusz Michałek and Bernd Sturmfels,** Invitation to Nonlinear Algebra, 2021
210 **Bruce E. Sagan,** Combinatorics: The Art of Counting, 2020
209 **Jessica S. Purcell,** Hyperbolic Knot Theory, 2020
208 **Vicente Muñoz, Ángel González-Prieto, and Juan Ángel Rojo,** Geometry and Topology of Manifolds, 2020
207 **Dmitry N. Kozlov,** Organized Collapse: An Introduction to Discrete Morse Theory, 2020
206 **Ben Andrews, Bennett Chow, Christine Guenther, and Mat Langford,** Extrinsic Geometric Flows, 2020
205 **Mikhail Shubin,** Invitation to Partial Differential Equations, 2020
204 **Sarah J. Witherspoon,** Hochschild Cohomology for Algebras, 2019
203 **Dimitris Koukoulopoulos,** The Distribution of Prime Numbers, 2019
202 **Michael E. Taylor,** Introduction to Complex Analysis, 2019
201 **Dan A. Lee,** Geometric Relativity, 2019
200 **Semyon Dyatlov and Maciej Zworski,** Mathematical Theory of Scattering Resonances, 2019
199 **Weinan E, Tiejun Li, and Eric Vanden-Eijnden,** Applied Stochastic Analysis, 2019
198 **Robert L. Benedetto,** Dynamics in One Non-Archimedean Variable, 2019
197 **Walter Craig,** A Course on Partial Differential Equations, 2018
196 **Martin Stynes and David Stynes,** Convection-Diffusion Problems, 2018
195 **Matthias Beck and Raman Sanyal,** Combinatorial Reciprocity Theorems, 2018
194 **Seth Sullivant,** Algebraic Statistics, 2018
193 **Martin Lorenz,** A Tour of Representation Theory, 2018
192 **Tai-Peng Tsai,** Lectures on Navier-Stokes Equations, 2018
191 **Theo Bühler and Dietmar A. Salamon,** Functional Analysis, 2018
190 **Xiang-dong Hou,** Lectures on Finite Fields, 2018
189 **I. Martin Isaacs,** Characters of Solvable Groups, 2018
188 **Steven Dale Cutkosky,** Introduction to Algebraic Geometry, 2018
187 **John Douglas Moore,** Introduction to Global Analysis, 2017
186 **Bjorn Poonen,** Rational Points on Varieties, 2017
185 **Douglas J. LaFountain and William W. Menasco,** Braid Foliations in Low-Dimensional Topology, 2017
184 **Harm Derksen and Jerzy Weyman,** An Introduction to Quiver Representations, 2017
183 **Timothy J. Ford,** Separable Algebras, 2017
182 **Guido Schneider and Hannes Uecker,** Nonlinear PDEs, 2017
181 **Giovanni Leoni,** A First Course in Sobolev Spaces, Second Edition, 2017
180 **Joseph J. Rotman,** Advanced Modern Algebra: Third Edition, Part 2, 2017
179 **Henri Cohen and Fredrik Strömberg,** Modular Forms, 2017
178 **Jeanne N. Clelland,** From Frenet to Cartan: The Method of Moving Frames, 2017
177 **Jacques Sauloy,** Differential Galois Theory through Riemann-Hilbert Correspondence, 2016
176 **Adam Clay and Dale Rolfsen,** Ordered Groups and Topology, 2016
175 **Thomas A. Ivey and Joseph M. Landsberg,** Cartan for Beginners: Differential Geometry via Moving Frames and Exterior Differential Systems, Second Edition, 2016
174 **Alexander Kirillov Jr.,** Quiver Representations and Quiver Varieties, 2016
173 **Lan Wen,** Differentiable Dynamical Systems, 2016

For a complete list of titles in this series, visit the
AMS Bookstore at **www.ams.org/bookstore/gsmseries/**.